Thermoelectric Materials

Thermoelectric Materials
Advances and Applications

Enrique Maciá-Barber

Pan Stanford Publishing

Published by

Pan Stanford Publishing Pte. Ltd.
Penthouse Level, Suntec Tower 3
8 Temasek Boulevard
Singapore 038988

Email: editorial@panstanford.com
Web: www.panstanford.com

British Library Cataloguing-in-Publication Data
A catalogue record for this book is available from the British Library.

Thermoelectric Materials: Advances and Applications

ISBN 978-981-4463-52-2 (Hardcover)
ISBN 978-981-4463-53-9 (eBook)

Printed in the USA

Contents

Preface

Environmental concerns regarding refrigerant fluids as well as the convenience of using non toxic and non expensive materials, have significantly spurred the interest in looking for novel, high- performance thermoelectric materials for energy conversion in small-scale power generation and refrigeration devices, including cooling electronic devices, or flat-panel solar thermoelectric generators.

This search has been mainly fueled by the introduction of new designs and the synthesis of new materials. In fact, the quest for good thermoelectric materials entails the search for solids simultaneously exhibiting extreme properties. On the one hand, they must have very low thermal- conductivity values. On the other hand, they must have both electrical conductivity and Seebeck coefficient high values as well. Since these transport coefficients are not independent among them, but are interrelated, the required task of optimization is a formidable one. Thus, thermoelectric materials provide a full-fledged example of the essential cores of solid state physics, materials science engineering, and structural chemistry working side by side towards the completion of a common goal, that is, interdisciplinary research at work.

Keeping these aspects in mind, the considerable lag between the discovery of the three main thermoelectric effects (Seebeck, Peltier and Thomson, spanning the period 1821–1851), and their first application in useful thermoelectric devices during the 1950s, is not surprising at all. In fact, such a delay can be understood as arising from the need of gaining a proper knowledge of the role played by the electronic structure in the thermal and electrical transport properties of solid matter. Thus, metals and most alloys (whose Fermi level falls in a partially filled allowed energy band) yield typically low thermoelectric conversion efficiencies, as compared to those observed in semiconducting materials (exhibiting a characteristic gap between valence and conduction bands).

According to this conceptual scheme, the first two chapters are devoted to present a general introduction to the field of thermoelectric materials, focusing on both basic notions and the main fundamental questions in the area. For the benefit of the non-acquainted readers, the contents of these chapters are presented in a tutorial way, recalling previous knowledge from solid state physics when required, and illustrating the abstract notions with suitable application examples.

In Chapter 1, we start by introducing the thermoelectric effects from a phenomenological perspective along with their related transport coefficients and the mutual relations among them. We also present a detailed description of the efficiency of thermoelectric devices working at different temperature ranges. Some more recent concepts, like the use of the compatibility factor to characterize segmented devices, or a formulation based on the use of the relative current density and the thermoelectric potential notions to derive the figure of merit and coefficient of performance expressions, are also treated in detail. Finally, several issues concerning the characterization of thermoelectric materials and some related industry standards will be presented.

In Chapter 2, we review the two basic strategies adopted in order to optimize the thermoelectric performance of different materials, namely, the control of the thermal conductivity and the power factor enhancement. The electronic structure engineering approach, nowadays intensively adopted, is introduced along with some useful theoretical notions related to the spectral conductivity function and its optimization.

Within a broad historical perspective, the next three chapters focus on the main developments in the field from the 1990s to the time being, highlighting the main approaches followed in order to enhance the resulting thermoelectric efficiency of different materials. In this way, the low thermal conductivity requirement has led to the consideration of complex enough lattice structures, generally including the presence of relatively heavy atoms within the unit cell, or to the consideration of nanostructured systems characterized by the emergence of low-dimensional effects. By fully adopting this structural complexity approach, in Chapter 3, we progressively introduce the different kinds of bulk materials which have been considered, starting from the main properties of the elemental solids of thermoelectric interest (bismuth, antimony and tellurium), going through a number of binary and ternary alloys of growing chemical and structural complexity, to finish with the promising large unit cell inclusion compounds, including half-Heusler alloys, skutterudites, clathrates and Chevrel phases.

By all indications, attaining large values of the electrical conductivity and Seebeck coefficient usually requires a precise doping control as well as an accurate tailoring of the sample's electronic structure close to the Fermi level. Thus, next generation thermoelectric materials will require more attention to be paid to enhancing their electronic properties, as the lattice thermal conductivity of most thermoelectric materials of interest has already been greatly reduced. To this end, a main goal focuses on obtaining a fundamental guiding principle, in terms of an electronic band structure tailoring process aimed at optimizing the thermoelectric performance of a given material. Following this route, in Chapter 4 we will analyze the role played by the electronic structure in the thermoelectric performance of the different materials described in Chapter 3, paying a special attention to the benefits resulting from a systematic recourse to the band engineering concept.

In Chapter 5, we take a step further along the structural complexity approach by considering materials able to possess atomic lattices which are *both* complex (low thermal conductivity) and highly symmetric (favorable electronic properties). This leads us beyond periodic order into the realm of aperiodic crystals characterized by either incommensurate structures or fully new lattice geometries based on scale-invariance symmetry and long-range aperiodic order, as it occurs in quasicrystals and their related phases.

The inorganic thermoelectric materials we have considered in the five previous chapters are hindered by issues like high cost of production, scarcity of constituting elements, or toxicity. Because of these problems associated with inorganic compounds, organic electronic materials have spurred a growing interest in thermoelectric community. Consequently, in Chapter 6 we consider novel materials based on organic semiconductors and conducting polymers. We also explore recent advances in the study of thermoelectric phenomena at the nanoscale, focusing on the transport properties through molecular junctions and analyzing the potential of DNA based thermoelectric devices.

The book contains **58** proposed **exercises** (highlighted in boldface through the text) accompanied by their detailed **solutions**. I have prepared the exercises mainly from results published and discussed in regular research papers during the last decade in order to provide a glimpse into the main current trends in the field. Although the exercises and their solutions are given at the end of each chapter for convenience, it must be understood that they are an integral part of the presentation, either motivating or illustrating the different concepts and notions. In the same way, most exercises of Chapters 5 and 6 assume the reader is well acquainted with the contents presented in the previous four chapters, and may serve as a control test. Accordingly, it is highly recommended to the reader that he/she try to solve the exercises in the sequence they appear in the text, then check his/her obtained result with those provided at the end of the chapter, and only then to resume the reading of the main text. In this way, the readers (who are intended to be both graduate students as well as senior scientists approaching this rapidly growing topic from other research fields) will be able to extract the maximum benefit from the materials contained in this book in the shortest time.

All the references are listed in the bibliography section at the end of the book. I have tried to avoid a heavily referenced main text by concentrating most references in the places where they are most convenient to properly credit results published in the literature, namely, in the figures and tables captions, in the footnotes, and in the exercises and their solutions. The references are arranged according to the following criteria: in the first place, some historical papers are given, followed by a series of reference textbooks covering different topics directly related to the materials treated in this book, then I list the reviews and monographs published on related issues during the last decade. Afterwards, a list of archival research papers is given in the order they appear in the text from Chapters 1 to 6.

I am gratefully indebted to Professors Esther Belin-Ferré, Jean Marie Dubois, Kaoru Kimura, Uichiro Mizutani, Tsunehiro Takeuchi, and Terry M. Tritt for their continued interest in my research activities during the last two decades. Their illuminating advice has significantly guided my scientific work in the field of thermoelectric materials.

It is a pleasure to thank Emilio Artacho, Janez Dolinsek, Roberto Escudero, G. Jeffrey Snyder, Oleg Mitrofanov, and José Reyes-Gasga for sharing very useful materials with me.

I am also grateful to Mr. Stanford Chong for giving me the opportunity to prepare this book and to Ms. Shivani Sharma for her continued help in dealing with editorial matters. Last, but not least, I warmly thank M. Victoria Hernández for her invaluable support, unfailing encouragement, and attention to detail.

Enrique Maciá-Barber
Madrid
Spring 2015

Chapter 1

Basic Notions

1.1 Thermoelectric Effects

During the nineteenth century, several phenomena linking thermal energy transport and electrical currents in solid materials were discovered within a time interval of 30 years, spanning from 1821 to 1851 (Fig. 1.1) These phenomena are collectively known as *thermoelectric effects*, and we will devote this section to briefly introducing them.[a]

Let us start by considering an elementary thermal effect: experience shows us that when a piece of matter is subjected to a temperature difference between its ends heat spontaneously flows from the region of higher temperature, T_H, to the region of lower temperature, T_C (Fig. 1.2a). This heat current is maintained over time until thermal equilibrium ($T_H = T_C \equiv T$) is reached and the temperature gradient vanishes (Fig. 1.2b). It was Jean Baptiste Joseph Fourier who first introduced the mathematical formulation describing this well-known fact in 1822. According to the so-called *Fourier's law*, the presence of a temperature gradient ∇T (measured

[a]In addition to the phenomena described in this section, we may also observe the so-called *galvanomagnetic* (when no temperature gradients are present) or *thermomagnetic* (when both thermal gradients and magnetic fields are present) effects. These phenomena, however, will not be covered in this book.

Thermoelectric Materials: Advances and Applications
Enrique Maciá-Barber

ISBN 978-981-4463-52-2 (Hardcover), 978-981-4463-53-9 (eBook)
www.panstanford.com

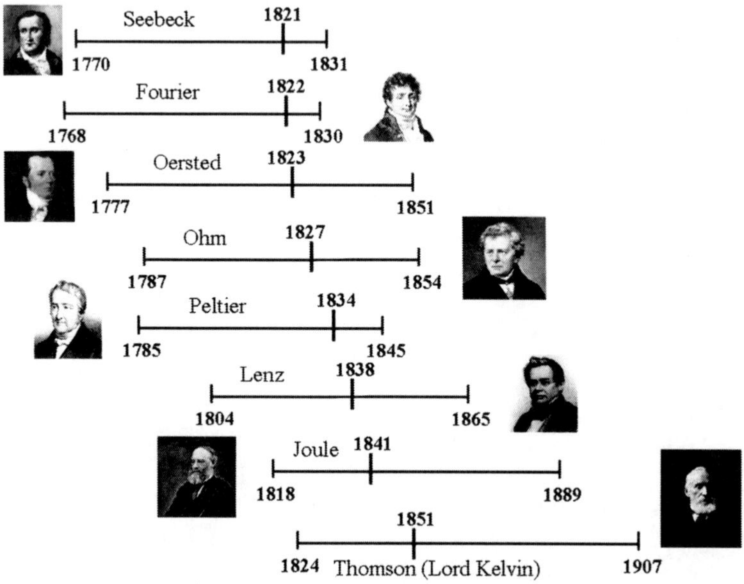

Figure 1.1 Chronogram showing the portraits and life span of the main characters in the origins of thermoelectric research. The ticks indicate the date when the corresponding thermoelectric phenomenon was first reported.

in Km^{-1}) induces in the material a *heat current density* $\mathbf{h}$ (measured in Wm^{-2} units) which is given by[a]

$$\mathbf{h} = -\kappa \,\nabla T, \tag{1.1}$$

where κ is a characteristic property of the considered material, referred to as its *thermal conductivity* (measured in $\mathrm{Wm}^{-1}\mathrm{K}^{-1}$ units). In general, the thermal conductivity depends on the temperature of the material, that is, $\kappa(T)$, and it always takes on positive values ($\kappa > 0$), so that the minus sign in Eq. (1.1) is introduced to properly describe the thermal current propagation sense. Indeed, if we reverse the temperature gradient ($\nabla T \to -\nabla T$) in Eq. (1.1) we get a heat flow reversal ($\mathbf{h} \to -\mathbf{h}$), so that heat always diffuses the same way: from the hot side to the cold one.

Five years after the publication of Fourier's work, Georg Simon Ohm reported that when a potential difference, ΔV (measured in V),

[a]Throughout this book boldface characters will denote vectorial magnitudes.

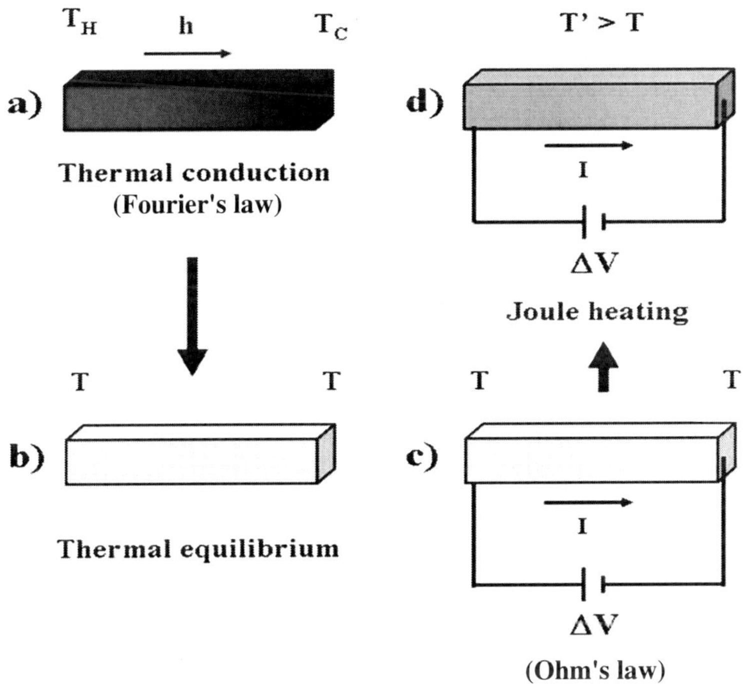

Figure 1.2 Thermal and electrical phenomena in homogeneous conductors.

is established between the end points of an isothermal conductor, an *electrical current intensity*, I (measured in A), flows through the material (Fig. 1.2c). Both magnitudes are linearly related according to the so-called *Ohm's law*

$$\Delta V = RI, \tag{1.2}$$

where $R > 0$ is a characteristic property of the considered material, referred to as its *electrical resistance* (measured in Ω units), which generally depends on the temperature, that is, $R(T)$. In this expression, $I > 0$ describes the motion of positive charge carriers moving from positive to negative electrodes.[a] In order to highlight the analogy between thermal and electrical currents, Ohm's law can

[a]We note that this convention was adopted before the electron, the main charge carrier in metallic conductors, was discovered by Joseph John Thomson (1856–1940) in 1897.

be expressed in terms of the *electrical current density* $\mathbf{j}$ (measured in Am^{-2} units) and the potential gradient ∇V (measured in Vm^{-1}) in the form

$$\mathbf{j} = -\sigma \nabla V, \tag{1.3}$$

where $\sigma(T) > 0$ (usually measured in $\Omega^{-1}\mathrm{cm}^{-1}$) is the *electrical conductivity*. This magnitude is the reciprocal (i.e., $\sigma = \rho^{-1}$) of the material's *electrical resistivity*

$$\rho = \frac{RA}{L}, \tag{1.4}$$

(measured in Ωcm units), where A is the cross-section of the material and L measures its length.

Using a calorimeter to measure heat and a galvanometer to measure electrical currents through a variety of resistive circuits, James Prescott Joule realized in 1841 that whenever an electrical current is flowing through a conductor, a certain amount of heat is released per unit time (Fig. 1.2d), according to the expression

$$W_J = RI^2 = \rho LA\, \mathbf{j} \cdot \mathbf{j}, \tag{1.5}$$

where W_J measures the heat power dissipated in the material in W units[a] (**Exercise 1.1**). Accordingly, an electrical current has an inherent thermal effect. The so-called *Joule effect* is an irreversible process, which means that if the sense of the current flowing through the conductor is reversed a heat liberation still occurs, instead of a heat absorption process leading to cooling down of the material. In modern scientific jargon, we say that Joule formula is invariant under the sign reversal operation $\mathbf{j} \to -\mathbf{j}$ in Eq. (1.5). A similar irreversible character is observed in the Fourier's heat current flow, as previously indicated.

Let us now consider what happens when an electric current passes through a homogeneous conductor along which a temperature gradient is also maintained. In this case, when charge carriers flow in the direction of the temperature gradient ∇T, both thermal and electrical currents are simultaneously present in the system and one may expect different behaviors to occur due to the coupling of these currents, depending upon whether $\mathbf{h}$ and $\mathbf{j}$ currents propagate in the same or the opposite sense (Fig. 1.3). This interesting issue

---

[a]Making use of Eqs. (1.2)–(1.5), one obtains the useful dimensional relation $[\mathrm{W}] = [\Omega^{-1}][\mathrm{V}^2] = [\Omega][\mathrm{A}]^2$ between mechanical and electrical magnitudes.

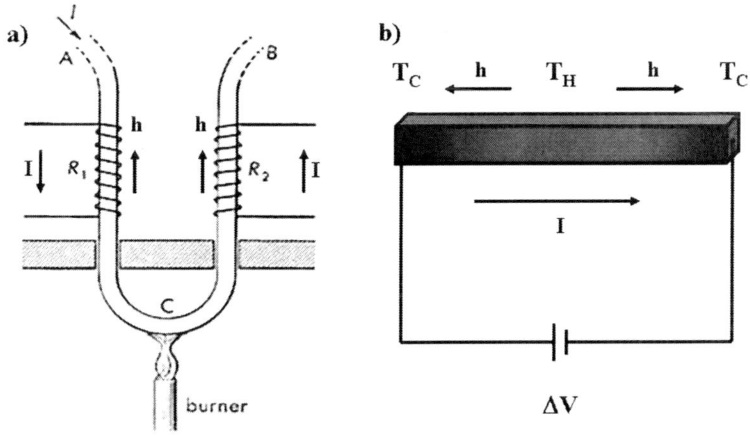

Figure 1.3 Experimental setup for a demonstration of the Thomson effect: (a) original drawing. (b) Schematic diagram.

was first addressed by William Thomson, first Baron Kelvin, who in 1851 proposed the existence of a specific thermal effect produced by the pass of an electrical current through an unequally heated conductor. This thermal effect results in the release or absorption of a certain amount of heat depending on the relative sense of the **h** and **j** currents, as well as on the material nature of the conductor.

In his original experimental setup, Thomson allowed an electrical current of intensity I to pass through an iron rod, which was bent into a U-shape (Fig. 1.3a). Two resistance coils, R_1 and R_2, were wound about the two sides and connected to an external electrical circuit known as a Wheatstone bridge. This extremely sensitive circuit was initially balanced in order to determine any possible variation of the resistivity of these coils. The bottom of the U-shaped conductor was then heated with a burner. This establishes two temperature gradients, a positive one extending from A to C and a negative one extending from C to B. Consequently, the thermal and electrical currents run parallel (anti-parallel) in the CB (AC) arms, respectively. By inspecting the behavior of the Wheatstone bridge, Thomson observed that it became unbalanced, indicating that the resistance R_1 has increased its value as a consequence of heat being liberated from the conductor.[a] On the contrary, at the position of resistance R_2 a certain amount of heat was absorbed by the conductor, so that some energy was supplied to the conductor at the expense of the thermal energy of the resistance. Therefore, carriers traversing the thermal gradient gain or release energy depending on their direction relative to ∇T.

The measured *Thomson heat* (in J units) is proportional to the current intensity passing during a time Δt, and to the temperature difference between the ends, according to the expression

$$\Delta Q_T = I\Delta t \int_{T_C}^{T_H} \tau(T)dT, \tag{1.6}$$

where the coefficient $\tau(T)$ is a temperature dependent property of the considered material called the *Thomson coefficient* and it is expressed in VK^{-1} units. Typically, Thomson coefficient values amount to a few $\mu\mathrm{VK}^{-1}$ for most metallic systems, for instance $\tau_{\mathrm{Cu}} = +1.4\ \mu\mathrm{VK}^{-1}$, $\tau_{\mathrm{Pt}} = -13\ \mu\mathrm{VK}^{-1}$ and $\tau_{\mathrm{Fe}} = -6.0\ \mu\mathrm{VK}^{-1}$ at room temperature. We must note that, for a given material choice, Thomson coefficient can take on either positive or negative values depending on the relative sense of propagation of thermal and electrical currents. The sign convention normally used is that $\tau > 0$ if heat is absorbed ($\Delta Q_T > 0$) when the electrical current flows toward the hotter region. Physically, when the electrical current moves from the hot to the cold end the conductor absorbs heat, making the cold end to get even colder, thereby preserving the original thermal temperature distribution.

In summary, when a current is flowing through a material both Joule (irreversible) and Thomson (reversible) effects are simultaneously taking place, though the magnitude of the latter is about two orders of magnitude smaller than the former.

Once we have considered the different behaviors of both thermal and electrical currents propagating through a homogeneous conductor, one may think of properly combining two or more conductors among them in order to construct a thermoelectric (TE) circuit made up of three different electrical conductors, say α, β, and γ, as it is illustrated in Fig. 1.4, where $T_H > T_0 > T_C$. It was observed

---

[a]The electrical resistivity usually increases with temperature in most metals according to the linear relationship $\rho = \rho_0 + \alpha T$, where ρ_0 and α take on characteristic values for each material.

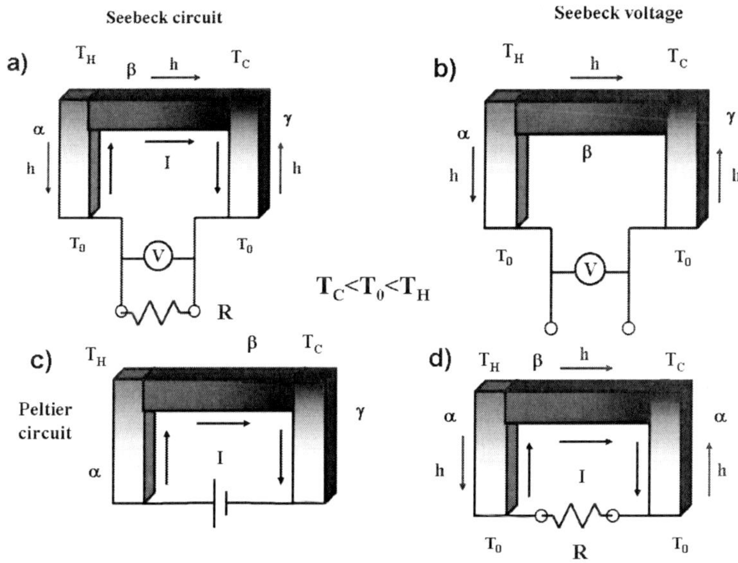

Figure 1.4 Seebeck and Peltier effects in thermoelectric circuits made of three homogeneous conductors (labeled α, β, and γ) connected in series.

by Thomas Johann Seebeck [1], that when the junctions between dissimilar conductors are subjected to different temperatures an electric current flows around the closed circuit (Fig. 1.4a). Thus, the *Seebeck effect* describes the conversion of thermal energy into electrical energy in the form of an electrical current. The magnitude of this effect can be expressed in terms of the *Seebeck voltage* related to the electromotive force set up under open-circuit conditions (Fig. 1.4b). Shortly after Seebeck's report, Oersted together with Fourier constructed the first pile based on the TE effect in 1823.[a]

For not too large temperature differences between the junctions, this voltage is found to be proportional to their temperature difference,

$$\Delta V_S = S_{\alpha,\beta}\,\Delta T, \tag{1.7}$$

where the coefficient of proportionality $S_{\alpha,\beta}(T)$ is a temperature dependent property of the junction materials called the *Seebeck coefficient* and it is expressed in VK^{-1} units. Thus, the Seebeck

[a]For the sake of comparison we recall that the Volta battery was introduced in 1799.

coefficient[a] measures the magnitude of an induced TE voltage in response to a temperature difference across the material. Its magnitude (usually comprised within the range from μVK^{-1} to mVK^{-1}) generally depends on the temperature of the junction and its sign is determined by the materials composing the circuit. The sign convention normally used is that $S_{\alpha,\beta} > 0$ if a clockwise electrical current is induced to flow from α to β at the hot junction (**h** and **j** are parallel through the β conductor in this case, see Fig. 1.4a).

Let us now consider that, instead of keeping the junctions at different temperatures, we allow them to reach thermal equilibrium and with the aid of an external battery we generate a relatively small electrical current around the circuit (Fig. 1.4c). It was reported by Jean Charles Peltier [2], that when the current flowed across the junction in one sense the junction was cooled, thereby absorbing heat from the surroundings ($\Delta Q_P > 0$), whereas when the current sense was reversed the junction was heated, thus releasing heat to the environment ($\Delta Q_P < 0$). This effect was nicely illustrated by Friedrich Emil Lenz, who placed a drop of water on the junction of bismuth and antimony wires. Passing an electrical current through the junction in one sense caused the water to freeze, whereas reversing the current caused the ice to quickly melt. In this way, the basic principle of TE refrigeration was first demonstrated in 1838.

The so-called *Peltier heat* (measured in J) is proportional to the magnitude (I) and duration (Δt) of the current applied,

$$\Delta Q_P = \Pi_{\alpha,\beta}(T) I \Delta t, \tag{1.8}$$

where the coefficient of proportionality is called the *Peltier coefficient* and it is expressed in V units. The origin of this effect resides in the transport of heat by an electrical current. Its magnitude (usually comprised within the range 30–0.1 mV at room temperature) generally depends on the temperature of the junction and its sign is determined by the materials making the circuit. The sign convention normally used is that $\Pi_{\alpha,\beta} > 0$ if a clockwise electrical current

[a]Also referred to as *thermopower* or *thermoelectric power*, though these terms are certainly misleading since this coefficient actually measures a voltage gradient, not an electric power. Nevertheless, they were generally adopted by the thermoelectric research community from the very beginning, and can be profusely found in the literature. Notwithstanding this, we will avoid the use of these terms as much as possible throughout the book.

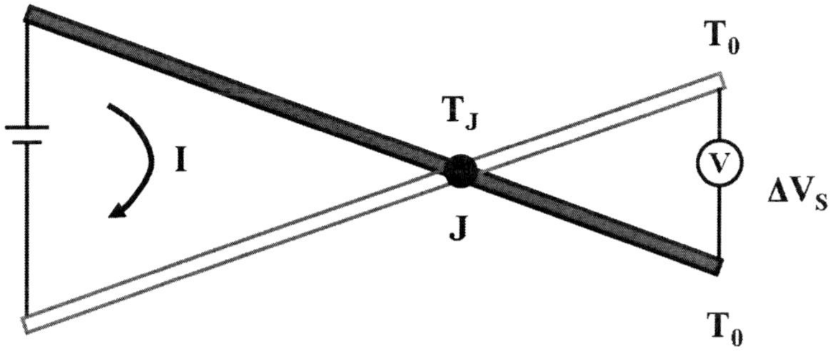

Figure 1.5 Peltier cross. The circuit consists of two different metallic wires contacting with one another at a single point, labeled J. The left part of the circuit is connected to a battery, whereas the right part contains a voltmeter.

induces a cooling effect at the hot junction (i.e., it absorbs heat) when flowing from α to β (see the circuit shown in Fig. 1.4c).

Attending to their phenomenological features, the Seebeck and Peltier effects are closely related to each other. To show the relation between the Seebeck effect and his new effect, Peltier used a circuit of his original design, known as the "Peltier cross" (Fig. 1.5). When the current flows through the left circuit, the junction is heated or cooled, depending on the current sense. In any case, this leads to a change in the temperature of the junction T_J, as compared to the temperature of the wires at the right ends, T_0. Accordingly, a Seebeck voltage can be measured among these ends, which is proportional to $|T_J - T_0|$ (**Exercise 1.2**). In this way, Peltier observed that, for a given applied current value, the *rate* of absorption or liberation of heat at a TE junction depended on the value of the Seebeck coefficient of the junction itself.

About two decades later, William Thomson disclosed the relationship between both coefficients by applying the first and second laws of thermodynamics to a TE circuit, assuming it to be a reversible system (hence neglecting Joule heating and Fourier heat conduction irreversible effects).[a] It is instructive to reproduce this

[a]The very possibility of transforming a certain amount of thermal energy into electrical energy through the presence of an electromotive force driving charge carriers motion in a metallic conductor was earlier proposed by W. Thomson, who referred this process as "*the convection of heat by electric currents*" [3].

derivation in order to gain a deeper understanding on the reversible TE effects we have just introduced. To this end, let us consider the situation depicted in Fig. 1.4d, where an electrical current is driven by the Seebeck voltage arising from the existence of a thermal gradient between the hot and cold junctions. This electrical current, in turn, gives rise to a Peltier heat at the contacts along with a Thomson heat through the homogeneous conductors composing the circuit. The first law of thermodynamics states that the variation in electrical energy equals the variation in thermal energy through the circuit, namely, $q\Delta V_S = \Delta Q = \Delta Q_P^{T_H} + \Delta Q_P^{T_C} + \Delta Q_T^{\alpha} + \Delta Q_T^{\beta} + \Delta Q_T^{\alpha}$, where $q = I\Delta t$ measures the charge flowing through the circuit, and the used notation is self-explanatory (note that, for the sake of simplicity, we have assumed $\gamma = \alpha$). By expressing Eq. (1.7) in the differential form $dV_S = S_{\alpha,\beta}dT$, and making use of Eqs. (1.6) and (1.8) we get

$$\int_{T_C}^{T_H} S_{\alpha,\beta}(T)dT = \Pi_{\alpha,\beta}(T_H) + \Pi_{\beta,\alpha}(T_C) + \int_{T_0}^{T_H} \tau_\alpha(T)dT + \int_{T_C}^{T_0} \tau_\alpha(T)dT - \int_{T_C}^{T_H} \tau_\beta(T)dT, \tag{1.9}$$

where the two first terms in the second member describe the Peltier cooling (heating) at the hot (cold) junctions, respectively, whereas the three remaining terms describe the Thomson cooling (heating) at the α (β) conductors, respectively. Eq. (1.9) can be grouped into the form

$$\int_{T_C}^{T_H} S_{\alpha,\beta}(T)dT = \Delta\Pi_{\alpha,\beta} + \int_{T_C}^{T_H} [\tau_\alpha(T) - \tau_\beta(T)]dT, \tag{1.10}$$

where we explicitly used the symmetry relation $\Pi_{\beta,\alpha}(T) = -\Pi_{\alpha,\beta}(T)$. Assuming the conductors are short enough, Eq. (1.10) can be expressed in the differential form

$$S_{\alpha,\beta}\, dT = d\Pi_{\alpha,\beta} + (\tau_\alpha - \tau_\beta)\, dT. \tag{1.11}$$

On the other hand, the second law of thermodynamics states that the entropy change vanishes in reversible processes, so that

$$\int \frac{\Delta Q}{T} = \frac{\Pi_{\alpha,\beta}(T_H)}{T_H} - \frac{\Pi_{\alpha,\beta}(T_C)}{T_C} + \int_{T_C}^{T_H} \frac{\tau_\alpha(T) - \tau_\beta(T)}{T} dT \equiv 0, \tag{1.12}$$

or, in differential form,

$$d\left(\frac{\Pi_{\alpha,\beta}}{T}\right) + \frac{\tau_\alpha - \tau_\beta}{T} dT = 0. \tag{1.13}$$

By properly relating Eqs. (1.13) and (1.11), one obtains

$$\Pi_{\alpha,\beta} = S_{\alpha,\beta} T, \tag{1.14}$$

hence indicating that Peltier and Seebeck coefficients are proportional to each other and have the same sign. Within this approach, we realize that the Seebeck coefficient provides a measure of the entropy associated with the Peltier electrical current.

Differentiating Eq. (1.14) and making use of Eq. (1.11), we get

$$\tau_\beta - \tau_\alpha = T\,\frac{dS_{\alpha,\beta}}{dT}, \tag{1.15}$$

so that we realize that Thomson effect is produced by the Seebeck coefficient variation induced by the temperature gradient present in the material, and it vanishes when the Seebeck coefficient is temperature independent.[a] Eqs. (1.14) and (1.15) are referred to as the first and second *Kelvin relations*, respectively, and they link the three TE coefficients among them. Thus, the knowledge of one of the Peltier, Thomson, or Seebeck coefficients leads to the knowledge of the two others.

Although the validity of separating the reversible TE effects from the irreversible processes may be questioned, the subsequent application of the theory of irreversible thermodynamics has resulted in the same relationships, which are known as the Onsager relations in this more general scenario [4]. In fact, the validity of Eq. (1.14) has been recently confirmed experimentally [12]. Thus, from Eqs. (1.14) and (1.15) one concludes that Peltier and Thomson effects can be regarded as different manifestations of a basic TE property, characterized by the magnitude $S_{\alpha,\beta}$ given by Eq. (1.7) (**Exercise 1.3**).

If we take a look at Eqs. (1.7) and (1.8), we see that the phenomenological expressions for the Seebeck and Peltier coefficients refer to junctions between dissimilar materials making a thermocouple, so that one cannot use these expressions in practice

---

[a]According to Eq. (1.15), a constant (non-null) value of the Thomson coefficient requires a logarithmic temperature dependence of the Seebeck coefficient of the form $S(T) = \tau \ln T$.

Table 1.1 Seebeck coefficient values of different materials at $T = 273$ K

Metal	S (μVK^{-1})	Metal	S (μVK^{-1})
Ni	−18.0	Pd	−9.00
Pt	−4.45	Pb	−1.15
V	+0.13	W	+0.13
Rh	+0.48	Ag	+1.38
Cu	+1.70	Au	+1.79
Mo	+4.71	Cr	+18.0

to measure the Seebeck and Peltier coefficients of each material in the couple. A convenient way of obtaining the Seebeck and Peltier coefficients values of a given material from experimental measurements relies on the following relationships $S_{\alpha,\beta} \equiv S_\alpha - S_\beta$, and $\Pi_{\alpha,\beta} \equiv \Pi_\alpha - \Pi_\beta$, between contact and bulk transport coefficient values, respectively. Then, to get the coefficients values for each component it is necessary to first measure the potential drop in the couple ΔV_S, divide it by the temperature difference to obtain $S_{\alpha,\beta}$, and then subtracting the absolute Seebeck coefficient of one of the components constituting the couple, which should be previously known. To this end, it is convenient to adopt as a suitable standard reference a material having $S = 0$ at the measurement temperature, a condition which is physically satisfied for superconducting materials below their critical temperatures. Thus, the Seebeck coefficient value for Pb-Nb_3Sn couples measured at low temperatures up to the critical temperature of Nb_3Sn (18 K) gives S_{Pb}, which has become a reference material.

For the sake of illustration, in Table 1.1, we list the Seebeck coefficient values of some representative metals. By convention, the sign of S represents the potential of the cold side with respect to the hot side. In metals the charge carriers are electrons, which diffuse from hot to cold end, then the cold side is negative with respect to the hot side and the Seebeck coefficient is negative. In a p-type semiconductor, on the other hand, charge carriers are holes diffusing from the hot to the cold side, so that the Seebeck coefficient is positive. This is not, however, the case for the metals exhibiting positive S values in Table 1.1. In this case, the Seebeck coefficient sign is determined by the energy dependence of the electrons concentration and their mean scattering time with metal lattice ions, as we will see in Chapter 4.

1.2 Transport Coefficients

The TE effects described in the previous section introduce in a natural way a number of characteristic coefficients of the material, namely the thermal conductivity κ, the electrical conductivity σ, and the Seebeck coefficient S. These coefficients relate thermal and electrical currents (effects) with thermal and electrical gradients (causes). In this section, we will consider these coefficients, generally referred to as *transport coefficients*, in more detail. In the first place, we will introduce a unified treatment of the electrical and thermal currents **j** and **h** in terms of the so-called TE *transport matrix*. Afterwards, we will present a microscopic description of the transport coefficients.

1.2.1 *Thermoelectric Transport Matrix*

In Section 1.1, we learnt that when a piece of matter is subjected to the simultaneous presence of thermal and electrical potential gradients a number of TE effects may occur, resulting in the presence of coupled thermal and electrical currents. Assuming, as a reasonable first approximation, a linear dependence between the electrical, **j**, and thermal, **h**, current densities, on the one side, and the electrical potential, ∇V, and temperature ∇T, gradients which originate them, on the other side, we obtain the following general expressions

$$\begin{aligned} \mathbf{j} &= -(\mathbf{L}_{11}\nabla V + \mathbf{L}_{12}\nabla T), \\ \mathbf{h} &= -(\mathbf{L}_{21}\nabla V + \mathbf{L}_{22}\nabla T), \end{aligned} \tag{1.16}$$

where the coefficients $\mathbf{L}_{ij}$ are tensors in the general case of materials exhibiting anisotropic physical properties. For materials endowed with a high structural symmetry degree, thereby showing an isotropic behavior, these tensor magnitudes reduce to scalar quantities. The minus sign is introduced in order to properly describe the phenomenological behavior reported for heat (Fourier's law) and electrical (Ohm's law) currents, as we will see below.

According to Eq. (1.16), the **j** and **h** current densities can be described in a unified way by introducing the matrix expression

$$\begin{pmatrix} \mathbf{j} \\ \mathbf{h} \end{pmatrix} = \begin{pmatrix} \mathbf{L}_{11} & \mathbf{L}_{12} \\ \mathbf{L}_{21} & \mathbf{L}_{22} \end{pmatrix} \begin{pmatrix} -\nabla V \\ -\nabla T \end{pmatrix}, \quad (1.17)$$

which can be, in turn, written in the more compact vectorial form

$$\mathbf{J} = -\tilde{\mathbf{L}}\nabla\mathcal{U} \quad (1.18)$$

where $\tilde{\mathbf{L}}$ is referred to as the TE *transport matrix* tensor, $\mathbf{J} \equiv (\mathbf{j}, \mathbf{h})^t$ is the *current* vector, and $\mathcal{U} \equiv (V, T)^t$, where the superscript t indicates vector transposition. Now, by recalling the main results presented in Section 1.1, we realize that, although conceptually straightforward, the transport matrix elements $\mathbf{L}_{ij}$ are not amenable to direct measurement. Instead, TE effects are naturally described in terms of a number of transport coefficients, namely, the thermal conductivity κ, the electrical conductivity $\sigma = \rho^{-1}$, and the mutually related Seebeck, S, Peltier, Π, and Thomson, τ, coefficients. Accordingly, it is convenient to express the transport matrix elements $\mathbf{L}_{ij}$ in terms of these transport coefficients. To this end, let us consider the following experimental setups:[a]

- The sample is kept at constant temperature ($\nabla T \equiv 0$) and an electrical current **j** is generated by applying an external voltage ∇V. Taking into account the Ohm's relation $\mathbf{j} = -\sigma \nabla V$, from Eq. (1.17) one gets

$$\sigma(T) = L_{11}. \quad (1.19)$$

- The sample is electrically insulated to prevent any electric current from flowing through it ($\mathbf{j} = \mathbf{0}$) and a thermal gradient ∇T is applied to generate the Seebeck potential $\nabla V = -S\,\nabla T$.[b] Hence, from Eq. (1.17) one gets

$$S(T) = L_{12}L_{11}^{-1}. \quad (1.20)$$

[a] For the sake of simplicity, in what follows we shall restrict ourselves to the consideration of isotropic materials, so that both the transport coefficients and the transport matrix elements are scalar magnitudes.

[b] As it is described in Section 1.2.2.2, the Seebeck electric field which opposes to the thermal drift of *positive* charge carriers is parallel to the thermal gradient, so that ∇V and ∇T are anti-parallel in the case of negative charge carriers.

- The sample is kept at constant temperature ($\nabla T \equiv 0$) as an electrical current $\mathbf{j}$ flows through the sample. Due to the Peltier effect (see Eq. (1.8)), we observe the presence of a thermal current density which is proportional to the electric current, that is, $\mathbf{h} \equiv \Pi\, \mathbf{j}$, so that from Eq. (1.17) one gets

$$\Pi(T) = L_{21} L_{11}^{-1}. \tag{1.21}$$

- The sample is electrically insulated to prevent any electric current from flowing through it ($\mathbf{j} = \mathbf{0}$) while a thermal gradient ∇T is maintained. According to Fourier's law, the measured heat current density is given by $\mathbf{h} = -\kappa \nabla \mathbf{T}$, so that from Eq. (1.17) one gets

$$\kappa(T) = L_{22} - L_{12} L_{21} L_{11}^{-1}. \tag{1.22}$$

By properly combining the nested relations given by Eqs. (1.19)–(1.22) and keeping in mind the first Kelvin relation $\Pi = ST$, one can finally express Eq. (1.17) in the form,[a]

$$\begin{pmatrix} \mathbf{j} \\ \mathbf{h} \end{pmatrix} = \begin{pmatrix} \sigma & \sigma S \\ \sigma S T & \kappa + \sigma S^2 T \end{pmatrix} \begin{pmatrix} -\nabla V \\ -\nabla T \end{pmatrix}. \tag{1.23}$$

Thus, measuring the transport coefficients $\sigma(T)$, $\kappa(T)$, and $S(T)$ we can completely determine the TE transport matrix describing the linear relations between currents and gradients. As we can see, in the limiting case $S = 0$ the transport matrix becomes diagonal and $\mathbf{j}$ and $\mathbf{h}$ are completely decoupled from each other. Thus, the Seebeck coefficient, appearing in the nondiagonal terms of the TE transport matrix, determines the coupled transport of electricity and heat through the considered sample (**Exercise 1.4**). We also see that the TE transport matrix given by Eq. (1.23) considerably simplifies when $\kappa \to 0$. This mathematical result indicates that materials exhibiting a very low thermal conductivity value may be of particular interest in TE research.

---

[a]We note the L_{22} element is closely related to an important parameter in thermoelectric research: the dimensionless figure of thermoelectric merit, ZT, which will be introduced in Section 1.4.2.

1.2.2 *Microscopic Description*

Once we have considered the phenomenological description of TE effects at a macroscopic scale, it is convenient to introduce now a microscopic description able to provide a physical picture of the main transport processes at work within the solid at the atomic scale. Indeed, at a microscopic level TE effects can be understood by considering that charge carriers inside solids, say electrons or holes, transport both electrical charge and kinetic energy when moving around interacting with the crystal lattice and among them.

1.2.2.1 Electrical conductivity

Let us consider a metallic conductor containing n electrons per unit volume. The electrical resistivity is defined to be the proportionality constant between the electric field **E** at a point in the metal and the current density **j** that it induces, namely $\mathbf{E} = \rho \mathbf{j}$. The current density can be expressed in the form $\mathbf{j} = -|e| n \mathbf{v}$, where e is the electron charge and **v** is the average velocity of the electrons. In fact, at any point in the metal, electrons are always moving in a variety of directions with different energies. Thus, in the absence of an electric field, all possible directions are equally probable and **v** averages to zero. The presence of an electric field, however, introduces a preferential direction of motion, so that the averaged velocity now reads [8],

$$\mathbf{v} = -\frac{|e|\tau}{m}\mathbf{E}, \tag{1.24}$$

where m is the electron mass and τ is the so-called *relaxation time*, which measures the average time elapsed between two successive collisions of a typical electron in the course of its motion throughout the solid. Thus, the electrical current density can be expressed as

$$\mathbf{j} = \frac{e^2 n\tau}{m}\mathbf{E} = -\frac{e^2 n\tau}{m}\nabla V, \tag{1.25}$$

and comparing with Eq. (1.3) we obtain

$$\sigma = \frac{e^2 n\tau}{m}. \tag{1.26}$$

Thus, the electrical conductivity coefficient will be always positive and it is related to the charge and mass of the carriers, to their volume concentration in the material and to one parameter measuring the role of scattering events in their overall dynamics.

1.2.2.2 Seebeck effect

Let us consider again the physical setup depicted in Fig. 1.4a, where a metallic conductor labeled β is heated at one end and cooled at the other end. The electrons at the hot region are more energetic and therefore have higher velocities than those in the cold region. Consequently, there is a net diffusion of electrons from the hot end toward the cold end resulting from the applied temperature gradient. This situation gives rise to the transport of heat in the form of a thermally induced heat current, **h**, along with a transport of charge in the form of an electrical current **j**. According to Eq. (1.1), in a system where both ends are kept at a constant temperature difference (i.e., $\nabla T = \text{cte}$), there is a constant diffusion of charges (i.e., $\mathbf{h} = \text{cte}$) from one end to the other. If the rate of diffusion of hot and cold carriers in opposite senses were equal, there would be no net change in charge at both ends. However, the diffusing charges are scattered by impurities, structural imperfections, and lattice vibrations. As far as these scattering processes are energy dependent, the hot and cold carriers will diffuse at different rates. This creates a higher density of carriers at one end of the material, and the resulting splitting between positive and negative charges gives rise to an electric field and a related potential difference: the Seebeck voltage.

Now, this electric field opposes the uneven scattering of carriers so that an equilibrium distribution is eventually reached when the net number of carriers diffusing in one sense is canceled out by the net number of carriers drifting back to the other side as a result of the induced electric field. Only an increase in the temperature difference between both sides can resume the building up of more charges on the cold side, thereby leading to a proportional increase in the TE voltage, as prescribed by Eq. (1.7). In this way, the physical meaning of the Seebeck coefficient can be understood in terms of processes taking place at the atomic scale.

1.2.2.3 Lattice thermal conductivity

When considered at a microscopic scale, the thermal conductivity transport coefficient appearing in Eq. (1.1) must be regarded as depending on two main contributions, namely, a contribution arising from the motion of charge carriers $\kappa_e(T)$, and a contribution due to the vibration of atoms around their equilibrium positions in the crystal lattice $\kappa_l(T)$. Therefore, $\kappa(T) = \kappa_e(T) + \kappa_l(T)$. The charge carrier contribution will be discussed in Section 1.2.3. In this section, we will consider the main features of the lattice contribution to the thermal conductivity [302].

We recall, from standard solid-state physics, that the dynamics of atoms in the crystal lattice can be properly described in terms of a number of collective oscillation modes characterized by their frequency values and their specific pattern of oscillation amplitudes. Within the framework of quantum mechanics, these oscillations are described in terms of the so-called phonons, which are elementary excitations characterized by an energy $\hbar\omega$, where $\hbar$ is the reduced Planck constant and ω is the mode frequency. By arranging the available phonons according to their energy value one obtains the vibrational density of states (DOS) $D(\omega)$, which express the number of modes per unit frequency (or energy) interval. For most solids, the vibrational DOS grows quadratically with the frequency for relatively small frequency, then displays a series of alternating maxima and minima for intermediate frequencies and finally decreases approaching zero at the upper limit cut-off frequency ω_D, referred to as the *Debye frequency* (Fig. 1.6). At any given temperature, the probability distribution of phonons able to contribute to heat transport is given by the Planck distribution function

$$p(\omega, T) = \frac{1}{e^{\hbar\omega/k_B T} - 1}, \tag{1.27}$$

where k_B is the Boltzmann constant. In terms of the vibrational DOS and the Planck distribution function, the lattice thermal conductivity can be expressed as [9],

$$\kappa_l(T) = \frac{v^2\hbar}{3V}\int_0^{\omega_D}\left(\frac{\partial p}{\partial T}\right)D(\omega)\tau(\omega, T)\omega d\omega, \tag{1.28}$$

where v is the sound velocity of the considered material, V is the sample's volume, and $\tau(\omega, T)$ is the average time between heat current degrading collisions involving phonons at a given temperature (the so-called phonon relaxation-time). In the simplest approach, the relaxation-time may be regarded as independent of

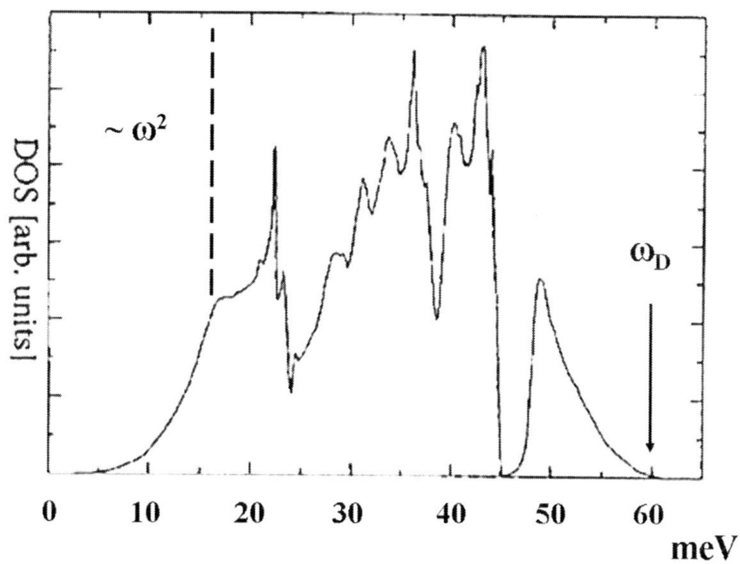

Figure 1.6 Phonon density of states as a function of their energy for a CaF_2 crystal obtained from numerical *ab initio* calculations. The dashed vertical line indicates the energy value limiting the ω^2 dependence interval. The energy value corresponding to the cut-off Debye frequency is marked with an arrow [39]. Reprinted with permission from Schmalzl K., Strauch D., and Schiber H., 2003 *Phys. Rev. B* **68** 144301, Copyright 2003, American Physical Society.

the phonon frequency and the temperature. In that case, Eq. (1.28) can be rewritten in the form

$$\kappa_l(T) = \frac{v^2\tau}{3V}\left[\frac{\partial}{\partial T}\int_0^{\omega_D} p(\omega, T)D(\omega)\hbar\omega d\omega\right], \tag{1.29}$$

where the expression in the brackets can be readily identified as the phonon contribution to the specific heat at constant volume [9], so that Eq. (1.29) reduces to the well-known formula

$$\kappa_l = \frac{1}{3}c_v v l, \tag{1.30}$$

where c_v is the sample's specific heat per unit volume and $l \equiv v\tau$ is the phonon mean-free-path [8–10]. Although the assumption of a constant relaxation-time value is too crude for most applications, in a first approximation this assumption allows for a rough experimental estimation of the phonon mean-free-path from the expression

$$\kappa_l = \frac{d}{3} C_p \bar{v} l, \tag{1.31}$$

where d is the density, C_p is the heat capacity at constant pressure, and the mean sound velocity is given by

$$\bar{v} = 3^{1/3} \left(v_l^{-3} + 2 v_t^{-3} \right)^{-1/3}, \tag{1.32}$$

where v_l and v_t are the longitudinal and transversal sound speed components, respectively (**Exercise 1.5**).

Introducing the dimensionless scaled energy variable $x_l \equiv \beta \hbar \omega$, where $\beta \equiv (k_B T)^{-1}$, and expressing the Planck distribution derivative in terms of hyperbolic functions (**Exercise 1.6**)

$$\frac{\partial p}{\partial T} = \frac{x_l}{4T} \operatorname{csch}^2 \left(\frac{x_l}{2} \right), \tag{1.33}$$

Eq. (1.28) can be rewritten in the form

$$\kappa_l(T) = \frac{v^2 k_B^2 T}{12 V \hbar} \int_0^{\Theta_D/T} x_l^2 \operatorname{csch}^2 \left(\frac{x_l}{2} \right) D(x_l) \tau(x_l, T) dx_l, \tag{1.34}$$

where we have introduced the so-called *Debye temperature*, which is defined from the relationship $\hbar \omega_D \equiv k_B \Theta_D$. In terms of parameters of the material, the Debye temperature is given by

$$\Theta_D = \frac{\hbar v}{k_B} \left(\frac{6\pi^2 N}{V} \right)^{1/3} = \frac{\hbar v}{k_B} \sqrt[3]{6\pi^2 n_a}, \tag{1.35}$$

where N is the number of atoms in the solid and $n_a \equiv N/V$ is the atomic density [9, 10]. The Debye temperature can be experimentally determined from a fitting analysis of the specific heat at low temperature using the formula

$$\Theta_D = \left(\frac{12\pi^4 R_g}{5\delta} \right)^{1/3}, \tag{1.36}$$

where R_g is the gas constant and δ is the coefficient of the T^3 term of the heat capacity curve.

Within the Debye model approximation, which assumes that the vibrational DOS adopts the parabolic form

$$D(\omega) = \frac{3V}{2\pi^2 v^3} \omega^2, \tag{1.37}$$

Eq. (1.34) can be written

$$\kappa_l(T) = \frac{3}{4}v^2 k_B n_a \left(\frac{T}{\Theta_D}\right)^3 \int_0^{\Theta_D/T} x_l^4 \operatorname{csch}^2\left(\frac{x_l}{2}\right) \tau(x_l, T) dx_l, \tag{1.38}$$

where we have made use of Eq. (1.35). As it is illustrated in Fig. 1.6, one reasonably expects the Debye model will be applicable in a relatively broad interval within the low frequency region of the phonon energy spectrum. Accordingly, Eq. (1.38) will hold as far as most phonons contributing to the thermal conductivity belong to this region of the spectrum as well.

The mean relaxation time of heat-carrying phonons is determined by the various scattering mechanisms phonons may encounter when propagating through the solid, such as grain boundaries, point defects (i.e., atomic isotopes, impurity atoms, or vacancies), phonon–phonon interactions, or resonant dynamical effects (e.g., rattling atoms, see Section 3.5.2). Thus, the overall phonon relaxation time can be expressed in the general form

$$\tau^{-1}(\omega, T) = \frac{v}{L} + A_1\omega^4 + A_2\omega^2 T \exp\left(-\frac{\Theta_D}{3T}\right) + \frac{A_3\omega^2}{(\omega_0^2 - \omega^2)^2}, \tag{1.39}$$

where L is the crystal size in a single-grained sample or measures the average size of grains in a poly-grained sample, A_1 (measured in s^3), A_2 (measured in sK^{-1}), and A_3 (measured in s^{-3}), are suitable constants and ω_0 is a resonance frequency. The first term on the right side of Eq. (1.39) describes the grain-boundary scattering, the second term describes scattering due to point defects, the third term describes anharmonic phonon–phonon Umklapp processes,[a] and the last term describes the possible coupling of phonons to localized modes present in the lattice via mechanical resonance.

The ω^4 dependence of the second term in Eq. (1.39) indicates that point defects are very effective in scattering short-wavelength phonons, and they have a lesser effect on longer wavelength phonons. Remarkably enough, short-wavelength phonons make the most important contribution to the thermal current. Then, a natural

[a]In the case of quasicrystals (see Section 5.1.3), the expression for the Umklapp processes must be modified to properly account for their characteristic self-similar symmetry, and the corresponding relaxation-time expression adopts a power law dependence with the temperature of the form $\tau^{-1} \sim \omega^2 T^n$ instead of an exponential one [40].

way of reducing the thermal conductivity of a substance, preserving its electronic properties, is by alloying it with an isoelectronic element. In that case, the phonon scattering by point defects is determined by the mass, size, and interatomic force differences between the substituted and the original atoms. As a general rule, in order to maximize the phonon scattering one should choose point defects having the largest mass and size differences with respect to the lattice main atoms. In this regard, an important type of point defects are the vacancies. Indeed, vacancies represent the ideal point defect for phonon scattering, as they provide the maximum mass contrast. However, vacancies can also act as electron acceptors, hence affecting the electronic transport properties.

In the absence of dynamical resonance effects,[a] Eq. (1.39) can be expressed in the form

$$\tau^{-1}(x_l, T) = \frac{v}{L} + c_0^2 x_l^2 \left[A_1 c_0^2 x_l^2 T + A_2 \exp\left(-\frac{\Theta_D}{3T}\right)\right] T^3, \quad (1.40)$$

where $c_0 = k_B/\hbar$. For most materials

$$A_1 = \frac{\bar{V}\Gamma_S}{4\pi \bar{v}_g},$$

where $\bar{V}$ is the average atomic volume, $\bar{v}_g$ is the average phonon group velocity, and Γ_S is the scattering parameter. For scattering processes dominated by mass fluctuations due to alloying, the scattering parameter reads

$$\Gamma_S = \frac{\sum_{i=1}^{N} c_i f_i^A f_i^B \left(\frac{M_i^A - M_i^B}{M}\right)^2}{\sum_{i=1}^{N} c_i},$$

where $M_i^{A,(B)}$ represents the mass of the substituting (substituted) atoms, c_i is the site degeneracy of the ith sublattice, and $f_i^{A,B}$ measures the fractional occupation of atoms A and B, respectively.

In the low-temperature regime, the average phonon frequency is low and only long-wavelength phonons will be available for heat transport, which are mostly unaffected by both point defects and phonon–phonon interactions. These long-wavelength phonons are

[a]These effects will be discussed in detail when studying thermal transport in skutterudites and clathrates compounds in Sections 3.5.2 and 3.5.3.

chiefly scattered by grain-boundaries (polycrystalline samples) and crystal dimensions (single crystals). Accordingly, $\tau \simeq L/v$ and Eq. (1.38) reads

$$\kappa_l(T) = \frac{3k_B}{4} Lvn_a \left(\frac{T}{\Theta_D}\right)^3 \mathcal{I}, \tag{1.41}$$

where

$$\mathcal{I} \equiv \int_0^\infty \left(\frac{x_l^2}{\sinh(x_l/2)}\right)^2 dx_l > 0, \tag{1.42}$$

since in the limit $T \to 0$ one gets $\Theta_D/T \to \infty$, and the integral in Eq. (1.38) reduces to a real positive number. Thus, in the low-temperature regime the thermal conductivity will show a cubic dependence with the temperature, as prescribed by the $(T/\Theta_D)^3$ factor in Eq. (1.41). From Eq. (1.41) we also see that at any given (low enough) temperature, the thermal conductivity takes on large values for those samples having larger (i) sizes, (ii) sound velocities, and (iii) atomic densities.

On the other hand, in the high temperature limit (i.e., $T > \Theta_D$), $\exp\left(-\frac{\Theta_D}{3T}\right) \to 1$ in Eq. (1.40), and the phonons wavelength is significantly shorter than the sample dimensions, so that it can be regarded as effectively infinite in size ($L \to \infty$). Thus, $v/L \to 0$ and Eq. (1.40) can be written

$$\tau^{-1}(x_l, T) = c_0^2 x_l^2 (A_1 c_0^2 x_l^2 T + A_2) T^3. \tag{1.43}$$

Plugging this relaxation time expression into Eq. (1.38) and making use of Eq. (1.35), we obtain

$$\kappa_l(T) = \frac{\hbar}{8\pi^2 v A_1 T} \int_0^{\Theta_D/T} \frac{x_l^2}{x_l^2 + A_4} \operatorname{csch}^2\left(\frac{x_l}{2}\right) dx_l, \tag{1.44}$$

where $A_4 \equiv (\hbar/k_B)^2 A_2 (A_1 T)^{-1}$ is a dimensionless constant. This expression can be further simplified by taking into account that at high enough temperatures ($x_l \ll 1$), we can approximate $\sinh(x_l/2) \simeq x_l/2$ in Eq. (1.44), which can then be explicitly integrated to get

$$\kappa_l(T) = \frac{k_B}{2\pi^2 v \sqrt{A_1 A_2 T}} \tan^{-1}\left(\frac{\Theta_D}{\sqrt{A_4} T}\right). \tag{1.45}$$

Finally, we must take into account that, at the high-temperature regime we are now considering, the phonon–phonon Umklapp processes generally overshadow the scattering due to impurities as a major mechanism degrading the thermal current, so that $A_1/A_2 \ll 1$. Therefore, one can make the approximation $\tan^{-1}\alpha \simeq \alpha$, and Eq. (1.45) can be rewritten in the form

$$\kappa_l(T) = \frac{k_B^2}{2\pi^2 v\hbar A_2}\frac{\Theta_D}{T} \sim T^{-1} \tag{1.46}$$

in agreement with experimental transport data obtained at high temperatures [8].

Making use of Eq. (1.35), we can express Eq. (1.46) in the form

$$\kappa_l(T) = \frac{k_B\sqrt[3]{6\pi^2 n_a}}{2\pi^2 A_2 T}. \tag{1.47}$$

We see that, for a given value of the parameter A_2, κ_l generally decreases as n_a decreases at a given temperature. Indeed, this property is exploited in TE generators based on materials characterized by complex structures with many atoms in their unit cells, as we will discuss in Chapters 3 and 4. On the other hand, by comparing Eqs. (1.41) and (1.46) we see that, whereas the thermal conductivity is improved by increasing the sound velocity at low enough temperatures, to have large v values leads to a poorer thermal conductivity in the high-temperature regime.

1.2.2.4 Phonon drag effect

When charge carriers diffuse in a solid driven by an applied thermal gradient they can experience scattering processes with the lattice vibrations, thereby exchanging momentum and energy. A rough estimation reveals that the wavelength of electrons is about 10^{-8} m at room temperature, which is about two orders of magnitude larger than the typical lattice periodicity in elemental solids, and about an order of magnitude larger than typical unit cell size in relatively structurally complex materials of TE interest, such as skutterudites (see Section 3.5.2) or clathrates (see Section 3.5.3). Accordingly, charge carriers will be more efficiently scattered by lattice vibration waves having a comparable long wavelength (the so-called acoustic phonons).

As a result of this interaction (usually referred to as electron–phonon interaction), phonons can exchange energy with electrons, so that the local energy carried by the phonon system is fed back to the electron system, resulting in an extra Peltier current source, namely, $\mathbf{h}_P = \Pi\mathbf{j} = (\Pi_e + \Pi_l)\mathbf{j}$, where Π_e indicates the contribution due to the charge carriers diffusion and Π_l gives the electron–phonon contribution. Taking into account the first Kelvin relation given by Eq. (1.14), the Seebeck coefficient can be properly expressed as the sum of two contributions, namely, a diffusion term arising from the charge carriers motion and the so-called *phonon-drag* term, due to interaction of those carriers with the crystal lattice. Thus, we have $S(T) = S_e(T) + S_l(T)$, where the first term accounts for the charge carriers and the second term gives the phonon-drag term. The phonon-drag contribution to the Seebeck coefficient is given by [8],

$$S_l(T) = \frac{k_B}{|e|}\frac{C_V(T)}{3nN_A k_B} = -\frac{k_B}{|e|}\frac{4\pi^4}{5n}\left(\frac{T}{\Theta_D}\right)^3, \tag{1.48}$$

and it was first observed in semiconducting germanium at low temperatures and subsequently identified in metals and alloys as well. The magnitude of S_l depends on the relative strength of phonon scattering by electrons compared to either phonon–phonon and phonon–defects interactions. Since these later scattering contributions dominate at temperatures comparable to the Debye one, one concludes that the phonon-drag effect is important at low temperatures only, say in the range $\Theta_D/10 \lesssim T \lesssim \Theta_D/5$, where it can make a significant contribution to the total Seebeck coefficient values. Therefore, since most applications of thermoelectric materials (TEMs) take place at temperatures comparable or above Θ_D, the contribution due to phonon-drag effects plays only a minor role in mainstream TE research.

1.2.3 *Transport Coefficients Coupling*

Once we have completed the description of transport coefficients of TE interest from a microscopic point of view, it is now convenient to consider their mutual relationships, which ultimately originate from the interaction between charge carriers and lattice vibrations, as well as due to the dual nature of charge carriers transport. Such a duality is nicely exemplified by metallic systems, whose thermal conductivity is mainly governed by the motion of electrons (i.e., $\kappa_l \ll \kappa_e$ at any temperature). Since this motion also determines their contribution to the resulting electrical conductivity, one should expect that the transport coefficients κ_e and σ will be tied up in these materials. Experimentally, the close interrelation between thermal and electrical currents in metals was disclosed by Gustav Heinrich Wiedemann (1826–1889) and Rudolf Franz in 1853. According to the so-called *Wiedemann–Franz's law* (WFL), the thermal and electrical conductivities of most metallic materials are mutually related through the relationship

$$\kappa_e(T) = L_0 T \sigma(T), \tag{1.49}$$

where $L_0 = (\pi k_B/e)^2/3 \simeq 2.44 \times 10^{-8}\ \mathrm{V^2K^{-2}}$ is the *Lorenz number*, named after Ludwig Valentin Lorenz (1829–1891). It was subsequently observed that Eq. (1.49) also holds for semiconducting materials, with L_0 being replaced by the somewhat smaller value $L_s = 2(k_B/e)^2 \simeq 1.48 \times 10^{-8}\ \mathrm{V^2K^{-2}}$ [10].

Strictly speaking, Eq. (1.49) only holds over certain temperature ranges, namely, as far as the motion of the charge carriers determines both the electrical and thermal currents. Accordingly, one expects some appreciable deviation from WFL when electron–phonon interactions, affecting in a dissimilar way to electrical and heat currents, start to play a significant role. Thus, WFL generally holds at low temperatures (say, as compared to the Debye temperature). As the temperature of the sample is progressively increased, the validity of WFL will depend on the nature of the interaction between the charge carriers and the different scattering sources present in the solid. In general, the WFL applies as far as elastic processes dominate the transport coefficients, and usually holds for a broad variety of materials, provided that the change in energy due to collisions is small as compared with $k_B T$ [8, 9]. Finally, at high enough temperatures the heat transfer is dominated by the charge carriers again, due to the onset of Umklapp phonon–phonon scattering processes, which reduce the number of phonons available for electron–phonon interactions. Accordingly, the WFL is expected to hold as well.

From a practical viewpoint, the importance of the WFL can be seen by considering that only the total thermal conductivity $\kappa(T)$ can be experimentally measured in a straightforward way, and the contributions $\kappa_e(T)$ and $\kappa_l(T)$ must be somehow separated. This is usually done by explicitly assuming the applicability of the WFL to the considered sample, so that the lattice contribution to the thermal conductivity is obtained from the expression

$$\kappa_l(T) = \kappa(T) - LT\sigma(T), \tag{1.50}$$

where $L = L_0$ for metallic systems and $L = L_s$ for semiconducting ones. Actually, this estimation of the lattice contribution should be regarded as a mere approximation, since one generally lacks a precise knowledge of the L value in real applications. On the one hand, as we have previously indicated, the Lorenz number is sample dependent and its value not only differs for metallic and semiconducting materials, but even in the case of semiconductors it can take on different values for different chemical compounds. For instance, the value $L = 2.0 \times 10^{-8}$ $\mathrm{V^2K^{-2}}$ is widely adopted in the study of skutterudites (see Section 3.5.2). On the other hand, even for a given material the L value usually varies with the temperature. Accordingly, the Lorenz number should more properly be evaluated, at any given temperature, from the ratio

$$L(T) \equiv \frac{\kappa_e(T)}{T\sigma(T)}, \tag{1.51}$$

which is referred to as the *Lorenz function*. This function can be experimentally determined is some cases, a topic we will discuss in more detail in Section 1.5 (**Exercise 1.7**).

Another important relationship between transport coefficients involves the electrical conductivity and the Seebeck coefficient. Indeed, in most materials the Seebeck coefficient decreases as the electrical conductivity increases and vice versa.[a] This is illustrated in Fig. 1.7 for the case of a clathrate compound (see Section 3.5.3). In Section 2.1, we will comment in detail on the important role played by this relationship in the TE performance of TEMs.

1.3 Thermoelectric Devices

Thermoelectric devices are small (a few mm thick by a few cm square), solid-state devices used in small-scale power generation

[a]Some noteworthy exceptions have been recently reported for unconventional materials [42, 43].

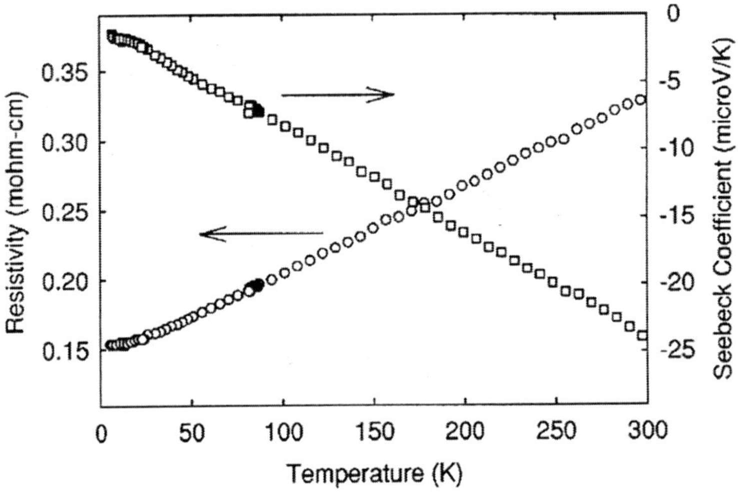

Figure 1.7 Temperature dependence of the Seebeck coefficient and the electrical resistivity for the SrZnGe clathrate [41]. Reprinted with permission from Qiu L., Swainson I. P., Nolas G. S., and White M. A. 2004 *Phys. Rev. B* **70**, 035208, Copyright 2004, American Physical Society.

and refrigeration applications, where a thermal gradient generates an electrical current flow (*TE generator*, TEG) or a DC current is applied to remove heat from the cold side (*TE cooler*, TEC). Thermoelectric devices generally consist of a relatively large number of *thermocouples* (Fig. 1.8) associated electrically in series and thermally in parallel, which can adopt a stacked configuration forming a multi-staged *thermoelectric module* (Fig. 1.9). Historically the interest in TE devices was significantly spurred by the intensive research work performed by the team led by Abram Fedorovich Ioffe (1880–1960, Fig. 1.10a) and his coworkers at the Physical-Technical Institute in Saint Petersburg, where they actively pursued TE research in USSR during the period 1930–1960, leading to some of the first commercial TE power generation and cooling devices [44]. Thus, one of the first TEGs was developed by Yuri Petrovich Maslakovets (1900–1967) during the late 1930s. The modules were based on 74 thermocouples of PbS (see Section 3.3.4) for the n-type leg and iron for the p-leg. Each leg was shaped as a four-sided truncated pyramid with a 2.1×2.2 cm^2 base and a 2.2 cm height.

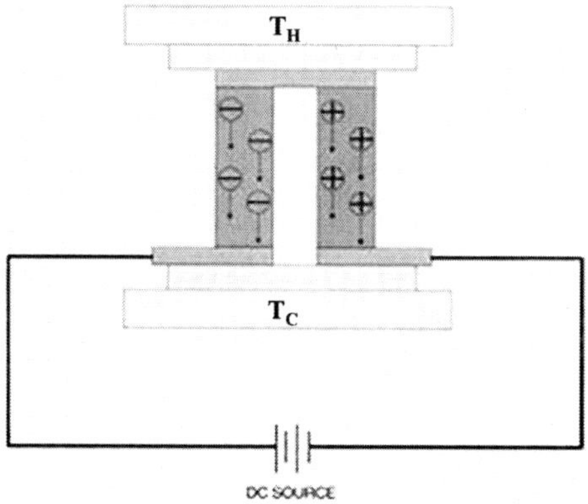

Figure 1.8 Sketch of a typical thermocouple composed of two ceramic substrates, that serve as foundation and electrical insulation for a n-type (p-type) semiconductor element on the left (right), respectively.

Subjected to a temperature difference of 300 C the TEG supplied 12 W of electrical power during 400 h. The most difficult problem in developing that TEG was the interconnection of legs with low enough contact resistance operating at relatively high temperatures for long times. The first contact material was metallic lead. After the Second World War, the ZnSb compound (see Section 3.3.3) was replaced by iron in the p-leg and the lead in the interconnections was replaced by strips of antimony, whose melting point is significantly higher than that of lead. Since 1948, the first commercial TEGs were produced in the URSS for the electrical supply of radio-receivers in rural areas. These ring-shaped TEGs were placed on a kerosene lamp, which served as the heat source (Fig. 1.10b). During the past several decades, TEGs have reliably provided power in remote terrestrial and extraterrestrial locations, mostly based on high temperature radioisotope TEGs on deep space probes such as Voyager 1 and Voyager 2 spacecrafts. Currently, a huge window of opportunity exists for thermoelectrics for low-grade waste heat recovery, such as in automobiles exhaust where TEGs working at intermediate temperatures (500–800 K) can be used to improve fuel economy and reduce greenhouse gas emission. Also, combined with photovoltaics, TEMs can be implemented in high temperature

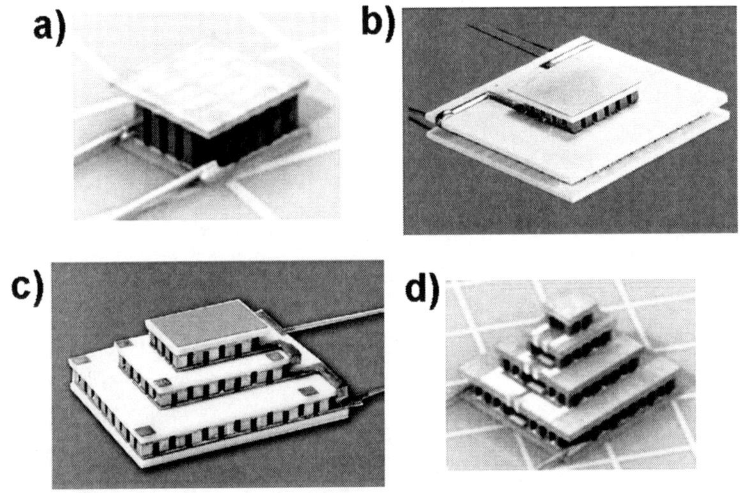

Figure 1.9 Thermoelectric cooling modules based on (a) single stage, (b) two stage, (c) three stage, and (d) four stage arrangements.

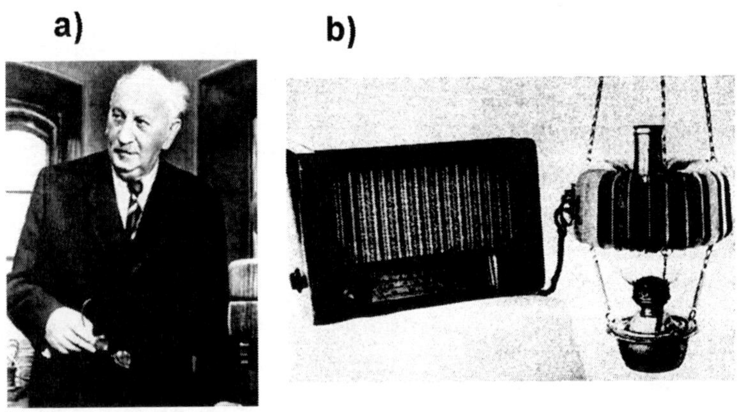

Figure 1.10 (a) Portrait of A. F. Ioffe; (b) Radio receiver powered by a thermolectric generator driven by the heat of a kerosene lamp.

solar TEGs [45], whereas textiles powered by body heat and IR solar energy can act as low temperature energy harvesters.

In a similar way, the practical uses of TECs are also wide-ranging. Starting at the 1950s, a number of TECs were made and successfully tested by Lazar Solomonovich Stil'bans (1917–1988). Their main

Table 1.2 Parameters characterizing a series of thermoelectric coolers developed during the period 1951–1954 at the Ioffe Physical-Technical Institute [44]. The last row device is a two-staged TE module. ΔT_m is the maximum cooling temperature difference and N_T is the number of thermocouples

n-leg	p-leg	ΔT_m (C)	N_T	Year
PbTe	ZnSb	10	16	1951
PbTe	BiSbTe			
PbTe	Bi_2Te_3	30		1952
PbTe	$(Bi,Sb)_2Te_3$	40		1953
PbTe:PbSe	$(Bi,Sb)_2Te_3$	60	336	1954

characteristics are summarized in Table 1.2. At the same time, a demonstration of 0°C cooling was given by H. Julian Goldsmid in 1954, using thermoelements based on Bi_2Te_3 [11, 46]. He also identified the importance of having a combination of large charge carriers mobility and effective masses along with low lattice thermal conductivities in semiconducting materials used for TE applications. Currently, TECs are commonly used for cooling electronic devices. Materials that provide efficient local cooling at temperatures below 200 K would greatly affect the electronics industry, since the performance of many semiconducting and other electronic devices is dramatically enhanced below room temperature. Indeed, Peltier coolers are the most widely used solid-state cooling devices, enabling a wide range of applications from thermal management of optoelectronics and infrared detector arrays to polymerase chain reaction instruments. They are also used in other applications that have modest cooling demands, such as camping and portable coolers, or cases in which the energy cost is not the main concern, such as military and aerospace applications. Thus, as the field of cryoelectronics and cold computing grows, the need for low temperature (100–200 K) TEMs has become more evident. Also, cellular phone technology using superconducting narrowband spectrum dividers to increase frequency band utilization is a reliable cooling technology calling for improved TEMs. In summary, while there are a considerable number of applications in the high-temperature regime, there is even greater potential device applications for refrigeration at temperatures between 80 and

400 K. The dearth of these potential applications makes the need for the investigation of entirely new materials for this regime even more acute [47].

Certainly, TE devices offer distinct advantages over other energy technologies, namely:

- they work with no moving parts, thereby minimizing vibration and noise related effects
- their overall volume is relatively small
- they exhibit very high reliability because of their solid-state construction
- they do not contain refrigerant fluids (i.e., chlorofluorocarbons), which may require periodic replenishment or may be harmful to environment
- they can be used for both heating and cooling
- precise temperature control within ± 0.1 K can be achieved using appropriate support circuitry
- they are scalable and no position dependent (as compared to photovoltaic or thermosolar devices)

Despite the numerous advantages of TE devices, their commercial use still remains quite limited [33]. One reason is that their current efficiency is excessively low for them to be economically competitive. Thus, enhancing the performance of such devices is very necessary. Another reason is that the cost of traditional TE modules per watt of cooling, heat or power generation is yet excessively high to enable the replacement of existing technologies, with the exception of a few applications in which the beneficial of being solid state outweighs cost and performance limitations [48]. Therefore, the main goals in current TE research are to improve the TE performance of suitable materials and to optimize the design of TE devices. These important issues will be fully discussed in the next sections.

1.4 Thermoelectric Efficiency

The efficiency of TE devices was early studied by Edmund Altenkirch (1880–1953) within a model framework based on the assumption

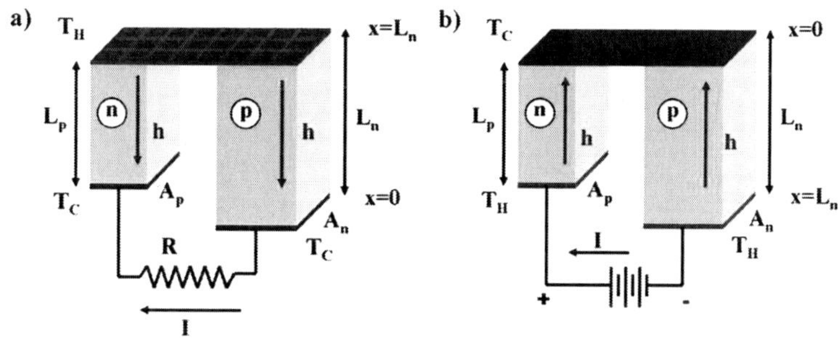

Figure 1.11 Basic design parameters for (a) a thermoelectric generator and (b) a thermoelectric cooler, both based on a n-type/p-type thermocouple.

that the transport properties of the considered materials remained essentially the same (i.e., the materials are homogeneous and the transport coefficients are temperature independent) under usual working temperature ranges. In this way, he properly derived mathematical expressions giving the maximum efficiency of a TEG in 1909 and the optimal performance of a TEC in 1911. Later on, Ioffe developed in 1949 a comprehensive theory (including both macroscopic and microscopic aspects) of TE energy conversion by semiconductor materials. In this seminal work, the efficiency for TE refrigeration and heating was explicitly made in terms of the celebrated *TE figure of merit*, which was introduced in order to characterize the TE performance of a given material. This theory was originally published by the Academy of Science of the URSS in 1950 as a classified edition and subsequently spread worldwide [7]. In what follows, we will present the main results obtained in these fundamental works.

1.4.1 *Power Factor*

Consider the two-element TEG (also referred to as thermopile or thermocouple) depicted in Fig. 1.11a. A couple formed by a n-type and a p-type thermoelements (also referred to as *legs* or *branches*) of length L_n and L_p and cross-section A_n and A_p, respectively, are connected by a conductor at the hot end (at temperature T_H), which is assumed to have negligible electrical and thermal resistances. A load resistor, with a resistance R, is connected between the cold ends of thermoelements (at temperature T_C), closing the circuit. Following Altenkirch's original approach, from now on we will assume that the thermoelements transport coefficients ρ, S, and κ are temperature independent and their values are those corresponding to the average working temperature $T_M = (T_H + T_C)/2$, that is, $\rho = \rho(T_M)$, $S = S(T_M)$, and $\kappa = \kappa(T_M)$ (**Exercise 1.8**). The temperature difference $T_H - T_C \equiv \Delta T > 0$ generates the Seebeck voltage $\Delta V_S = (S_p - S_n)\Delta T$ at the hot junction, where $S_p > 0$ and $S_n < 0$ are the Seebeck coefficients of the p-type and n-type thermoelements, respectively. The internal electrical resistance of the thermopile is

$$r = \frac{\rho_n L_n}{A_n} + \frac{\rho_p L_p}{A_p}, \tag{1.52}$$

where ρ_n and ρ_p are the legs resistivities. According to Ohm's law (Eq. (1.2)), the intensity of the electrical current through the overall circuit is

$$I = \frac{\Delta V_S}{r + R}. \tag{1.53}$$

The power delivered to the external load is the result of subtracting to the electric power due to the Seebeck effect the power lost by Joule heating due to the internal resistance, and it is given by

$$W = I \Delta V_S - r I^2 = \frac{R(\Delta V_S)^2}{(r + R)^2}. \tag{1.54}$$

By imposing the extremum condition $dW/dR = 0$, we get that maximum output power is obtained when the load resistance equals internal resistance, that is $R^* = r$. Making use of this condition in Eq. (1.54), we obtain

$$W^* = \frac{(\Delta V_S)^2}{4r} = \frac{(S_p - S_n)^2 \Delta T^2}{4(\rho_n L_n A_n^{-1} + \rho_p L_p A_p^{-1})}. \tag{1.55}$$

This expression can be conveniently simplified if one assumes that the thermoelements: (1) have the same dimensions (i.e., $L_n = L_p = L$ and $A_n = A_p = A$), and (2) they share the same transport coefficients value at the considered temperature T_M (i.e., $\rho_n = \rho_p = \rho$ and $S_p = -S_n = S$). In this case, Eq. (1.55) simplifies to

$$W^* = \frac{A}{2L} \sigma S^2 \Delta T^2, \tag{1.56}$$

where $\sigma = \sigma(T_M)$ is the electrical conductivity. The first factor in Eq. (1.56) includes the thermopile design parameters. Thus, power generation can be increased by either increasing the cross-section or decreasing the length of the legs (a quite convenient feature in order to reduce the device volume). The second factor measures the influence of the transport properties of the legs materials in terms of the product

$$P(T) = \sigma(T)S^2(T), \tag{1.57}$$

which is commonly referred to as the TEM *power factor* (usually measured in $\mu\text{Wcm}^{-1}\text{K}^{-2}$ units). Finally, the third factor in Eq. (1.56) expresses the optimal power dependence upon the temperature difference across the thermoelements.

1.4.2 *Figure of Merit*

The efficiency of a TEG is defined by the ratio $\eta = W/\dot{Q}_H$, where W is the power delivered to the external load and $\dot{Q}_H$ is the heat power (measured in W) entering the hot junction (source) to the cold junction (sink). For the thermopile depicted in Fig. 1.11a W is given by Eq. (1.54) and $\dot{Q}_H$ at the hot junction is determined by the heat loses stemming from two main contributions: the diffusive (Fourier) heat flow to the sink and the Peltier cooling effect driven by the current I (given by Eq. (1.53)) crossing the junction. This heat loss will lead to a progressive lowering of T_H, thereby reducing the temperature gradient through the device, which must be constantly fed with an external thermal energy supply at the hot junction in order to keep it running.

Making use of Eqs. (1.1) and (1.8), the heat power supplied to the p- and n-type legs can be respectively written as

$$\dot{Q}_p = A_p h + \frac{\Delta Q_P}{\Delta t} = \kappa_p(T_H)A_p(\nabla T)_{L_p} + \Pi_p(T_H)I, \tag{1.58}$$

$$\dot{Q}_n = A_n h + \frac{\Delta Q_P}{\Delta t} = \kappa_n(T_H)A_n(\nabla T)_{L_n} - \Pi_n(T_H)I, \tag{1.59}$$

where we have explicitly taken into account that the Peltier coefficient takes on negative values for n-type materials, so that $-\Pi_n$ is a positive quantity. Although the Peltier cooling effect taking place at the hot junction is naturally described in terms of the Peltier coefficients $\Pi_p(T_H)$ and $\Pi_n(T_H)$, in practice it is more practical to characterize TEMs in terms of their Seebeck coefficient values, which are easier to determine experimentally. Accordingly, it is convenient to replace Π's by S's in Eqs. (1.58)–(1.59) making use of the first Kelvin relation given by Eq. (1.14). In this way, the overall heat power, $\dot{Q}_H = \dot{Q}_p + \dot{Q}_n$, can be expressed in the form

$$\dot{Q}_H = \kappa_p A_p \left(\frac{dT}{dx}\right)_{L_p} + \kappa_n A_n \left(\frac{dT}{dx}\right)_{L_n} + (S_p - S_n) T_H I, \quad (1.60)$$

where all transport coefficients are evaluated at T_H and x is measured along the thermal gradient direction, with $x = 0$ at the sink (T_C) and $x = L_{n,p}$ at the source (T_H) junctions, respectively.

In order to explicitly determine the thermal gradient value at the hot ends we must derive the temperature profile along each leg, which is obtained by solving the heat equation

$$\frac{d}{dx}(\kappa_\alpha \nabla T) = -T \frac{dS_\alpha}{dT} j_\alpha \nabla T - \rho_\alpha j_\alpha^2, \quad (1.61)$$

respectively, where j_α measures the electrical current density (in Wm^{-2}) through each leg in the thermopile. The first term on the right hand of Eq. (1.61) describes the Thomson heat release and the second term gives the Joule heating. Since we have assumed that the transport coefficients take on constant values which are independent of both position and temperature, the Thomson contribution vanishes,[a] and Eq. (1.61) reduces to

$$\frac{d^2 T}{dx^2} = -\frac{j_\alpha^2 \rho_\alpha}{\kappa_\alpha}. \quad (1.62)$$

Making use of the boundary conditions $(dT/dx)_{L_\alpha/2} = \Delta T / L_\alpha$ and $T(x = 0) = T_C$, one obtains

$$T(x) = T_C + \frac{\Delta T}{L_\alpha} x + \frac{j_\alpha^2 \rho_\alpha}{2\kappa_\alpha} x (L_\alpha - x), \quad (1.63)$$

and using this parabolic temperature profile in Eq. (1.60) one finally gets

$$\dot{Q}_H = I T_H (S_p - S_n) + \mathcal{K} \Delta T - \frac{I^2}{2} r, \quad (1.64)$$

[a] Some recent works have considered in detail the possible role of Thomson effect in the performance of TE devices. In particular, it has been confirmed that the standard approach (assuming constant values for the transport coefficient, thereby neglecting the Thomson heat contribution) provides a reasonably accurate estimation of TE module efficiency, despite its limiting assumptions [49].

where we have used the relation $I = j_n A_n = j_p A_p$ and introduced the internal thermal conductance

$$\mathcal{K} = \frac{\kappa_p A_p}{L_p} + \frac{\kappa_n A_n}{L_n}. \tag{1.65}$$

Thus, the heat power is determined by the Peltier reversible heat release and two irreversible contributions coming from the Fourier diffusion and the Joule effect (we note that exactly one-half of the total Joule heating is detracted from the resulting heat power). Making use of Eqs. (1.54) and (1.64), we have

$$\eta = \frac{W}{\dot{Q}_H} = \frac{I\Delta V_S - rI^2}{I\frac{\Delta V_S}{\Delta T}T_H + \mathcal{K}\Delta T - \frac{I^2}{2}r}. \tag{1.66}$$

If there were no irreversible effects, this expression will reduce to the Carnot limit value $\eta_c = \Delta T/T_H$. Accordingly, from very first principles, we realize that a good TE material must have both low electrical resistivity (minimizing Joule heating) and low thermal conductivity (minimizing thermal conduction).

In order to gain additional insight on the role of the transport coefficients of the material, it is convenient to rearrange Eq. (1.66) in the form (**Exercise 1.9**)

$$\eta = \eta_c \frac{m}{1 + m + \frac{(1+m)^2}{\mathcal{Z}T_H} - \frac{\Delta T}{2T_H}}, \tag{1.67}$$

where $m \equiv R/r$ measures the ratio between the load and internal resistances and we have introduced the *thermocouple* figure of merit

$$\mathcal{Z} \equiv \frac{(S_p - S_n)^2}{\mathcal{K}r} \tag{1.68}$$

which is determined by the transport properties of the legs. For a given material, $\mathcal{Z}$ also depends on the relative dimensions of the legs. Its value is maximized when the product $\mathcal{K}r$ is minimized in Eq. (1.68) denominator, which occurs when

$$\frac{L_n A_p}{L_p A_n} = \sqrt{\frac{\rho_p \kappa_n}{\rho_n \kappa_p}}. \tag{1.69}$$

In that case, Eq. (1.68) becomes

$$\mathcal{Z} \equiv \frac{(S_p - S_n)^2}{(\sqrt{\rho_n \kappa_n} + \sqrt{\rho_p \kappa_p})^2}. \tag{1.70}$$

In selecting thermopile materials, one aims to get the largest possible value for the thermocouple parameter $\mathcal{Z}$. In general, this condition must not necessarily coincide with choosing the best p-type and n-type materials according to their individual Z_α value. At most temperatures of interest, the TE properties of the best available p-type and n-type materials are similar, so that $\mathcal{Z}$ for a thermocouple is approximately the average of Z_p and Z_n (**Exercise 1.10**). If one assumes the transport properties are exactly the same for each leg, as we did to obtain Eq. (1.56), then Eq. (1.68) reduces to the so-called *material figure of merit* (FOM)

$$Z \equiv \frac{\sigma S^2}{\kappa}, \tag{1.71}$$

which is the most commonly used TE performance metric. Accordingly, materials for TE applications are evaluated in terms of their FOM value, which is, in turn, determined by their thermal conductivities (appearing in the FOM denominator) and the power factor $P \equiv \sigma S^2$, which appears in the FOM numerator. Thus, Eq. (1.71) indicates that large FOM values require both small thermal conductivity values and large Seebeck coefficient values. Therefore, in searching for promising TEMs one must focus on materials exhibiting strong couplings between the electrical and thermal currents.

Now, lets us come back to Eq. (1.67), which expresses the efficiency as a function of the hot side temperature, the cold side temperature, and the device parameters m and $\mathcal{Z}$. As we saw in Section 1.4.1, for any given temperature gradient, the maximum output power for a given TEG is delivered when the load is matched to the generator internal resistance (i.e., $R = r$). In that case, $m = 1$, and Eq. (1.67) reads

$$\eta_R^* = \frac{\eta_C}{2}\left(1 + \frac{2}{\mathcal{Z}T_H} - \frac{\Delta T}{4T_H}\right)^{-1}. \tag{1.72}$$

Alternatively, one can keep m as a free parameter and impose the extremum condition $d\eta/dm = 0$ to Eq. (1.67) to get $m = \sqrt{1 + \mathcal{Z}T_M}$. Plugging this resistance ratio into Eq. (1.67), one obtains (**Exercise 1.11**)

$$\eta^* = \eta_C \frac{\sqrt{1 + \mathcal{Z}T_M} - 1}{\sqrt{1 + \mathcal{Z}T_M} + \frac{T_C}{T_H}}, \tag{1.73}$$

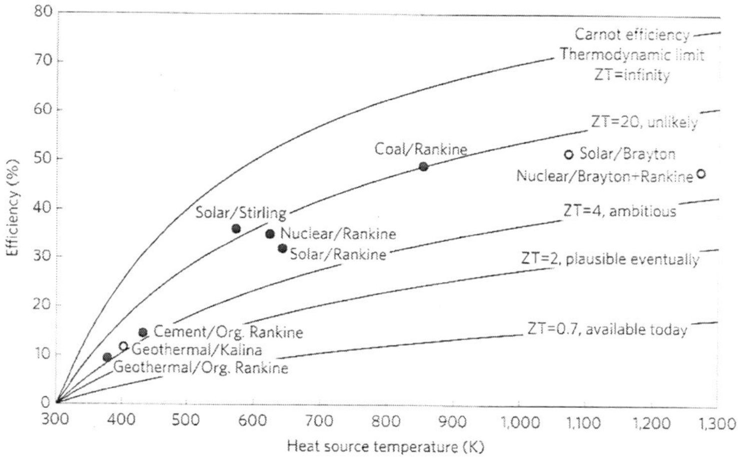

Figure 1.12 Thermoelectric efficiency curves as a function of the hot junction temperature with $T_C = 300$ K for different ZT values. The circles indicate the efficiency values for several heat engines of current use [50]. Reprinted by permission from Macmillan Publishers Ltd: *Nature Materials* (C. B. Vining, *Nature Mater.* **8**, 83), Copyright 2009.

where $\mathcal{Z}T_M$ is the *dimensionless* thermocouple FOM. Therefore, the efficiency of a TEG is governed by the TE properties of the generator materials and the temperature drop across the generator. The lowest efficiency value $\eta^* = 0$ is obtained for $\mathcal{Z}T_M = 0$, as expected. On the other hand, the temperature difference between the hot and the cold junctions sets the upper limit efficiency, which is given by the Carnot efficiency. Since Eq. (1.73) is a monotonously growing function of $\mathcal{Z}T_M$, the highest efficiency value $\eta^* = \eta_C$ can only be obtained in the limit $\mathcal{Z}T_M \to \infty$. In this way, the TEM transport properties determine how close the efficiency can be to η_C, in terms of their FOM value. As it is illustrated in Fig. 1.12, the Carnot limit is very slowly attained in practice. For instance, according to Eq. (1.73), an efficiency of approximately 8.4% can be reached by imposing a temperature difference of 200°C across good TEMs with $\mathcal{Z}T_M = 1$ when $T_C = 20$°C. Since the efficiency improves only slowly with increasing $\mathcal{Z}T_M$, in order to be competitive with conventional small-scale power generation and refrigeration devices working at mid (500–900 K) to high (1000–2000 K) temperatures, $\mathcal{Z}T_M \sim$ 2–4 values will be necessary (**Exercise 1.12**). For example, a value $\mathcal{Z}T_M = 3$ would make TE home refrigerators economically competitive with traditionally compressor-based ones [36, 50].

If we consider a TE module composed of a number, say N_T, of identical thermocouples linked electrically in series and in thermal parallel, the electrical power generated by the module will be

$$W_T = I N_T \Delta V_S - N_T r I^2 = N_T W, \tag{1.74}$$

and the heat power entering the hot junction will be

$$\dot{Q}_H^T = I T_H N_T (S_p - S_n) + N_T \mathcal{K} \Delta T - I^2 N_T r/2 = N_T \dot{Q}_H. \tag{1.75}$$

Thus, the efficiency of a TE module composed of identical thermocouples reduces to that of a single thermocouple, that is, $\eta_T \equiv W_T/\dot{Q}_H^T = W/Q_H = \eta$. Indeed, Eq. (1.66) is invariant upon a simultaneous scaling of the parameters $\Delta V_S, r$, and $\mathcal{K}$, by a constant factor N_T. This result supports the common practice of considering that $\mathcal{Z}$ is the main parameter one must consider in the design of an efficient TEG. In practice, however, it is observed that TEGs having the same $\mathcal{Z}$ value usually display different efficiency curves. This behavior has been related to the existence of temperature drops across contacts and interfaces, and also to the nonideal nature of heat sources and sinks (**Exercises 1.13 and 1.14**) [51, 52].

Furthermore, at low temperatures around 200 K, it is usually possible to tailor the p-type material for operation to a greater extent than it is possible for the n-type material (see Chapter 3), making the properties of the two legs grossly different. At even lower temperatures, as the liquid nitrogen temperature (77 K) is approached, the best thermocouples are composed of n-type BiSb alloys and a passive (i.e., $S = \sigma = 0$) superconducting leg. The superconducting thermoelement cross-section must be large enough to keep the current density below the critical value and the transport properties and dimensions of the BiSb leg are then adjusted to maximize the thermocouple $\mathcal{Z}$ value.

1.4.3 *Coefficient of Performance*

Thermoelectric coolers have been traditionally based on the Peltier effect, which describes the reversible heat transported by an electric current (Section 1.1). The cooling performance of the thermocouple shown in Fig. 1.11b can be analyzed in a way completely analogous to the efficiency study presented for TEGs in the previous section. In a TEC, electrons in the n-type leg and holes in the p-type leg both move away from the cold junction to the hot one, driven by the external battery, thus carrying heat from the cold junction to the hot one. By inspecting Fig. 1.11b we see that the heat power is now determined by the *opposite* contributions stemming from the Fourier flow and the heat flow due to the Peltier effect, now taking place at the cold junction ($x = 0$). Accordingly, the overall heat power $\dot{Q}_C = \dot{Q}_p + \dot{Q}_n$, can be expressed in the form

$$\dot{Q}_C = (S_p - S_n)IT_C - \kappa_p A_p \left(\frac{dT}{dx}\right)_{0,p} - \kappa_n A_n \left(\frac{dT}{dx}\right)_{0,n}, \quad (1.76)$$

where the temperature profile along the branches is given by Eq. (1.63). Therefore, the total heat power extracted from the cold junction is given by (**Exercise 1.15**)

$$\dot{Q}_C = IT_C(S_p - S_n) - \mathcal{K}\Delta T - \frac{I^2}{2}r, \quad (1.77)$$

where the first term on the right describes the Peltier cooling contribution, the second term describes the heat leakage from the hot junction to the cold junction due to Fourier diffusion (note that the sign of this contribution is reversed as compared to that in Eq. (1.64)), and the last term accounts for the fraction of Joule heat generated inside the legs which is conducted back to the cold junction.

The electric power consumed by the battery feeding the thermocouple includes the work *against* the Seebeck voltage and the Joule heating due to the legs resistance and is given by

$$W_C = I\Delta V_S + rI^2 \quad (1.78)$$

where we note that both contributions are additive (in contrast to their subtracting nature in Eq. (1.54)). The efficiency of a TEC is expressed in terms of the so-called *thermocouple coefficient of performance* (COP) defined as

$$\phi \equiv \frac{\dot{Q}_C}{W_C} = \frac{IT_C(S_p - S_n) - \mathcal{K}\Delta T - \frac{I^2}{2}r}{I\Delta V_S + rI^2}, \quad (1.79)$$

which relates the rate of heat extraction at the cold end $\dot{Q}_C$ to the power consumption W_C in the device. If there were no irreversible transport effects, one would find $\phi = \phi_C \equiv T_C/\Delta T$, which is the Carnot efficiency for cooling. The maximum cooling power is determined from the extremum condition $d\dot{Q}_C/dI$ on Eq. (1.77) and leads to the optimal current value $I_C^* = (S_p - S_n)T_C/r$. Plugging it into Eq. (1.79), we get

$$\phi_I^* = \frac{\frac{1}{2}\mathcal{Z}T_C^2 - \Delta T}{\mathcal{Z}T_C T_H}, \tag{1.80}$$

where $\mathcal{Z}$ is given by Eq. (1.68). On the other hand, the current leading to the maximum COP is determined by the condition $d\phi/dI = 0$, which leads to

$$I_\phi^* = \frac{(S_p - S_n)\Delta T}{r(\sqrt{1 + \mathcal{Z}T_M} - 1)}, \tag{1.81}$$

and substituting into Eq. (1.79) we obtain

$$\phi^* = \phi_C \frac{\sqrt{1 + \mathcal{Z}T_M} - \frac{T_H}{T_C}}{\sqrt{1 + \mathcal{Z}T_M} + 1}. \tag{1.82}$$

The optimized COP of a thermocouple strongly depends on the $\mathcal{Z}T_M$ of the device (Fig. 1.13). On the other hand, ϕ_C diverges in the limit $T_C \to T_H$, since no heat flows across the legs in this case and $\dot{Q}_C = 0$. Thus, in the physical temperature range $0 < T < T_H$, ϕ^* approaches ϕ_C as $\mathcal{Z}$ goes to infinity. For the sake of illustration, TE cooling devices based on Bi_2Te_3 alloys (with $ZT \simeq 1$) have an efficiency of about 10% of Carnot efficiency. The lowest possible COP value $\phi^* = 0$ determines the maximum cooling temperature difference, ΔT_m, that a TEC can reach. This maximum temperature difference is obtained by equating Eq. (1.82) to zero as $\Delta T_m = \mathcal{Z}T_C^2/2$. For the best current commercial materials, one gets $\Delta T_m \simeq 60$–70 K in single-stage devices (compare this figure with earlier values listed in Table 1.2). Further cooling can be achieved in multi-staged modules, such as those shown in Fig. 1.9, where each lower stage is larger than the upper one, hence adopting a pyramidal shape. This geometry arises from the fact that the increased cooling power required for the higher temperature stages

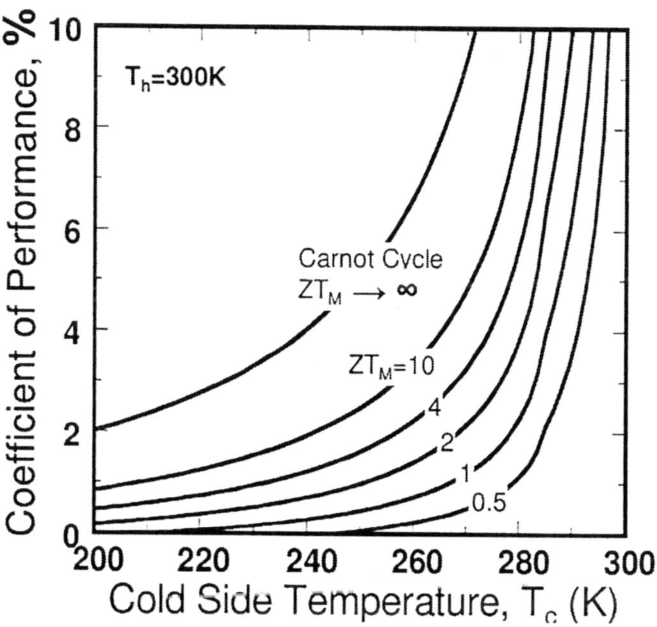

Figure 1.13 Coefficient of performance of thermoelectric coolers with different values of $\mathcal{Z}T_M$. The hot junction is at $T_H = 300$ K.

of a multi-staged module is usually achieved by increasing N_T rather than by reducing the thermocouple height L. Indeed, in principle each stage can produce additional cooling to lower temperatures, regardless of the $\mathcal{Z}$ value of the TEMs in the stage. In practice, however, the unavoidable thermal losses and complications of fabrication limit the performance of multi-stage devices. In Table 1.3, we list the maximum cooling temperature differences for multi-stage TECs as reported by commercial manufacturer's brochures (as May 2014). The transport properties across the thermocouple legs can be manipulated to improve cooling performance, although this approach has proved less effective in reducing ΔT_m than the multi-stage one. Recently, the very possibility of designing TEC based on the Thomson effect instead of the Peltier effect has been analyzed [53].

Table 1.3 Maximum cooling temperature difference ΔT_m with $T_H = 298$ K for different multi-staged thermoelectric coolers available in the market distributed by Huimao Cooling Equipment Co. Ltd. and RMT Ltd., respectively

# - stages	ΔT_m (°C)	ΔT_m (°C)
1	-	70–72
2	75–106	94–104
3	90–109	107–122
4	100–114	117–125
5	118	-
6	128–131	-

1.4.4 *Compatibility Factor*

According to Eq. (1.73), high efficiency TEGs require materials having large FOM values working subjected to large temperature gradients. Now, we know that the TE transport coefficients determining the ZT value will generally depend on temperature, and such a dependence becomes more significant as the temperature difference drop across the thermoelements is progressively increased in order to reduce the T_C/T_H ratio in the Eq. (1.73) denominator. As a consequence, it usually occurs that a material, well performing within a relatively narrow temperature range, gets worse when the working temperature interval of interest is widened, so that it is no longer convenient to use just one type of material in the TE device.

To cope with this technical problem, novel designs based on the concept of segmentation have been introduced. Thus, in the so-called *segmented* devices two or more different materials are joined together in each TEG leg in such a way that a material with high TE efficiency at high temperatures (located close to the hot junction) is combined with a different material (located close to the cold end) with high TE efficiency at lower temperatures. According to this design (Fig. 1.14a), both materials are operating in their most efficient temperature ranges, thereby improving the resulting overall performance of the TE device. In a segmented TEG, all segments are thermally and electrically in series. An alternative design is provided by the so-called *cascaded* devices,

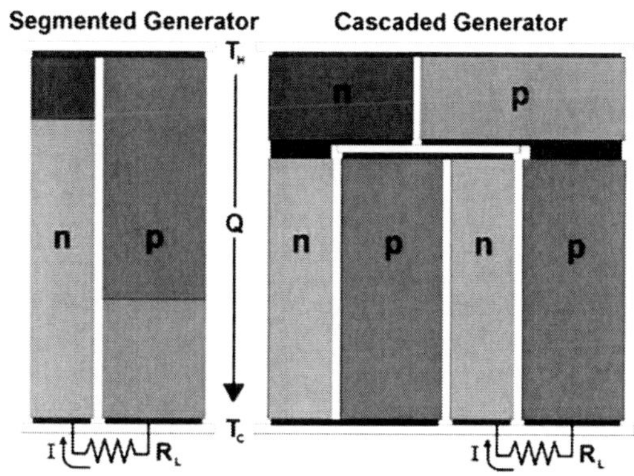

Figure 1.14 Schematic diagram comparing segmented and cascaded designs in thermoelectric generators [54]. Reprinted with kind permission from G. Jeffrey Snyder, *Appl. Phys. Lett.*, **84**, 2436 (2004), Copyright 2004, American Institute of Physics.

which contain an independent electrical circuit for each segmented stage (Fig. 1.14b). Though cascaded arrangements may render a better performance, they are generally much more difficult to implement than segmented ones.

As an illustrative example of the segmented design, we can mention a TE generator that has been successfully used on several NASA space missions. This TEG is composed of a p-type leg based on the alloy $(AgSbTe_2)_{0.15}(GeTe)_{0.85}$ (a representative of the so-called TAGS material, see Section 3.4.2), which must be maintained below 525°C for long term applications, while the other leg is composed of a PbTe n-type element, which operates efficiently at 600°C. In order to be able to achieve a 600°C working temperature for the device as a whole, a segment of p-type SnTe material is added to the TAGS p-type leg to operate between 525°C and 600°C.

In the previous sections, we have derived closed expressions for the TEG TE efficiency and TEC coefficient of performance under the assumption that the legs' transport coefficients are independent of both position and temperature. In particular, this assumption was explicitly used when deriving Eqs. (1.66) and (1.79) making use of the heat power flux related to the temperature profile given by Eq. (1.63). In turn, this temperature profile was obtained by solving the heat equation (1.61). This equation cannot be so easily solved in the case of inhomogeneous materials with a temperature dependent Seebeck coefficient. Therefore, the TE efficiency calculation becomes rather involved in the more general case and usually requires finite elements methods using averaged material parameters [304].

Quite remarkably, a new concept has been recently introduced by Jeffrey Snyder that allows one to also derive closed analytical expressions for the efficiency formula in the study of inhomogeneous materials [54, 55]. Let us consider that the legs of a TEG can be regarded as composed of a series of extremely thin layers of resistivity ρ, Seebeck coefficient S, thermal conductivity κ, cross-section A, and width $\Delta L \ll 1$. Each inhomogeneous leg in the thermocouple can then be treated as a continuous series of infinitesimal homogeneous layers in thermal contact to each other operating locally with a given FOM value each. The power delivered by an arbitrary layer when subjected to a local temperature gradient ΔT is given by (see Eq. (1.54))

$$\Delta W = IS\Delta T - I^2 \frac{\rho \Delta L}{A} \tag{1.83}$$

and the heat power entering the layer hotter surface is (see Eq. (1.58))

$$\dot{Q}_H = IS_H T_H + \kappa_H A \frac{\Delta T}{\Delta L} \tag{1.84}$$

where $S_H \equiv S(T_H)$, and $\kappa_H \equiv \kappa(T_H)$. The TE efficiency of an arbitrary elemental layer under steady-state conditions is given by

$$\Delta \eta = \frac{\Delta W}{\dot{Q}_H} = \frac{jS\Delta T - j^2 \rho \Delta L}{jS_H T_H + \kappa_H \frac{\Delta T}{\Delta L}}, \tag{1.85}$$

where $j = I/A$ is the electrical current density. Dividing Eq. (1.85) by the Carnot efficiency η_c and taking the limit $\Delta T \to 0$, we obtain the *reduced* efficiency of an elemental segment of the thermoelement as

$$\eta_r \equiv \lim_{\Delta T \to 0} \frac{\Delta \eta}{\eta_c} = \frac{jS - j^2 \rho (\nabla T)^{-1}}{jS_H + \kappa_H \frac{\nabla T}{T_H}}. \tag{1.86}$$

At this stage, the explicit knowledge of the position variable dependence in the transport coefficients appearing in Eq. (1.86) will be eliminated, hence allowing for a more general treatment of TE efficiency in inhomogeneous systems. To this end, it is convenient to introduce the so-called *relative current* density, defined as the ratio of the electrical current density to the heat flux due to thermal diffusion,

$$u \equiv \frac{\mathbf{j} \cdot \mathbf{j}}{\mathbf{h} \cdot \mathbf{j}} = -\frac{j^2}{\kappa \nabla T . \mathbf{j}} \tag{1.87}$$

(measured in V^{-1}), where we have used Eq. (1.1) in the right-hand expression. The relative current density is positive when the thermal, **h**, and electrical, **j**, currents are both parallel, whereas it is negative when they run antiparallel. For the sake of clarity, we shall consider parallel currents in what follows (**Exercise 1.16**), so that Eq. (1.87) can be expressed as $j = u\kappa\nabla T$. Making use of this expression into Eq. (1.86), we have

$$\eta_r = \frac{w(1-w)}{w + \frac{1}{ZT}}, \tag{1.88}$$

where $w \equiv uS/Z$, and we have reasonably assumed that the transport coefficients are temperature independent through a given layer (i.e., $S_H = S$, $\kappa_H = \kappa$, and $T_H = T$). Thus, the reduced efficiency can be expressed as a function of the variables u, S, Z, and T, whereas explicit reference to the spatial geometric parameters A and ΔL has been eliminated. In terms of the variable u, the reduced efficiency increases from zero ($u = 0$) to a maximum value and then decreases to zero again at $u = Z/S$. Accordingly, we will consider the value of the reduced current which maximizes η_r at a given T value. From the condition $d\eta_r/dw = 0$, one has

$$u_+^* = \frac{\sqrt{1+ZT}-1}{ST} \equiv s, \tag{1.89}$$

where the subscript '+' here indicates that the currents **h** and **j** are parallel. The just obtained u_+^* value maximizing the reduced efficiency is defined as the *compatibility factor* s, since materials with dissimilar s values cannot be efficiently combined in a single device. Indeed, from Eq. (1.89) it is clear that s is, like Z, a temperature dependent material property, which depends on the transport coefficients $\sigma(T)$, $S(T)$, and $\kappa(T)$. Likewise, the value of s will change discontinuously between different materials composing a segmented leg. Accordingly, the optimum u_+^* current of one segment can be significantly different from the optimum value for the other, so that there may be no suitable current value where both parts are operating efficiently. This is the physical basis for TE compatibility. For instance, if the compatibly factors differ by a factor of two or more, a given u cannot be suitable for both materials and segmentation will not be efficient. Thus, the closer the compatibility factor values are for two materials, the higher the combined efficiency will be when they are used in a segmented leg. Therefore, the ultimate goal in TEG design is to select high FOM materials that have similar compatibility factors as well.

In terms of the relative current density u the heat equation (1.61) can be expressed in the form

$$\frac{du^{-1}}{dx} = -T\frac{dS_\alpha}{dT}\nabla T - u\rho_\alpha\kappa_\alpha\nabla T. \tag{1.90}$$

Now, we eliminate the spatial variable x dependence through the relation

$$\frac{du^{-1}}{dx} = \frac{du^{-1}}{dT}\frac{dT}{dx} = \frac{du^{-1}}{dT}\nabla T \tag{1.91}$$

so that Eq. (1.90) reads,

$$\frac{du^{-1}}{dT} + T\frac{dS_\alpha}{dT} + u\kappa_\alpha\rho_\alpha = 0, \tag{1.92}$$

which can be further simplified by introducing the auxiliary function

$$\Phi_\alpha \equiv u^{-1} + S_\alpha T \tag{1.93}$$

to obtain

$$\frac{d\Phi_\alpha}{dT} - S_\alpha + u\kappa_\alpha\rho_\alpha = 0. \tag{1.94}$$

Making explicit use of the FOM given by Eq. (1.71), the heat equation (1.94) can be expressed in the form

$$\frac{uS_\alpha}{Z_\alpha} = w_\alpha = 1 - \frac{\Phi'_\alpha}{S_\alpha}, \tag{1.95}$$

where the prime indicates derivation with respect to the temperature. Plugging Eq. (1.95) into Eq. (1.88), and taking into account Eq. (1.93), we obtain a very compact expression for the elemental reduced efficiency in terms of the function Φ_α alone, namely, (**Exercise 1.17**)

$$\eta_r = T\frac{d(\ln\Phi_\alpha)}{dT}. \tag{1.96}$$

Finally, from the knowledge of this elemental reduced efficiency, the overall efficiency of the finite leg can be obtained from the expression (we drop the subscript α for the sake of clarity thereafter) [56],

$$\eta = 1 - \exp\left(-\int_{T_C}^{T_H} \frac{\eta_r}{T} dT\right), \tag{1.97}$$

so that, making use of Eq. (1.96) in Eq. (1.97), one finally obtains

$$\eta = \frac{\Delta\Phi}{\Phi_H} = 1 - \frac{S_C T_C + u_C^{-1}}{S_H T_H + u_H^{-1}}, \tag{1.98}$$

where $\Phi_H \equiv \Phi(T_H)$, $\Phi_C \equiv \Phi(T_C)$, and $\Delta\Phi \equiv \Phi_H - \Phi_C$. We note that function Φ plays in Eq. (1.98) a role similar to that played by the temperature in the Carnot's efficiency formula $\eta_c = \Delta T / T_H$.

According to Eq. (1.98), the efficiency of a thermoelement can be obtained from properties evaluated at the junctions, which play a role similar to that played by the initial and final states in usual thermodynamic state functions. Thus, the auxiliary function defined by Eq. (1.93) can be properly regarded as a *thermoelectric potential*, and the TE efficiency given by Eq. (1.98) becomes an intensive thermodynamic quantity (**Exercise 1.18**). Under the optimal current condition $u = s$, we can use Eq. (1.89) to express Eq. (1.93) as $\Phi = s^{-1}\sqrt{1 + ZT}$, so that Eq. (1.98) reads

$$\eta = 1 - \frac{s_H}{s_C}\frac{\sqrt{1 + (ZT)_C}}{\sqrt{1 + (ZT)_H}}. \tag{1.99}$$

Therefore, both ZT and s must be simultaneously optimized to obtain a large η value. The TE compatibility of several materials of current technological interest has been reviewed during the last decade [54]. For the sake of illustration, let us consider the case of the TAGS-SnTe based segmented TEG previously mentioned. By inspecting the FOM temperature dependence curves shown in Fig. 1.15b it appears that p-type PbTe would be a better choice than SnTe within the 525–600 °C temperature interval. However, we see that the compatibility factor values shown in Fig. 1.15d for p-type PbTe drop much lower than those of SnTe in the same temperature interval, and segmentation actually results in a decrease in efficiency (see Table 1.4). On the contrary, SnTe has a closer compatibility factor to TAGS material, and thus produces a higher efficiency

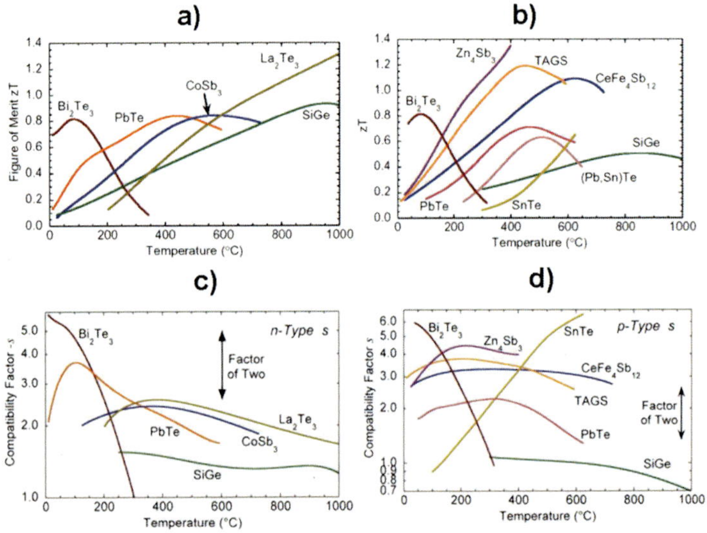

Figure 1.15 Temperature variation of the figure of merit and the compatibility factor for several n-type (a) and (c), and p-type (b) and (d) materials of current interest in thermoelectric research [54]. Reprinted with kind permission from G. Jeffrey Snyder, *Appl. Phys. Lett.*, **84**, 2436 (2004), Copyright 2004, American Institute of Physics.

Table 1.4 Maximum efficiencies for segmented legs in TEGs with $T_C = 100$ C. u^* is the relative current density that gives the maximum efficiency η^*. T_I is the temperature at the interface joining both materials

p-leg material	η^* (%)	u^* (V^{-1})	T_I (C)	T_H (C)
TAGS/PbTe	10.33	2.33	525	600
TAGS/SnTe	11.09	2.84	525	600
TAGS/$Ce_{0.85}Fe_{3.5}Co_{0.5}Sb_{12}$	11.87	2.94	525	600
TAGS/$Ce_{0.85}Fe_{3.5}Co_{0.5}Sb_{12}$	13.56	2.88	525	700

Source: Adapted from Ref. [54].

segmented generator, despite its lower ZT compared to PbTe. An even better material to segment with TAGS are p-type filled skutterudites such as $Ce_{0.85}Fe_{3.5}Co_{0.5}Sb_{12}$ (see Section 3.5.2), which produce more than twice increase in efficiency as compared to SnTe (see Table 1.4).

It can be readily checked that, when the transport properties are constant with respect to the temperature, the expressions obtained making use of the TE potential approach coincide with those derived within the constant transport coefficients approach described in the previous sections (**Exercise 1.19**). On the other hand, the expressions just obtained for the TEG case can be straightforwardly extended to the TEC study by properly replacing Eq. (1.97) by the equivalent expression [53],

$$\phi^{-1} = \exp\left(\int_{T_C}^{T_H} \phi_r^{-1} \frac{dT}{T}\right) - 1, \tag{1.100}$$

where ϕ_r is the local reduced coefficient of performance given by

$$\phi_r = \frac{w + \frac{1}{ZT}}{w(1-w)}, \tag{1.101}$$

which can be optimized through the condition $d\phi_r/dw$ to lead to the reduced current density[a]

$$u_C^* = -\frac{\sqrt{1+ZT}+1}{ST} \equiv s_C, \tag{1.102}$$

defining the TE compatibility factor for coolers s_C [53]. Plugging Eq. (1.102) into Eq. (1.101), we get the optimal reduced coefficient of performance given by

$$\phi_r^* = \frac{\sqrt{1+ZT}-1}{\sqrt{1+ZT}+1}. \tag{1.103}$$

It has been recently shown that if one considers materials with a constant Z value but allowing $\sigma(T)$, $S(T)$, and $\kappa(T)$ to vary with temperature, one can obtain a substantial improvement in cooling. In fact, in that case one has (**Exercise 1.20**)

$$\frac{1}{\phi^*} = \left(\frac{\chi_H}{\chi_C}\right)^2 \exp\left(\frac{2(\chi_H - \chi_C)}{\chi_H \chi_C}\right) - 1, \tag{1.104}$$

where $\chi_i \equiv \sqrt{1+ZT_i} - 1$. At variance with the case of constant transport coefficients, where ϕ can take on negative values over certain temperature intervals, one realizes that Eq. (1.104) guarantees that ϕ^* is always greater than zero, thereby rendering a better cooling operation.

---

[a]Note that in cooling devices **h** and **j** run in opposite sense, so that $u < 0$.

1.5 Thermoelectric Materials Characterization

The characterization of TEMs requires accurate enough measurements of the transport coefficients $\sigma(T)$, $S(T)$, and $\kappa(T)$ determining their corresponding ZT curves. Although the mutual relations among different transport coefficients in the ZT expression make the design of good TEMs a relatively complex engineering task, this does not imply that the resulting $ZT(T)$ curve must be particularly complex at all. In fact, the $ZT(T)$ curve is characterized by a relatively smooth topology, typically vanishing at very low temperatures, then uniformly increasing as the temperature is progressively increased, until it reaches a relatively broad peak, then slowly decreasing to zero as the temperature is further increased (see Fig. 1.15a–b). The location of the peak significantly differs among diverse materials, which determines their respective potential for low-, mid-, or high-temperature TE applications. Thus, Bi_3Te_2 based alloys are the materials of choice for low-temperature applications, whereas SiGe alloys are commonly used for high-temperature ones.

To properly determine a $ZT(T)$ curve, it is highly desirable to be able to measure all the involved transport coefficients simultaneously on a single sample, in order to avoid side effects arising from spatial inhomogeneities or material changes due to thermal cycling. The electrical conductivity and Seebeck coefficients can be determined with high accuracy and the obtained results generally show consistency between the measured values obtained by different teams. However, some inconsistencies arise in the ZT values obtained in different laboratories for the same materials, which are mainly attributed to differences in the observed thermal conductivities of the corresponding samples, since accurate enough $\kappa(T)$ curves are more difficult to obtain. Indeed, thermal conductivity is generally determined from the expression $\kappa = D_T C_p d$, where D_T is the thermal diffusivity (measured in mm^2s^{-1}), C_p is the heat capacity (measured in $Jg^{-1}K^{-1}$), and d is the density (in gcm^{-3}). Therefore, the measurement of ZT requires six distinct measurements, each of them implying a sizable uncertainty, leading to a total uncertainty which can be expressed as

$$\frac{\Delta(ZT)}{ZT} = 2\frac{\Delta S}{S} + \frac{\Delta\sigma}{\sigma} + \frac{\Delta C_p}{C_p} + \frac{\Delta D_T}{D_T} + \frac{\Delta d}{d} + \frac{\Delta T}{T}. \quad (1.105)$$

As the uncertainty in each of these measurements at high temperature is frequently as large as 10–20%, the total uncertainty on ZT usually amounts >50% [36].

For instance, early measurements have systematically overestimated the thermal conductivity, which has led to an underestimate of ZT. This overestimation of thermal conductivity can be traced back to the unavailability of certain data or the technique used for high temperature κ measurement. Today the most used technique to this end is the flash diffusivity method as opposed to the steady-state method because of inaccuracies resulting from radiation losses during measurements. Thus, the thermal conductivity determined by the steady-state method is significantly higher than that measured using the diffusivity technique.

Certainly, the best way to avoid the uncertainties prescribed by Eq. (1.105), specially those regarding the thermal conductivity measurement, is to determine the FOM from a single, direct measurement. Indeed, although it is quite common to obtain the FOM by measuring the three transport coefficients σ, S, and κ, separately, a method was introduced some time ago which allows one to obtain a quantitative evaluation of ZT in a straightforward way. This method is referred to as the *Harman* method [13, 58], and its physical principle consists in employing the Peltier effect to establish a temperature gradient along the sample, as it is sketched in Fig. 1.16. The material to be characterized is subjected to a direct current I, driven by an external battery, so that the cold end of the sample gets colder. To this end, the battery polarity is chosen in such a way that the resulting electrical and thermal currents will respectively flow in opposite sense to each other. Neglecting irreversible Joule heating effects and assuming that there are no heat losses (adiabatic conditions), the stationary state is attained when the Peltier heat power driven by the electrical current equals the heat power delivered by the thermal gradient $\Delta T = T_H - T_C$. Thus, making use of Eqs. (1.1) and (1.8), we have

$$\Delta T = \frac{\Pi I \Delta L}{\kappa A}, \quad (1.106)$$

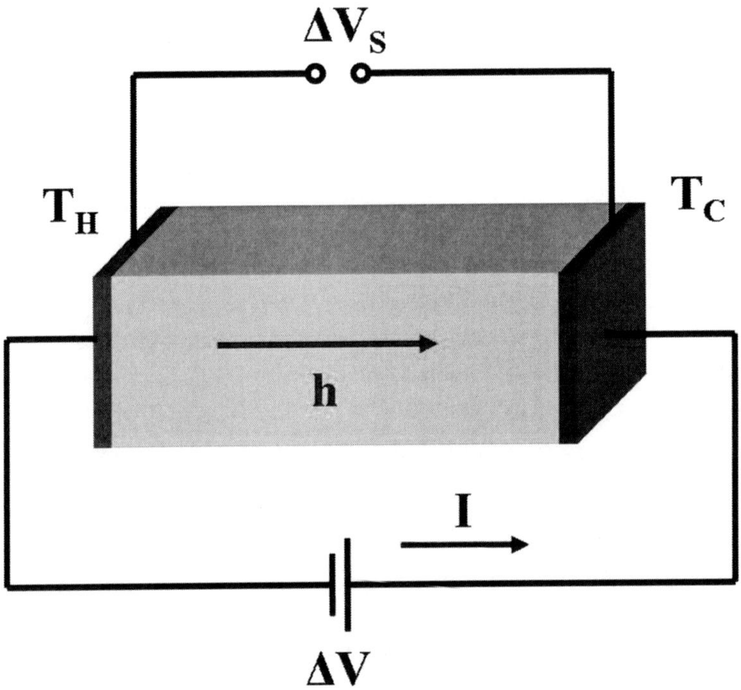

Figure 1.16 Sketch showing the Harman's method experimental setup.

where Π is the Peltier coefficient, κ is the thermal conductivity, A is the cross-section, and L measures the sample length. At the same time, the temperature difference between the contacts gives rise to the TE potential drop $\Delta V_S = S\Delta T$, where S is the Seebeck coefficient. The total potential difference across the sample then reads $\Delta V_T = \Delta V + \Delta V_S$. Making use of the first Kelvin relation and the Ohm's law, respectively given by Eqs. (1.2) and (1.14), into Eq. (1.106), we can write

$$\Delta V_T = \Delta V \left(1 + \frac{S^2 T}{\kappa \rho}\right) = \Delta V (1 + ZT). \tag{1.107}$$

Therefore, the FOM can be directly obtained as the ratio of two voltage measurements through the formula

$$ZT = \frac{\Delta V_S}{\Delta V}. \tag{1.108}$$

Albeit its simple appearance, the application of Harman method is severely limited by the practical difficulty in experimentally achieving perfect adiabatic conditions. Accordingly, alternative TEM characterization approaches based on the simultaneous measuring of the material transport properties are being progressively refined by experimental researchers nowadays [59, 60].

In fact, besides the inaccuracy related to the thermal conductivity coefficient measurement itself, current TE research is particularly interested in properly disclosing the lattice contribution to the thermal conductivity, since the control of this magnitude is one of the most important ways to optimize the overall efficiency in TEMs (see Section 2.2.1). As a first approximation, the lattice contribution can be estimated from the knowledge of the total thermal conductivity making use of Eq. (1.50). Nevertheless, this procedure relies on a number of assumptions, namely, (1) the validity of the WFL within the considered temperature range, (2) the constancy of the Lorenz function L within such temperature interval, and (3) the knowledge of the L value itself, which is difficult to measure, so that it is usually tentatively assumed to lie between L_0 and L_S for most materials. Notwithstanding this, experience demonstrates that Eq. (1.50) works reasonably well as far as the charge carrier and lattice contributions considerably differ in magnitude. However, when κ_l and κ_e become comparable to each other, one must look for other methods to differentiate between the two contributions.

A suitable experimental assessment on the WFL validity over a given temperature range can be obtained from the study of the magnitude $\kappa(T)/\sigma(T) = T\,L(T) + B(T)$, where the Lorenz function is given by Eq. (1.51) and $B(T) \equiv \kappa_l(T)/\sigma(T)$ accounts for the phonon contribution to the heat transport. For instance, a study of the temperature variation of the measured κ/σ ratio in several intermetallic compounds showed that the experimental data may be fitted by a linear temperature dependence of the form $\kappa/\sigma = LT + B_0$, over the temperature range 350–800 K, thereby confirming the constancy of the Lorenz number, which otherwise differs from both L_0 and L_S theoretical values [21, 61]. In a similar way, room temperature L values larger than L_0 were experimentally reported for other complex metallic alloys (see Section 5.1.4), suggesting the convenience of replacing Eq. (1.49) by a slightly modified expression of the form

$$\kappa_e(T) = (1 + \varepsilon)L_0 T \sigma(T), \tag{1.109}$$

in order to accurately describe experimental data, where the enhancement parameter ε is related to the structural complexity of the underlying lattice [62].

A direct measurement of the Lorenz function value can be experimentally done by utilizing a transverse magnetic field in order to suppress the charge carrier contribution to the thermal conductivity. In this way, the Lorenz number value has been recently measured for two compounds of interest in TE research. The reported values, $L = 2.16 \times 10^{-8}$ V^{2}K^{-2} and $L = 2.33 \times 10^{-8}$ V^{2}K^{-2} at 100 K , are within the interval of commonly used values, being relatively close to the Sommerfeld's value $L_0 = 2.44 \times 10^{-8}$ V^{2}K^{-2}, derived within the framework of the free-electron model for metallic systems [63]. One of the limitations of this technique, however, is the requirement of high mobility values for the considered sample charge carriers, along with the necessity of applying intense magnetic fields. Another way of experimentally determining the Lorenz number, not subjected to these requirements, is through the controlled introduction of impurities in the sample, thereby inducing a correlated change in the electrical conductivity and the κ_e contribution, from which one can determine κ_l [64].

1.6 Industrial Requirements

It is well known that the measured ZT value of a TE device is smaller than theoretically expected because in actual devices there are many parasitic losses (some of them stemming from packing, soldering, or electrical circuitry-related effects) that lower the efficiency below the predicted values as determined from their composing materials ZT values. Accordingly, efficient TE energy conversion imposes a series of constraints, and it requires a good average TE efficiency of the involved TEMs over a wide temperature range. For example, according to Eq. (1.73) a $ZT = 2$ would result in $\eta = 28.7\%$ when the heat sink is at room temperature and the heat source is at 1300 K. In practice, to achieve this efficiency figure one would need both p-type and n-type materials with an *average* $ZT = 2$ value each over the entire 1000 K working temperature interval (not merely a $ZT = 2$ peak somewhere in between T_C and T_H). Nevertheless, large ZT values are obtained over pretty narrow temperature intervals only (see Fig. 1.15a–b), which results in significantly lower efficiency values than expected in the best possible temperature conditions, even when segmented designs are fully implemented (see Section 1.4.4).

On the other hand, some reflection on the conditions and assumptions under which Eqs. (1.73), (1.82), (1.99), and (1.103) were derived reveals that there are several factors playing a role in the device TE efficiency that were not explicitly taken into account. For instance, in obtaining the aforementioned expressions, electrical and thermal resistances between the contacts and the electrodes are neglected. However, making good contacts is challenging for several reasons. On the one hand, TE devices usually operate over relatively broad temperature ranges, naturally leading to the presence of thermally induced mechanical stresses, the onset of atomic diffusion processes, and the activation of chemical reactions at the interfaces. Accordingly, the materials used as electrodes should have thermal expansion coefficients matching those of leg's TEMs and should also be chemically inert and stable enough over the lifetime operation of the TE module. For instance, it has been reported that the temperature dependence of the coefficient of thermal expansion does not significantly alter the ZT value of TE devices based on $Ba_8Ga_{16}Ge_{30}$ clathrates (Section 3.5.3) from room temperature up to 1173 K [65].

Furthermore, since in order to yield large FOM values, the TEMs composing the device must have high electrical conductivity values, very low contact resistances are required between these materials and the electrodes, which limits the choice for possible electrode materials.

TE device efficiency measurements require quantifying the heat inputs and the electrical output. If the transport properties were not dependent on temperature the maximum efficiency can be calculated from Eq. (1.67) and occurs when the external load is matched with the internal resistance by a factor $\sqrt{1 + \mathcal{Z}T_M}$. In reality, this load matching is best achieved by using a programmable current source running in reverse to the current direction driven by the temperature gradient. The most difficult part of the efficiency measurements is in precisely determining the heat balance, as the electrodes can lead to additional thermal leakage. Thus, careful calibration of the various heat losses is essential, and, by all indications, the TE community presently lacks a standard in efficiency control [36]. To this end, an important issue for TE community to address is to establish standard TE measurement protocols in order to properly assess the *reproducibility* of the transport coefficients temperature dependence curves and the resulting reported FOM values. Today, this seems to be a key factor for the future development of a growing TE device-based industry. In fact, the wide distribution of transport properties data reported in the literature for similar (or even identical) materials is likely due to differences incurred by the various fabrication and manufacturing processes used to prepare them. Thus, establishing reproducible synthesis parameters for the growth of the samples would pave the way for possible large-scale production of high-quality TEMs.

To conclude, some words are in order regarding the main criteria guiding the choice of materials for TE applications, which, in addition to exhibiting common large ZT values for both n-type and p-type legs over a broad temperature range, as well as comparable compatibility factors, should include thermal and chemical stability, reduced heat losses, durability, nontoxicity,[a] reliability, longevity, scalability, easy processing requirements, availability, and a small cost-efficiency ratio.

Since the seminal work by Ioffe, degenerated semiconductors or semimetals have traditionally been considered as the most appropriate materials for TE applications, because metals exhibit very small Seebeck coefficients and too large thermal conductivity values, hence yielding very small FOM values. In fact, since the lattice contribution to the thermal conductivity is almost negligible as compared to the electronic one in metals at room temperature

---

[a]Being made of nontoxic elements does not necessarily make a compound nontoxic. For instance, Zn_3P_2 is made of nontoxic elements but itself is highly toxic (rat poison). Conversely, being made of toxic elements does not necessarily make a compound toxic. Rather it is the actual chemical form and bioavailability that makes it so. Thus, the lead in the PbS compound of TE interest (see Section 3.3.4) is extremely strongly bound, not bioavailable and therefore nontoxic [66].

(i.e., $\kappa_l \ll \kappa_e$), making use of the WFL Eq. (1.71) reduces to $ZT \simeq S^2/L_0$. By considering $|S| = 10\ \mu VK^{-1}$ as a representative figure for metallic alloys at room temperature, typical values of $ZT \simeq 10^{-3}$ are then obtained. On the other hand, inorganic insulators simply exhibit too low electrical conductivity values to be of any practical interest ($\sigma \to 0$ in Eq. (1.71)), though they generally exhibit large Seebeck coefficient values and relatively low thermal conductivities. For instance, typical room temperature transport coefficient values for ionic crystals are $\sigma = 10^{-7} - 10^{-6}\ \Omega^{-1}cm^{-1}$, $S = 400 - 800$ μVK^{-1}, and $\kappa = 2 - 10\ Wm^{-1}K^{-1}$ (depending on doping conditions) [30, 67], yielding an average FOM value of about $ZT \simeq 10^{-9}$.

The case of polymers, which exhibit very low thermal conductivity values within the range $\kappa \simeq 0.1$–$0.5\ Wm^{-1}K^{-1}$ at room temperature, deserve an special attention (see Section 6.2.1). Indeed, these figures are more than one order of magnitude smaller than those usually measured in inorganic semiconductors currently used in TE devices. Quite interestingly, during the last two decades several classes of polymers, such as polyaniline, polypyrrole, or polythiophene, were found to have high electrical conductivities when chemically doped in a proper way. On the other hand, Seebeck coefficient values within the range from $+5$ to $+25$ μVK^{-1} have also been reported in the literature for polyacetylene, polyanilines, polypyrroles, and polythiophenes, suggesting their possible consideration as potential TEMs. The best conjugated polymer to this end up to now is polyacetylene ($ZT \simeq 0.6$ at room temperature) due to its good electrical conductivity and relatively high Seebeck coefficient. Nevertheless, it has poor stability in the (required) doped state, even in an inert atmosphere. In contrast, semiconductors can be doped with excess electrons or holes, and thus can have large negative or positive values of S, respectively. Presently, TEMs for current generation at moderate temperature gradients applications are based on PbTe. The most commonly used TEM for refrigeration in the temperature range from -120 to $230^{\circ}C$ is the alloy, $(Bi,Sb)_2(Te,Se)_3$, usually referred to as "bismuth telluride." In Chapters 3 and 4, we will present a detailed account on the most promising bulk materials currently being considered in the field of TE research.

1.7 Exercises

(1) Making use of Eqs. (1.1)–(1.4) derive the second term of Eq. (1.5).

(2) A Peltier cross made of copper and iron wires is initially in thermal equilibrium with the environment at 0°C. Determine the Seebeck voltage measured by the voltmeter when the battery is switched on and a current of 1 A flows through the circuit during 41 s. Assume there are not heat losses and consider the junction's volume is $V = 1$ mm^3 and it is made of copper ($\rho_M = 8.96$ gcm^3; $C_P = 0.093$ cal g^{-1}°C^{-1}). $\Pi_{\text{Cu,Fe}} =$ 2.3 mV at $T = 300$ K.

(3) In differential form the spatial and temporal rate variation of Thomson heat through a conductor can be expressed as

$$\frac{d^2 Q_T}{dxdt} = \tau I \nabla T. \tag{1.110}$$

Making use of suitable expressions introduced in Section 1.1 show that Eq. (1.110) can be rewritten in the form

$$\frac{d^2 Q_T}{dxdt} = -\frac{dS^{-1}}{dT} E_S\, h_P,$$

where $E_S = \frac{dV_S}{dx}$ is the Seebeck electric field and $h_P = \frac{dQ_P}{dt}$ is the Peltier heat current.

(4) Show that the TE transport matrix introduced in Eq. (1.23) can be expressed in a symmetric form.

(5) Estimate the phonon mean free path for the $Ba_8Ga_{16}Ge_{30}$ clathrate (see Section 3.5.3) at room temperature making use of the following physical data: $\kappa_l = 1.0$ Wm^{-1}K^{-1}, $v_l = 4500$ ms^{-1}, $v_t = 2690$ ms^{-1}, $C_p = 0.3065$ Jg^{-1}K^{-1}, and $d = 5.59$ gcm^{-3} [68].

(6) Derive the relationship given by Eq. (1.33).

(7) Show that for metals obeying the Wiedemann–Franz law Eq. (1.23) can be expressed in terms of a symmetric transport matrix.

(8) In the so-called standard integral-averaged model, the Seebeck coefficient value of a thermoelement is given by

$$\bar{S} = \frac{1}{\Delta T} \int_{T_C}^{T_H} S(T) dT,$$

where a functional form of $S(T)$ is assumed to be known. Show that in the case of the linear temperature dependence given by Eq. (2.51) one has $\bar{S} = S(T_M)$.

(9) Making use of Eq. (1.53) show that Eq. (1.66) can be written in the form given by Eq. (1.67).

(10) Consider a thermopile whose branches are made of elemental bismuth and antimony crystals. Making use of the data listed in Table 3.1 into Eqs. (1.70) and (1.71) determine the figure of merit of each branch and the resulting FOM for the thermopile as a whole. Comment on the obtained results.

(11) Derive Eq. (1.73).

(12) Making use of Eq. (1.73), determine the efficiency of a TEG with $ZT = 1$ and working at the temperatures (a) $T_C = 300$ K to $T_H = 1300$ K, (b) $T_C = 200$ K to $T_H = 300$ K. Comment on the obtained results.

(13) The average transport coefficients values of a series of filled skutterudites (Section 3.5.2), within the temperature range $T_C = 300$ K to $T_H = 800$ K, are listed in Table 1.5, along with the ZT_H values measured at T_H. Making use of these data, determine the TE efficiency for the different samples at T_M and T_H, and compare the resulting plots $\eta(ZT_M)$ and $\eta(ZT_H)$.

(14) Thermoelectric generation from window glasses utilizing the temperature difference between the solar heated outdoor and the relatively cold indoor of a building was recently demonstrated by integrating TE pellets through drilled holes (5 mm ∅) inside a 5 mm thick glass. The TE module legs were manufactured from $Bi_{1.75}Te_{3.25}$ and Sb_2Te_3 commercial powders, whose transport coefficient values, measured at 20°C read, $\sigma = 154\ \Omega^{-1}cm^{-1}$, $S = -125\ \mu VK^{-1}$, $\kappa = 1.0\ Wm^{-1}K^{-1}$ and $\sigma = 429\ \Omega^{-1}cm^{-1}$, $S = +118\ \mu VK^{-1}$, $\kappa = 0.8\ Wm^{-1}K^{-1}$, respectively [70]. Determine (a) the figure of merit of each

Table 1.5 Average temperature $T_M = (T_H + T_C)/2$ transport coefficients and high temperature FOM values for skutterudite samples. DD stands for the so-called didymium metal, a natural rare-earth mixture consisting of 4.76 mass Pr and 95.24 mass Nd [69]

Sample	ρ ($\mu\Omega$cm)	S (μVK^{-1})	κ (Wm^{-1}K^{-1})	ZT_H
$Ca_{0.91}Fe_4Sb_{12}$	454	92	4.3	0.42
$Ca_{0.93}Fe_4Sb_{12}$	452	124	3.4	0.77
$Ba_{0.87}Fe_4Sb_{12}$	468	117	3.2	0.83
$Ca_{0.74}Fe_3CoSb_{12}$	435	129	3.7	0.90
$Ba_{0.74}Fe_3CoSb_{12}$	635	136	2.8	1.02
$DD_{0.65}Fe_3CoSb_{12}$	667	156	2.4	1.20

branch and the resulting FOM for the thermopile as a whole, (b) the respective TE efficiencies in the working temperature range $T_C = 20^\circ$C to $T_H = 43.5^\circ$C, and (c) the average power factor of a pellet and the electrical power delivered under optimal conditions.

(15) Derive Eq. (1.77).

(16) Obtain the reduced efficiency and the value of the reduced current which optimizes it when the thermal **h** and electrical **j** currents are antiparallel.

(17) Derive Eq. (1.96).

(18) Making use of Eq. (1.23) and the relative current density given by Eq. (1.87) derive the relationship $\mathbf{h} = \Phi\mathbf{j}$, where Φ is the TE potential introduced in Section 1.4.4.

(19) Making use of the TE potential approach introduced in Section 1.4.4, derive the optimum efficiency expression for a TE generator when the transport coefficients are temperature independent. Compare the obtained expression with Eq. (1.73) in Section 1.4.2.

(20) Derive the optimum coefficient of performance for a thermoelectric cooler making use of the TE potential approach introduced in Section 1.4.4 when the transport coefficients depend on the temperature subjected to the constraint $\sigma(T)S^2(T)/\kappa(T) = Z =$ cte.

1.8 Solutions

(1) Assuming an homogeneous conductor we can express $\nabla V = \Delta V/L$, where ΔV is the potential drop through the conductor and L is its length. After Eq. (1.3) we have

$$\mathbf{j}\cdot\mathbf{j} = \sigma^2(\nabla V)^2 = \left(\frac{\Delta V}{RA}\right)^2,$$

where we have used Eq. (1.4). Thus, making use of Eq. (1.2) we get $\mathbf{j}\cdot\mathbf{j} = (I/A)^2$, so that $I^2 = A^2\,\mathbf{j}\cdot\mathbf{j}$. Therefore, $W_J = RI^2 = RA^2\,\mathbf{j}\cdot\mathbf{j} = \rho LA\,\mathbf{j}\cdot\mathbf{j}$.

(2) Assuming that neither heat losses by conduction nor radiation are present, the Peltier heat released at the junction increases its temperature an amount given by

$$\Delta Q_P = \Pi I\,\Delta t \equiv \rho_M V C_P \Delta T$$

which yields

$$\Delta T = \frac{2.3\times10^{-3}\ \mathrm{V}\times 1\ \mathrm{A}\times 41\ \mathrm{s}}{8.96\ \mathrm{g\,cm^{-3}}\times10^{-3}\ \mathrm{cm^3}\times 0.093\ \mathrm{cal\,g^{-1}\,{}^\circ C^{-1}}\times 4.18\ \mathrm{J\,cal^{-1}}} = 27.07^\circ\mathrm{C}$$

Plugging the obtained temperature difference into the Seebeck voltage expression, we obtain

$$\Delta V_S = S_{\mathrm{Cu,Fe}}\Delta T = \Pi_{\mathrm{Cu,Fe}}\frac{\Delta T}{T} = 2.3\times10^{-3}\ \mathrm{V}\,\frac{27\ \mathrm{K}}{300\ \mathrm{K}} = 0.21\ \mathrm{mV},$$

where we have made use of the first Kelvin relation given by Eq. (1.14).

(3) Equations (1.7) and (1.8) can be expressed in the differential form $dV_S = SdT$ and $dQ_P = \Pi I dt$. Dividing them by dx and dt, respectively, we have

$$\frac{dV_S}{dx} = S\frac{dT}{dx} = S\nabla T,$$
$$\frac{dQ_P}{dt} = \Pi I = STI,$$

where we made use of the first Kelvin relation. Therefore,

$$E_S h_P \equiv \frac{dV_S}{dx}\frac{dQ_P}{dt} = IS^2T\nabla T.$$

Thus, Eq. (1.110) can be written in the form

$$\frac{d^2 Q_T}{dxdt} = \frac{\tau}{S^2 T} E_S\, h_P,$$

and making use of the second Kelvin relation, we finally have

$$\frac{d^2 Q_T}{dxdt} = \frac{1}{S^2}\frac{dS}{dT} E_S\, h_P = -\frac{dS^{-1}}{dT} E_S\, h_P.$$

This expression shows that three different physical effects, characterized by the magnitudes Q_T, h_P and E_S, are mutually related in terms of the Seebeck coefficient.

(4) Making use of the first Kelvin relation, the nondiagonal terms in the TE transport matrix can be symmetrized, so that Eq. (1.23) can be expressed as,

$$\begin{pmatrix} \mathbf{j} \\ \mathbf{h} \end{pmatrix} = \sigma \Pi \begin{pmatrix} \Pi^{-1} & 1 \\ 1 & \Pi\left(1 + \frac{\kappa}{\sigma S^2 T}\right) \end{pmatrix} \begin{pmatrix} -\nabla V \\ -\nabla \ln T \end{pmatrix}. \quad (1.111)$$

Since the resulting transport matrix is real and symmetric it can be diagonalized. However, the form of the electrical and thermal currents is invariant under the similarity transformation which diagonalizes the TE transport matrix. Therefore, in the study of TE effects one cannot simplify the physical description by choosing a more convenient basis to express the magnitudes of interest, as it occurs in other areas of physics.

(5) $\bar{v} = 3^{1/3}\left(v_l^{-3} + 2v_t^{-3}\right)^{-1/3} = 3^{1/3}\left(4500^{-3} + 2 \times 2690^{-3}\right)^{-1/3}$
$= 2977\ \mathrm{ms}^{-1}$

$$l = \frac{3k_l}{\bar{v} C_p d} = \frac{3\ \mathrm{Wm^{-1}K^{-1}}}{2977\ \mathrm{ms^{-1}}\ 5590\ \mathrm{kg\,m^{-3}}\ 0.3065 \times 10^3\ \mathrm{J\,kg^{-1}K^{-1}}}$$
$$= 5.88 \times 10^{-10}\ \mathrm{m} \simeq 6\ \text{Å}$$

(6) After expressing Eq. (1.27) in terms of the dimensionless variable $x_l \equiv \beta\hbar\omega$ one has

$$\frac{\partial p}{\partial T} = \frac{x_l}{T}\frac{e^{x_l}}{(e^{x_l} - 1)^2} = \frac{x_l}{T}\frac{e^{x_l}}{e^{2x_l} + 1 - 2e^{x_l}} = \frac{x_l}{T}\frac{1}{e^{x_l} + e^{-x_l} - 2},$$

and making use of the relations $\cosh x_l \equiv (e^{x_l} + e^{-x_l})/2$ and $\cosh x_l - 1 = 2\sinh^2(x_l/2)$ one readily obtains Eq. (1.33).

(7) For metals we have $\kappa_l \ll \kappa_e$ over a wide temperature range, so that $\kappa \simeq \kappa_e = \sigma L_0 T$, according to the Wiedemann–Franz law. Making use of this expression for the thermal conductivity into Eq. (1.23), one gets

$$\begin{pmatrix} \mathbf{j} \\ \mathbf{h} \end{pmatrix} = \sigma \begin{pmatrix} 1 & S \\ ST & L_0 T + S^2 T \end{pmatrix} \begin{pmatrix} -\nabla V \\ -\nabla T \end{pmatrix}$$
$$= \sigma \begin{pmatrix} 1 & S \\ S & L_0 + S^2 \end{pmatrix} \begin{pmatrix} -\nabla V \\ -\frac{1}{2}\nabla T^2 \end{pmatrix}.$$

We note that the transport matrix determinant reads σL_0 in this case.

(8) According to Eq. (2.51) $S(T) = AT$, where A only depends on the electronic structure of the considered material. Thus,

$$\bar{S} = \frac{A}{\Delta T}\int_{T_C}^{T_H} T\,dT = \frac{A(T_H^2 - T_C^2)}{2(T_H - T_C)} = A\frac{T_H + T_C}{2}$$
$$= AT_M = S(T_M).$$

Therefore, the standard integral-averaged model will only differ from the standard Altenkirch's one when $S(T)$ is a nonlinear function.

(9) Making use of Eq. (1.53), we can express Eq. (1.66) in the form

$$\eta = \frac{RI^2}{I^2(r+R)\frac{T_H}{\Delta T} + \mathcal{K}\Delta T - \frac{I^2}{2}r}.$$

Simplifying the term I^2 and substituting the relation $R = mr$, we have

$$\eta = \frac{mr}{r(1+m)\frac{T_H}{\Delta T} + \frac{\mathcal{K}\Delta T}{I^2} - \frac{r}{2}}. \qquad (1.112)$$

Making use of Eq. (1.53) and the Seebeck voltage $\Delta V_S = (S_p - S_n)\Delta T$ expression the term $\frac{\mathcal{K}\Delta T}{I^2}$ in Eq. (1.112) reads $\frac{\mathcal{K}\Delta T}{(S_p - S_n)^2(\Delta T)^2}(1+m)^2 r^2$. Plugging this expression into Eq. (1.112), simplifying r, and multiplying denominator and numerator by $\eta_C = \Delta T / T_H$, we get

$$\eta = \eta_C \frac{m}{1 + m + \frac{\mathcal{K}r}{(S_p - S_n)^2 T_H}(1+m)^2 - \frac{\Delta T}{2T_H}}$$

which reduces to Eq. (1.67) making use of Eq. (1.68).

(10) Plugging the data listed in Table 3.1 into Eq. (1.71), we have

$$Z_{\text{Bi}} = 7.4 \times 10^3\ \Omega^{-1}\text{cm}^{-1}\ 10^2\ \text{m}^{-1}\text{cm}\ \frac{(-100)^2 \times 10^{-12}\ \text{V}^2\text{K}^{-2}}{6.1\ \text{Wm}^{-1}\text{K}^{-1}}$$
$$= 1.21 \times 10^{-3}\ \text{K}^{-1},$$
$$Z_{\text{Sb}} = 4.1 \times 10^4\ \Omega^{-1}\text{cm}^{-1}\ 10^2\ \text{m}^{-1}\text{cm}\ \frac{(40)^2 \times 10^{-12}\ \text{V}^2\text{K}^{-2}}{24.2\ \text{Wm}^{-1}\text{K}^{-1}}$$
$$= 2.71 \times 10^{-4}\ \text{K}^{-1},$$

and the figure of merit mean value is $\langle Z \rangle = (Z_{\text{Bi}} + Z_{\text{Sb}})/2 = 7.4 \times 10^{-4}\ \text{K}^{-1}$. On the other hand, making use of the transport coefficients data listed in Table 3.1 into Eq. (1.70) we obtain

$$\mathcal{Z} = \frac{(40+100)^2 \times 10^{-12}\ \text{V}^2\text{K}^{-2}}{\left(\sqrt{\frac{6.1\ \text{Wm}^{-1}\text{K}^{-1}}{7400\ \Omega^{-1}\text{cm}^{-1}10^2\ \text{m}^{-1}\ \text{cm}}} + \sqrt{\frac{24\ \text{Wm}^{-1}\text{K}^{-1}}{41000\ \Omega^{-1}\text{cm}^{-1}10^2\ \text{m}^{-1}\ \text{cm}}}\right)^2}$$
$$= 7.0 \times 10^{-4}\ \text{K}^{-1}.$$

Therefore, the mean value and the device FOM values are very similar, though the device FOM value is somewhat smaller by about 5.4%.

(11) Plugging the obtained optimal load ratio $m = \sqrt{1+\mathcal{Z}T_M}$ into Eq. (1.67) we get

$$\eta = \eta_C \frac{\sqrt{1+\mathcal{Z}T_M}}{1+\sqrt{1+\mathcal{Z}T_M} + \frac{(1+\sqrt{1+\mathcal{Z}T_M})^2}{\mathcal{Z}T_H} - \frac{\Delta T}{2T_H}}, \qquad (1.113)$$

now, we note that

$$\frac{(1+\sqrt{1+\mathcal{Z}T_M})^2}{\mathcal{Z}T_H} - \frac{\Delta T}{2T_H} = \frac{2(1+\sqrt{1+\mathcal{Z}T_M})}{\mathcal{Z}T_H} + \frac{T_C}{T_H}, \qquad (1.114)$$

and substituting Eq. (1.114) into Eq. (1.113) we have

$$\eta = \eta_C \frac{\sqrt{1+\mathcal{Z}T_M}}{(1+\sqrt{1+\mathcal{Z}T_M})\left(1+\frac{2}{\mathcal{Z}T_H}\right) + \frac{T_C}{T_H}}, \qquad (1.115)$$

Multiplying the numerator and denominator of Eq. (1.115) by $\sqrt{1+\mathcal{Z}T_M} - 1 > 0$, we obtain

$$\eta = \eta_C \frac{\sqrt{1+\mathcal{Z}T_M}(\sqrt{1+\mathcal{Z}T_M} - 1)}{\mathcal{Z}T_M + 1 + \sqrt{1+\mathcal{Z}T_M}\frac{T_C}{T_H}}, \qquad (1.116)$$

and expressing $1+\mathcal{Z}T_M = \sqrt{1+\mathcal{Z}T_M}\sqrt{1+\mathcal{Z}T_M}$, Eq. (1.116) can be further simplified to the form given by Eq. (1.73).

(12) (a) Plugging the data $ZT = 1$, $T_C = 300$ K, and $T_H = 1300$ K in Eq. (1.73), we have

$$\eta^* = \frac{1000}{1300} \frac{\sqrt{2} - 1}{\sqrt{2} + \frac{300}{1300}} = 0.1936 \rightarrow 19.4\%$$

when the device is working in the regime of high temperatures (note that the Carnot efficiency is $\eta_c = 10/13 = 0.769 \rightarrow 76.9\%$).

(b) Plugging the data $ZT = 1$, $T_C = 200$ K, and $T_H = 300$ K in Eq. (1.73), we get

$$\eta^* = \frac{100}{300} \frac{\sqrt{2} - 1}{\sqrt{2} + \frac{200}{300}} = 0.0663 \rightarrow 6.63\%$$

when the device is working in the regime of low temperatures (note that $\eta_c = 1/3 = 0.33 \rightarrow 33.3\%$). Two main conclusions can be drawn from the obtained results: (i) the efficiency of TE devices is substantially lower than that corresponding to the Carnot efficiency at any temperature, and (ii) the difference between η^* and the ideal η_c increases as the working temperature increases (see Fig. 1.12). Accordingly, low temperature (cooling) TE applications are the most efficient ones, whereas one must substantially improve the FOM of the module materials in order to be competitive with current power generation alternative technologies.

(13) Making use of data listed in Table 1.5 into Eqs. (1.71) and (1.73), we obtain the results listed in Table 1.6. In Fig. 1.17, we plot the efficiency curves derived from the data listed in Table 1.6. As we see, the efficiency values corresponding to the more realistic average temperature transport coefficients are significantly smaller than those corresponding to the high temperature data.

(14) (a) Plugging the transport coefficient data into Eq. (1.71), we have

$$\begin{aligned} ZT_{\mathrm{BiTe}} &= 154\ \Omega^{-1}\mathrm{cm}^{-1}\ 10^2\ \mathrm{m}^{-1}\mathrm{cm}\ \frac{125^2 \times 10^{-12}\ \mathrm{V}^2\mathrm{K}^{-2} \times 293\ \mathrm{K}}{1\ \mathrm{Wm}^{-1}\mathrm{K}^{-1}} \\ &= 0.07, \\ ZT_{\mathrm{SbTe}} &= 429\ \Omega^{-1}\mathrm{cm}^{-1}\ 10^2\ \mathrm{m}^{-1}\mathrm{cm}\ \frac{118^2 \times 10^{-12}\ \mathrm{V}^2\mathrm{K}^{-2} \times 293\ \mathrm{K}}{0.8\mathrm{Wm}^{-1}\mathrm{K}^{-1}} \\ &= 0.219, \end{aligned}$$

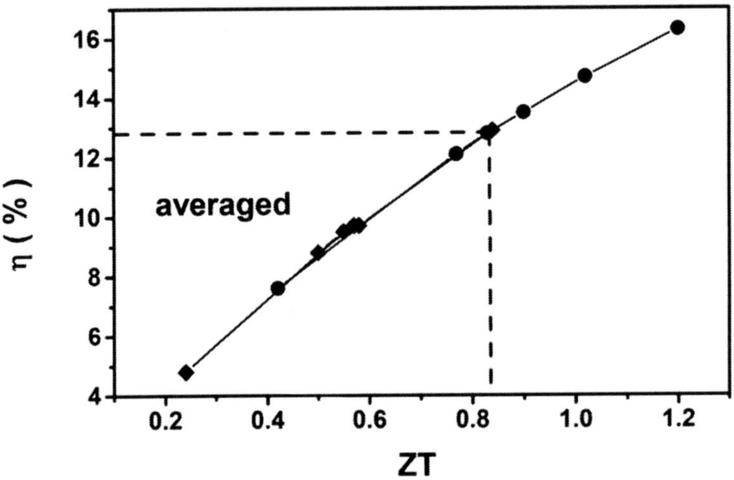

Figure 1.17 Efficiency as a function of the ZT value for the skutterudite compounds listed in Table 1.6. Circles denote high temperature data and diamonds denote averaged temperature data.

and the figure of merit mean value is $\langle ZT \rangle = (ZT_{\text{BiTe}} + ZT_{\text{SbTe}})/2 = 0.145$. On the other hand, making use of the same transport coefficients data into Eq. (1.70), we obtain

$$\mathcal{Z}T = \frac{(118+125)^2 \times 10^{-12}\ \text{V}^2\text{K}^{-2} \times 293\ \text{K}}{\left(\sqrt{\dfrac{1\ \text{Wm}^{-1}\text{K}^{-1}}{154\ \Omega^{-1}\ \text{cm}^{-1}10^2\text{m}^{-1}\text{cm}}} + \sqrt{\dfrac{0.8\ \text{Wm}^{-1}\text{K}^{-1}}{429\ \Omega^{-1}\text{cm}^{-1}10^2\ \text{m}^{-1}\text{cm}}}\right)^2}$$
$$= 0.113,$$

Table 1.6 Average temperature figure of merit ZT_M value and thermoelectric generator efficiency values corresponding to the high temperature $\eta(T_H)$ and average working temperature $\eta(T_M)$ for the skutterudite samples listed in Table 1.5

Sample	ZT_M	$\eta(T_H)$	$\eta(T_M)$
$Ca_{0.91}Fe_4Sb_{12}$	0.24	7.6	4.8
$Ca_{0.93}Fe_4Sb_{12}$	0.55	12.1	9.5
$Ba_{0.87}Fe_4Sb_{12}$	0.50	12.8	8.8
$Ca_{0.74}Fe_3CoSb_{12}$	0.57	13.5	9.7
$Ba_{0.74}Fe_3CoSb_{12}$	0.57	14.7	9.7
$DD_{0.65}Fe_3CoSb_{12}$	0.84	16.3	12.9

so that the FOM derived from the legs mean value is larger than that corresponding to the device as a whole by about 22%.

(b) Making use of the values $\langle ZT \rangle = 0.145$, $\mathcal{Z}T = 0.113$, $T_C = 293$ K, and $T_H = 316.5$ K in Eq. (1.73), we get

$$\eta^* = \frac{23.5}{316.5} \frac{\sqrt{1.145} - 1}{\sqrt{1.145} + \frac{293}{316.5}} = 0.0026 \rightarrow 0.26\%$$

when considering the mean FOM value, and

$$\eta^* = \frac{23.5}{316.5} \frac{\sqrt{1.113} - 1}{\sqrt{1.113} + \frac{293}{316.5}} = 0.0020 \rightarrow 0.20\%$$

when considering the more accurate device FOM value (note that $\eta_C = 23.5/316.5 = 0.074 \rightarrow 7.4\%$ in both cases). We see that the efficiency of the considered TE modules is quite low.

(c) The average power factor is given by

$$\langle P \rangle = \frac{154\,\Omega^{-1}\text{cm}^{-1}\,125^2 \times 10^{-6}\,\text{V}^2\text{K}^{-2} + 429\,\Omega^{-1}\text{cm}^{-1}\,118^2 \times 10^{-6}\,\text{V}^2\text{K}^{-2}}{2}$$
$$= 4.2\,\mu\text{Wcm}^{-1}\,\text{K}^{-2}$$

and making use of Eq. (1.56), we get

$$\begin{aligned} W^* &= \frac{A}{2L} \langle P \rangle \Delta T^2 \\ &= \frac{\pi \times (2.5 \times 10^{-3})^2\,\text{m}^2}{2 \times 5 \times 10^{-3}\,\text{m}} \times 4.19\,\mu\text{Wcm}^{-1}\,\text{K}^{-2} \times 10^2\,\text{m}^{-1}\,\text{cm} \times 23.5^2\,\text{K}^2 \\ &= 454\,\mu\text{W} \end{aligned}$$

This figure is more than three orders of magnitude larger than the measured output $W^* = 0.112\,\mu$W delivered by an assembly of four pellets in electrical series and thermal parallel [70]. The difference between the measured and estimated values has been attributed to significant thermal losses and large contact resistances.

(15) Making use of the temperature profile given by Eq. (1.63), we have

$$\left(\frac{dT}{dx}\right)_{0,\alpha} = \frac{\Delta T}{L_\alpha} + \frac{j_\alpha^2 \rho_\alpha}{2\kappa_\alpha} L_\alpha, \tag{1.117}$$

and substituting Eq. (1.117) into Eq. (1.76), keeping into account Eqs. (1.65) and (1.52), we finally obtain Eq. (1.77).

(16) When the thermal and electrical currents are antiparallel, Eq. (1.87) leads to $j = -u\kappa\nabla T$. Plugging this density current value into Eq. (1.86), and assuming $S_H = S$, $\kappa_H = \kappa$, and $T_H = T$, we have

$$\eta_r = -\frac{uS + u^2\kappa\rho}{-uS + T^{-1}} = \frac{uS}{Z}\frac{1 + \frac{uS}{Z}}{\frac{1}{ZT} - \frac{uS}{Z}} = \frac{w(1+w)}{\frac{1}{ZT} - w}, \qquad (1.118)$$

so that the reduced efficiency vanishes for $w = 0$ $(u = 0)$ and $w = -1$ $(u = -Z/S)$. Imposing the extremum condition $d\eta_r/dw = 0$ to Eq. (1.118), we get

$$w^* = \frac{1 \pm \sqrt{1+ZT}}{ZT}.$$

Since $u < 0$ when the electrical and thermal currents are antiparallel, we finally obtain

$$u_-^* = \frac{1 - \sqrt{1+ZT}}{ST} = -u_+^*$$

(17) Making use of Eq. (1.95) into Eq. (1.88), we have

$$\eta_r = \frac{\left(1 - \frac{\Phi'_\alpha}{S_\alpha}\right)\frac{\Phi'_\alpha}{S_\alpha}}{\left(1 - \frac{\Phi'_\alpha}{S_\alpha}\right) + \frac{1}{ZT}} = \frac{\frac{\Phi'_\alpha}{S_\alpha}}{1 + \frac{\frac{1}{ZT}}{1 - \frac{\Phi'_\alpha}{S_\alpha}}}, \qquad (1.119)$$

now, after Eq. (1.95) the denominator of Eq. (1.119) can be rearranged in the form

$$1 + \frac{\frac{1}{ZT}}{\frac{uS_\alpha}{Z}} = 1 + \frac{1}{uS_\alpha T},$$

so that Eq. (1.119) reads

$$\eta_r = \frac{\Phi'_\alpha T}{S_\alpha T + u^{-1}} = \frac{\Phi'_\alpha T}{\Phi_\alpha} = T\frac{d\ln\Phi_\alpha}{dT},$$

where we have made use of Eq. (1.93).

(18) Making use of the heat current density expression given by Eq. (1.23), we explicitly evaluate the scalar product

$$\begin{aligned}\mathbf{h}\cdot\mathbf{j} &= (-\sigma ST\nabla V - \sigma S^2 T\nabla\mathbf{T})\cdot\mathbf{j} - \kappa\nabla\mathbf{T}\cdot\mathbf{j} \\ &= ST(-\sigma\nabla\mathbf{V} - \sigma S\nabla\mathbf{T})\cdot\mathbf{j} - \kappa\nabla\mathbf{T}\cdot\mathbf{j}. \qquad (1.120)\end{aligned}$$

Now, according to Eq. (1.23) we can identify $-\sigma\nabla\mathbf{V}-\sigma S\nabla\mathbf{T}=\mathbf{j}$ in the first term, whereas the second term can be expressed as $\kappa\nabla\mathbf{T}\cdot\mathbf{j}=-u^{-1}\mathbf{j}\cdot\mathbf{j}$ after Eq. (1.87). Therefore, Eq. (1.120) takes the form

$$\mathbf{h}\cdot\mathbf{j}=ST\mathbf{j}\cdot\mathbf{j}+u^{-1}\mathbf{j}\cdot\mathbf{j}=(ST+u^{-1})\mathbf{j}\cdot\mathbf{j}\equiv\Phi\mathbf{j}\cdot\mathbf{j}\Rightarrow\mathbf{h}=\Phi\mathbf{j}$$

Thus, we realize that electrical and heat current densities can be directly connected by means of the TE potential given by Eq. (1.93), which plays a very fundamental role in TEMs theory [55, 56].

(19) When the transport coefficients are temperature independent, the heat equation (1.92) reduces to

$$\frac{du^{-1}}{dT}=-\frac{1}{u^2}\frac{du}{dT}=-u\kappa\rho, \tag{1.121}$$

where $\kappa\rho$ is a constant. Eq. (1.121) can be immediately integrated between the hot and cold junction temperatures to obtain

$$\frac{1}{u_C^2}-\frac{1}{u_H^2}=2\kappa\rho\Delta T=2S^2\frac{\Delta T}{Z}\equiv\Lambda. \tag{1.122}$$

Making use of Eq. (1.122) into Eq. (1.98), one has ($S_C=S_H=S$)

$$\eta=\frac{S\Delta T+u_H^{-1}-\sqrt{\Lambda+u_H^{-2}}}{ST_H+u_H^{-1}}. \tag{1.123}$$

Imposing the extremum condition $d\eta/du_H^{-1}=0$, we get

$$(u_H^{-1})^*=\frac{S}{ZT_M}\left(T_H+T_C\sqrt{1+ZT_M}\right), \tag{1.124}$$

and plugging this value into Eq. (1.123), after some algebra, we finally obtain

$$\eta^*=\eta_C\frac{\sqrt{1+ZT_M}-1}{\sqrt{1+ZT_M}+\frac{T_C}{T_H}}, \tag{1.125}$$

which exactly coincides with Eq. (1.73).

(20) Making use of Eq. (1.103) into Eq. (1.100), we have

$$1+\phi_*^{-1} = \exp\left(+\int_{T_C}^{T_H} \frac{dT}{T}\frac{\sqrt{1+ZT}+1}{\sqrt{1+ZT}-1}\right). \tag{1.126}$$

Since Z does not depend on the temperature, the change of variable $\chi \equiv \sqrt{1+ZT}-1$ (note that $\chi > 0\,\forall ZT \neq 0$) reduces the integral in Eq. (1.126) to the form

$$2\int_{\chi_C}^{\chi_H} \frac{1+\chi}{\chi^2} d\chi = 2\int_{\chi_C}^{\chi_H} \chi^{-2} d\chi + 2\int_{\chi_C}^{\chi_H} \frac{d\chi}{\chi}, \tag{1.127}$$

which can be readily integrated to obtain Eq. (1.104).

Chapter 2

Fundamental Aspects

2.1 Efficiency Upper Limit

Certainly, to get efficient TEMs, having high enough ZT values, would significantly increase the economical competitiveness of TE devices, considerably expanding their range of possible applications. Consequently, during the past seven decades a broad collection of different materials has been systematically explored in order to look for more efficient TEMs. Unfortunately, according to the measured transport data, an empirical barrier soon appeared in the way to achieve this goal: the recorded FOM values for all of these materials were always constrained to take on values below an upper limit threshold.

In the first research period, spanning from the 1940s to the 1990s, the upper limit was set at about $ZT \approx 1$. Subsequently, during the last two decades, the upper limit has been shifted up to $ZT \approx 2$–3 by considering, on the one hand, bulk materials with very *complex* crystalline structures, and, on the other hand, *nanostructured* materials, such as BiTe/SbTe superlattices, or PbTe/PbTeSe quantum dot heterostructures.

In Fig. 2.1, we present a ZT versus temperature chart showing several materials with $ZT > 1$ which have been reported during the

Thermoelectric Materials: Advances and Applications
Enrique Maciá-Barber

ISBN 978-981-4463-52-2 (Hardcover), 978-981-4463-53-9 (eBook)
www.panstanford.com

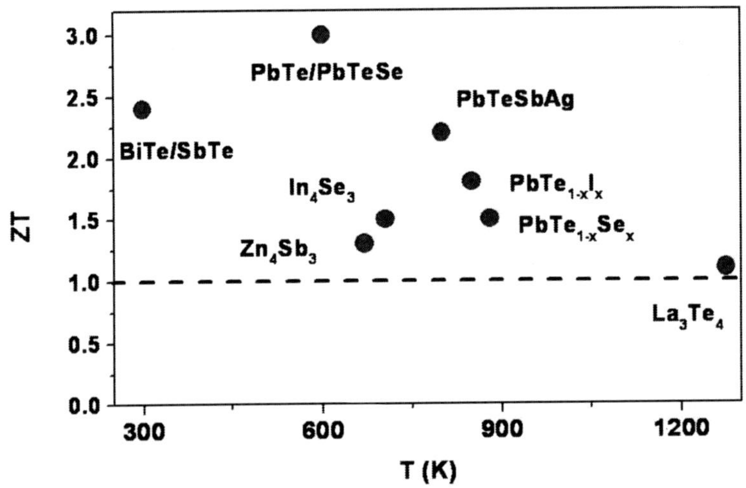

Figure 2.1 ZT–T chart showing the largest reported FOM values for different bulk materials, including: rare-earth chalcogenides, such as La_3Te_4 [72], β-Zn_4Sb_3 alloys [73], $In_4Se_{3-\delta}$ crystals [74], $PbTe_{1-x}I_x$ alloys [75], $PbTe_{1-x}Se_x$ alloys [76], and PbSbAgTe materials [77], as well as some nanostructured materials, such as BiTe/SbTe and PbTe/PbTeSe superlattices (left upper corner) [78, 79].

last decade. These materials are arranged according to their best working temperature performance. As we see, good enough TEMs are still missing within both the 350–550 K and 900–1200 K broad temperature ranges. We should also note that the largest $ZT > 2$ values have been obtained in research laboratories only and are not yet attainable for practical applications due to severe technical shortcomings (e.g., too brittle materials, nonreproducible results for industrial standards).

In the light of these empirical results, the question regarding the possible existence of some physical reason constraining the FOM value arises in a natural way. Thermodynamical analyses of TEGs, relying on either entropy second law or the Carnot limit efficiency criteria, indicate that apparently it does not exist any physical upper bound on ZT [80]. However, general thermodynamic arguments alone do not suffice to properly establish any realistic bound by their own, since they ignore many relevant aspects related to the atomic and electronic structures of actual materials. In fact, in order to precisely compute and optimize the material parameters entering into the FOM expression, it is necessary to assume a specific physical model of the considered solid. On this basis, no definitive theoretical justification accounting for the relatively small ZT values obtained to date has yet been put forward.

A physical property generally limiting the FOM value in most systems has to do with the inverse correlation between the electrical conductivity and the Seebeck coefficient (see Fig. 1.7 in Section 1.2.3). Indeed, the opposite trends observed in the $\sigma(T)$ and $S(T)$ curves for most materials naturally limit their resulting power factor value, which shows an upper limit stemming from a balance between both trends, thereby bounding above the ZT value (**Exercises 2.1** and **2.2**).

For the sake of illustration, let us consider the case of certain semiconductors that exhibit the following relationship between the Seebeck and electrical conductivity coefficients (Fig. 2.2)

$$S(T) = \frac{k_B}{e}\left[A(T) - \ln \sigma(T)\right], \tag{2.1}$$

where $A(T)$ includes scattering effects, the carriers mobility, and their effective mass [83]. In this case, the power factor can be expressed as an explicit function of the electrical conductivity, given by

$$P = \left(\frac{k_B}{e}\right)^2 \sigma(A - \ln \sigma)^2. \tag{2.2}$$

The extreme condition $dP/d\sigma \equiv 0$, then leads to $A = \ln \sigma$ (which corresponds to the trivial minimum $P = 0$) and $\ln \sigma = A - 2$, which characterizes the electrical conductivity value yielding the maximum power factor $P_{\max} = (2k_B/e)^2 \sigma_*$, with $\sigma_* = \exp(A - 2)$. In this way, one obtains a universal value for the ratio of the maximum electrical power factor at optimum electrical conductivity which depends on fundamental constants only.[a] On the other hand, plugging the obtained $P_{\max}$ value into Eq. (1.71), and assuming that the WFL (see Section 1.2.3) holds in the form $\kappa_e = L_0 \sigma_* T$, one gets

$$(ZT)_{\max} = \left(\frac{2k_B}{e}\right)^2 \frac{T}{L_0 T + \kappa_l(T) e^{2-A}}. \tag{2.3}$$

---

[a]The value $P_{\max}/\sigma_* \simeq 2.44 \times 10^{-8}$ V^{2}K^{-2} was reported from a linear regression analysis of suitable transport data for several established TEMs, in reasonable agreement with the theoretical figure $P_{\max}/\sigma_* = (2k_B/e)^2 \simeq 2.97 \times 10^{-8}$ V^{2}K^{-2} [82].

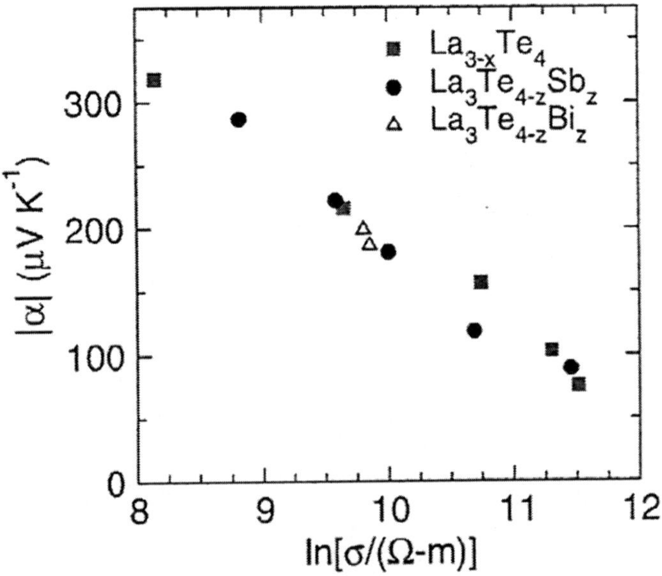

Figure 2.2 Seebeck coefficient as a function of the natural logarithm of the electrical conductivity (measured at 1000 K) for different rare-earth tellurides (see Section 4.3.7) [81]. Reprinted with permission from May A. F., Falge-Larsen E., and Snyder G. J., 2010 *Phys. Rev. B* **81**, 125205, Copyright 2010, American Physical Society.

At moderate to high temperature regimes, the lattice contribution to the thermal conductivity progressively decreases, due to the onset of the phonon–phonon scattering processes (see Section 1.2.2.4), so that the maximum FOM value approaches the limiting value $(ZT)_{\max} \sim \left(\frac{2k_B}{e}\right)^2 L_0^{-1} = 12/\pi^2 \simeq 1.2$.

2.2 ZT Optimization Strategies

The existence of a FOM upper limit can then be understood as stemming from the necessity of simultaneously tuning three interrelated transport coefficients (see Section 1.2.3): the electrical conductivity, the Seebeck coefficient, and the thermal conductivity. As a consequence of this three-way balance, FOM optimization becomes a difficult task, posing a very appealing challenge from the materials engineering viewpoint.[a] In order to properly address this fundamental issue, it is convenient to start by recalling the ZT mathematical expression obtained in Section 1.4.2, that is,

$$ZT = \frac{\sigma S^2 T}{\kappa_e + \kappa_l}, \qquad (2.4)$$

where the thermal conductivity coefficient is explicitly expressed in the form $\kappa(T) = \kappa_e(T) + \kappa_l(T)$, describing the contributions from both charge carriers, κ_e, and collective lattice oscillations, κ_l. This expression naturally suggests two complementary approaches to optimize ZT, namely, to first reduce the thermal conductivity terms appearing in the denominator and then to enhance the power factor contribution σS^2, appearing in the numerator. In the following sections, we will consider each procedure separately.

2.2.1 *Thermal Conductivity Control*

The quest for promising TEMs should reasonably start by considering materials having a very small contribution due to the lattice thermal conductivity κ_l. Indeed, according to Eqs. (1.49) and (2.4), the TE performance of a *hypothetical* material, having a vanishing lattice contribution to the thermal conductivity and obeying the WFL, will be completely determined by the Seebeck coefficient, hence significantly simplifying the FOM optimization task. At this point, it is convenient to recall that, even in the case of electrical insulators, where $\kappa_e \simeq 0$, there exists a non-negligible contribution to the thermal conductivity coming from the lattice dynamics, hence preventing the mathematical condition $\kappa \to 0$ from being attained in practice.[b]

Notwithstanding this, the idea of improving the FOM value by reducing the lattice contribution to the thermal conductivity as much as possible, while keeping relatively high values for both the electrical conductivity and the Seebeck coefficient, provides a useful guideline, referred to as the *electron crystal-phonon glass* concept

[a] Not surprisingly, the number of good TEMs is relatively small, as we will see through Chapters 3 and 4.

[b] The limit $\kappa_l(T \to 0) = 0$ is attained for all materials at very low temperatures. But this extreme temperature regime is of no practical interest in TE research, since $S(T \to 0) = 0$ as well, thereby leading to a vanishing ZT.

[17]. According to this view, promising TEMs should conduct charge carriers like long-range ordered crystalline materials (having large σ values) and should conduct thermal current like amorphous ones (having small κ values). Indeed, an efficient and economic way to reduce the thermal conductivity of bulk TEMs, without degrading their electronic properties, is to exploit inelastic resonant scattering between heat-carrying acoustic phonons and different kinds of scattering sources in the solid.

Inspired by this idea, several approaches have been proposed in order to properly control the thermal conductivity, such as:

- Starting from parent elemental solids exhibiting good electronic transport properties, but too large κ values, one can obtain compounds with substantially lower κ coefficients (say, by an order of magnitude or more) by simply *alloying* them, still preserving their main electronic structure features (see Section 1.2.2.3). Certainly, this alloying effect is more pronounced when involving heavy atoms. This is the case, for instance, of BiSb, ZnSb, and SiGe alloys of extensive use in current TE devices (see Sections 3.3.1, 3.3.3, and 3.3.5, respectively). In the same vein, the possible presence of additional disorder sources, such as vacancies or isotopic substitutional disorder, may contribute to further reducing the thermal conductivity value. Actually, not only atomic disorder (typically introduced by alloying) but also the disorder arising from the valence value of atoms occupying certain atomic sites (the so-called *valence fluctuation*) can strongly reduce the thermal conductivity (see Section 3.4.3).
- Another approach seeks for structurally *complex lattice structures*, characterized by atomic frameworks showing voids, cages, channels, or layers which can be conveniently filled with heavy, small ions able to loosely rattle about inside the framework, thereby degrading the thermal current propagation. Well-known examples of this scattering mechanisms are provided by rattling motion of atoms inside cages in skutterudites (Section 3.5.2) and clathrates (Section 3.5.3) or the complex dynamics of phonons in hierarchical cluster-based aggregates in quasicrystals (Section 5.3.1).

- The reduction of the phonon mean free path due to enhanced layer scattering effects in *nanostructured* semiconductor superlattices and *low-dimensional systems* has been intensively explored during the last decade, although these low-dimensional systems also scatter charge carriers, which implies an undesired electrical conductivity decrease.

More subtle mechanisms of reducing the thermal conductivity in order to boost TE performance have been recently studied too. Thus, it has been reported that certain crystal structures may have quite low thermal conductivity values due to their strong *lattice anharmonicity* [84, 85]. The so-called Peierls distortion mechanism (see Section 4.2.1) has been also considered to this end. Thus, in In_4Se_3 crystals,[a] characterized by the presence of covalently bonded In-Se chain-like structures, the Peierls mechanism gives rise to the emergence of a charge density wave, which originates a lattice distortion disturbing phonon propagation.

Anyway, it seems unlikely to reduce the thermal conductivity below a certain limit (close to $\sim 0.1\ \mathrm{Wm^{-1}K^{-1}}$ at room temperature), which can be estimated from the expression [87–89],

$$\kappa_m = 0.4 k_B n_a^{2/3} (v_l + 2v_t), \tag{2.5}$$

where $n_a = N/V$ is the number atomic density and v_l and v_t are the longitudinal and transverse speeds of sound, respectively (**Exercises 2.3** and **2.4**). A more accurate expression, allowing for an explicit determination of the thermal conductivity temperature dependence is given by [87],

$$\kappa_m(T) = \left(\frac{\pi}{6}\right)^{1/3} k_B n_a^{2/3} \sum_i v_i \left(\frac{T}{\theta_i}\right)^2 \int_0^{\theta_i/T} \frac{x_l^3 e^{x_l}}{(e^{x_l}-1)^2} dx_l, \tag{2.6}$$

where the summation is over the one longitudinal and the two transverse modes, v_i is the respective sound velocity for each mode, $x_l = \hbar\omega\beta$, and

$$\theta_i \equiv \frac{\hbar}{k_B} \left(6\pi^2 n_a\right)^{1/3} v_i. \tag{2.7}$$

plays the role of a mode-related effective Debye temperature.

[a]Reaching $ZT = 1.48$ at 705 K [86] (See Fig. 2.1).

2.2.2 *Power Factor Enhancement*

Once a material with a relatively low thermal conductivity value has been selected, the next step is to enhance its power factor in order to further increase the FOM. To this end, two main routes have been thoroughly described in the literature during the last decade.

In the first route, one exploits low-dimensional quantum confinement effects in nanostructured materials. In this case, the FOM improvement was originally attributed to an increase in the DOS close to the Fermi level, a mechanism that works well for single conduction band models within the constant relaxation time approximation [90, 91]. Detailed experimental studies, however, indicated that, in most cases, the reported ZT enhancement may be mainly due to a parallel substantial reduction of the thermal conductivity in nanostructured materials [78, 92], with no significant improvement in their electronic transport properties. In fact, recent theoretical studies have claimed that the power factor may even decrease below the corresponding bulk values when the energy dependence of the charge carrier's relaxation time is fully taken into account [93, 94].[a] Furthermore, while the results so far obtained for quantum-confined systems in laboratory conditions are promising, such structures may be of limited practical use due to difficulties in manufacturability.

An alternative trend entirely focuses on bulk materials. On the basis of an exhaustive comparison with experimental data of well-characterized bulk semiconducting TEMs, it has been shown that the power factor is maximized when the Seebeck coefficient is in the range $130 \leq |S^*| \leq 190\ \mu\text{VK}^{-1}$ [96]. Accordingly, efforts aimed at further improving the FOM should be addressed toward increasing the electrical conductivity at the given S^*. In doing so, however, one must keep control of the charge carrier contribution to both the electrical and thermal conductivities as well. Indeed, by only maximizing the power factor, one is ignoring the parallel decremental effect that any improvement of σ will have on the resulting FOM, as prescribed by the WFL (see Section 1.2.3). Thus,

[a] A possible competition between quantum size effects and degeneracy in multisub-band models may lead to a power factor reduction as well [95].

ZT is typically optimized at smaller charge carriers concentration values than required to optimize the power factor alone.

In addition, in many published works, typically metallic Lorenz numbers, close to L_0, are regularly used to determine the electronic contribution to the thermal conductivity. Such a choice may overestimate the electronic contribution in semiconductors which are not heavily doped, leading to over reduced κ_l values (see Section 1.2.3). Therefore, the literature values of the lattice contribution to the thermal conductivity are often underestimated, which can lead to improper conclusions regarding the TE efficiency of the corresponding materials.

2.3 The Spectral Conductivity Function

A deeper insight into the fundamental issue concerning the possible existence of a ZT upper limit value (see Section 2.1) can only be gained by explicitly considering TE effects at a microscopic level, as we did in Section 1.2.2. Indeed, the transport properties of a given material greatly depend on the scattering processes determined by the mutual interactions between the carriers and the lattice, as well as by the presence of impurities and other structural defects, which often vary themselves with both temperature and external force fields. In order to properly include the physical processes determining the temperature dependence of the transport coefficients appearing in Eq. (2.4), it is convenient to introduce a characteristic function, which entails detailed information on the electronic structure of the material. This function is referred to as the *spectral conductivity function*, $\sigma(E) > 0$ (measured in $\Omega^{-1}\text{cm}^{-1}$ units), which is defined as the $T \to 0$ conductivity with the Fermi level at energy E.

For systems which can be described within the Boltzmann approach, one has

$$\sigma(E) = e^2\tau(E)n(E)v^2(E), \tag{2.8}$$

where $\tau(E)$ is the relaxation time, $n(E)$ measures the charge carriers density (in cm^{-3}), and $v(E)$ is the group velocity of the carriers. On the other hand, in the case of systems for which the applicability of the Boltzmann approach is not guaranteed, one can consider the more general relationship

$$\sigma(E) = \frac{e^2}{V} N(E) D(E), \tag{2.9}$$

where $N(E)$ is the DOS (measured in states $(\text{eV})^{-1}$ per atom), $D(E)$ measures the diffusivity of the states (in cm^2s^{-1}), and V is the volume of the system.

From the knowledge of the spectral conductivity function, one can derive the kinetic coefficients,[a]

$$\mathcal{L}_{ij}(T) = (-1)^{i+j} \int_{-\infty}^{+\infty} \sigma(E)\,(E-\mu)^{i+j-2} \left(-\frac{\partial f}{\partial E}\right) dE, \tag{2.10}$$

where $i,\, j = 1,\, 2$, $\mu(T)$ is the chemical potential (which equates the Fermi energy, E_F, at $T = 0$), and

$$f(E, T) = \frac{1}{1 + e^{(E-\mu)\beta}}, \tag{2.11}$$

is the Fermi–Dirac distribution function, where $\beta \equiv (k_B T)^{-1}$. In this formulation, all the microscopic details of the system are included in the spectral conductivity function $\sigma(E)$. Therefore, the temperature dependence of the transport coefficients appears in the Fermi–Dirac distribution, whereas all peculiarities of the scattering processes are incorporated in $\sigma(E)$. The kinetic coefficients derived from Eq. (2.10) are valid for both extended and localized states. In fact, as we will see below, this formalism has been applied to study the TE properties of disordered systems undergoing a disorder-driven metal–insulator transition [97], as well as to study transport phenomena in quasicrystals and related approximants characterized by critical wavefunctions exhibiting spatial fluctuations at all scales (see Chapter 5).

The macroscopic transport coefficients introduced in Section 1.1 are related to the kinetic coefficients given by Eq. (2.10) through the

---

[a]This approach is known as the Chester–Thellung–Kubo–Greenwood formulation. [14, 16] This description is valid provided that the charge carriers are noninteracting among them and the scattering with impurities and lattice phonons is elastic. No assumption is made about the strength of disorder and the nature of the states.

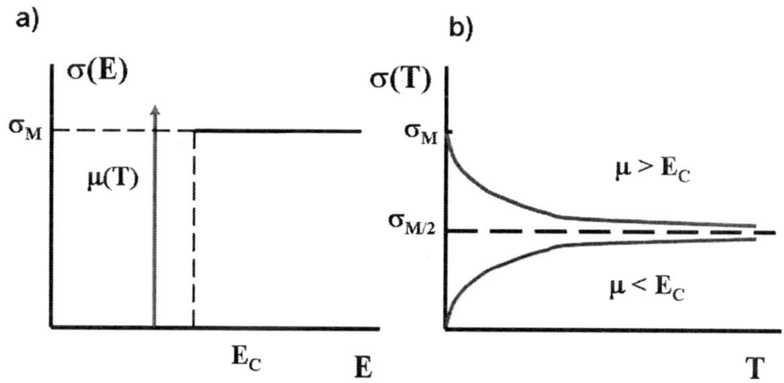

Figure 2.3 (a) Diagram showing the spectral conductivity function for the Mott model, (b) temperature dependence of the electrical conductivity for a system undergoing a metal–insulator transition. More details in the text

expressions [16],

$$\sigma(T) = \mathcal{L}_{11}(T), \tag{2.12}$$

$$S(T) = \frac{1}{|e|T}\frac{\mathcal{L}_{12}(T)}{\mathcal{L}_{11}(T)}, \tag{2.13}$$

$$\kappa_e(T) = \frac{1}{e^2 T}\left(\mathcal{L}_{22}(T) - \frac{\mathcal{L}_{12}^2(T)}{\mathcal{L}_{11}(T)}\right). \tag{2.14}$$

For the sake of illustration, let us consider the spectral conductivity function describing an insulating-metal transition which takes place in certain disordered materials whose electronic structure can be described in terms of the so-called Mott spectral model. The corresponding spectral conductivity function is sketched in Fig. 2.3a and consists of two separate regions given by the step function

$$\sigma(E) = \begin{cases} 0, & \text{if } E < E_C \\ \sigma_M, & \text{if } E \geq E_C \end{cases} \tag{2.15}$$

where σ_M is a constant and the energy E_C is referred to as the *mobility edge* separating the region where all wave functions are localized (i.e., $E < E_C$) from that having extended wave functions in the energy spectrum (i.e., $E \geq E_C$). Accordingly, an insulator–metal transition is expected to occur when the Fermi level crosses from the insulating to the conducting region.

The existence of this phase transition can be readily confirmed by plugging Eq. (2.15) into Eqs. (2.10) and (2.12) to obtain

$$\sigma(T) = \sigma_M \int_{E_C}^{+\infty} \left(-\frac{\partial f}{\partial E}\right) dE = \sigma_M \left[f(E_C) - \lim_{E\to+\infty} \frac{1}{1+e^{(E-\mu)\beta}} \right]$$
$$= \frac{\sigma_M}{1+e^{(E_C-\mu)\beta}}. \quad (2.16)$$

Thus, depending on the relative positions of the chemical potential and the mobility edge, three possible $\sigma(T)$ curves can be observed in the system, as it is illustrated in Fig. 2.3b. In the case $\mu < E_C$, the electrical conductivity vanishes in the limit $T \to 0$ (insulator) and it progressively increases as the temperature increases due to a thermally activated transport involving hopping of charge carriers among localized states. On the contrary, in the case $\mu > E_C$ one gets $\sigma(T \to 0) = \sigma_M$ (residual conductivity) and the electrical conductivity progressively decreases as T increases (metallic behavior). Both the thermally activated and the metallic transport curves asymptotically approach the value $\sigma(T \to \infty) = \sigma_M/2$ in the high temperature limit, which corresponds to the particular case where the chemical potential is exactly located at the mobility edge (i.e., $\mu = E_C$).

Let us now consider the Seebeck coefficient temperature dependence for this spectral conductivity model. According to Eq. (2.13), we must calculate the kinetic coefficient $\mathcal{L}_{12}(T)$ making use of Eqs. (2.10) and (2.15). In doing so, after an integration by parts, we get[a]

$$\mathcal{L}_{12}(T) = -\sigma_M(E_C-\mu) f(E_C) - \sigma_M \int_{E_C}^{+\infty} f(E) dE. \quad (2.17)$$

To perform the integral appearing in Eq. (2.17), it is convenient to introduce the dimensionless scaled energy variable $x \equiv (E-\mu)\beta$ and to express the Fermi–Dirac distribution function in the form

$$f(x) = \frac{1}{1+e^x} = \frac{1}{2}\left(1 - \tanh\frac{x}{2}\right), \quad (2.18)$$

[a]Making use of the L'Hôpital rule we have

$$\lim_{E\to\infty} \frac{E-\mu}{1+e^{(E-\mu)\beta}} = \lim_{E\to\infty} \frac{1}{\beta e^{(E-\mu)\beta}} = 0.$$

so that,

$$\int_{E_C}^{+\infty} f(E)dE = \frac{\beta^{-1}}{2}\int_{x_C}^{+\infty}\left(1-\tanh\frac{x}{2}\right)dx$$
$$= \beta^{-1}\left[\ln\left(2\cosh\frac{x_C}{2}\right)-\frac{x_C}{2}\right], \qquad (2.19)$$

where, in order to evaluate the improper integral limit, we take into account the relation $\lim_{x\to+\infty}\ln\cosh(x/2) \simeq x/2 - \ln 2$. Making use of Eq. (2.19) in Eq. (2.17) and plugging the resulting $\mathcal{L}_{12}(T)$ coefficient along with Eq. (2.16) into Eq. (2.13), we finally obtain

$$S(T) = -\frac{k_B}{|e|}\left[x_C + (1+e^{x_C})\left[\ln\left(2\cosh\frac{x_C}{2}\right)-\frac{x_C}{2}\right]\right]. \qquad (2.20)$$

As it occurred for the electrical conductivity, the temperature dependence of the Seebeck coefficient strongly depends on the relative positions of the chemical potential and the mobility edge. In the particular case $\mu = E_C$ (i.e., $x_C = 0$), we have a temperature independent Seebeck coefficient value $S(T) \equiv S_M = -k_B \ln 4/|e| \simeq -121\ \mu\text{VK}^{-1}$, which defines a model reference value. Note that this value is close to the proposed lower limit value $|S_*| \simeq 130\ \mu\text{VK}^{-1}$ yielding an optimal power factor (see Section 2.2.2). On the other hand, in the case $\mu < E_C$ (thermally activated regime) the Seebeck coefficient diverges to $-\infty$ in the limit $T \to 0$ (i.e., $x_C \to +\infty$), whereas it asymptotically approaches the value $S = S_M$ as the temperature is progressively increased. In a similar way, in the case $\mu > E_C$ (metallic regime) one gets $S(T \to 0) \to 0$ and $S(T \to \infty) = S_M$ (**Exercise 2.5**).

In the description of realistic systems, one generally finds rather involved spectral conductivity functions, naturally leading to difficult to solve integrals when obtaining the transport coefficients. In order to simplify the mathematical treatment, it is convenient to use Eq. (2.18) in order to express the Fermi distribution derivative in terms of hyperbolic functions as

$$-\frac{\partial f}{\partial E} = \frac{\beta}{4}\,\text{sech}^2\left(\frac{x}{2}\right). \qquad (2.21)$$

This is an even function with respect to the variable x exhibiting a narrow, pronounced peak around the chemical potential value ($x = 0$). Making use of Eq. (2.21), one can then rewrite Eq. (2.10) in terms of the scaled variable x to obtain $\mathcal{L}_{ij} = (-1)^n\beta^{-n}J_n/4$, where $n \equiv$

$i + j - 2$, and we have introduced the *reduced* kinetic coefficients (measured in $\Omega^{-1}\text{cm}^{-1}$ units),

$$J_n(\beta) = \int_{-\infty}^{\infty} x^n \sigma(x, \beta)\,\text{sech}^2(x/2)dx. \tag{2.22}$$

According to Eq. (2.22), $J_0 \geq 0$ and $J_2 \geq 0$, whereas the sign of J_1 will depend on the chemical potential relative position, and allows for both n-type ($\mu < E_F$) and p-type ($\mu > E_F$) behaviors. Making use of Eq. (2.22), one can express Eqs. (2.12)–(2.14) in the form

$$\sigma(T) = \frac{J_0}{4}, \tag{2.23}$$

$$S(T) = -\frac{k_B}{|e|}\frac{J_1}{J_0}, \tag{2.24}$$

$$\kappa_e(T) = \left(\frac{k_B}{2e}\right)^2 \frac{T}{J_0} \begin{vmatrix} J_0 & J_1 \\ J_1 & J_2 \end{vmatrix}, \tag{2.25}$$

and, for the sake of completeness, one can also express the Lorenz function in the form

$$L(T) \equiv \frac{\kappa_e}{T\sigma} = \left(\frac{k_B}{e}\right)^2 J_0^{-2} \begin{vmatrix} J_0 & J_1 \\ J_1 & J_2 \end{vmatrix}. \tag{2.26}$$

Some general conclusions about the importance of the spectral conductivity function *parity* on the resulting TE performance can be drawn from Eqs. (2.22)–(2.24). To this end, without loss of generality, let us express $\sigma(x) \equiv \sigma_+(x) + \sigma_-(x)$, where $\sigma_+(x)$ is an even function in the variable x (i.e., $\sigma_+(-x) = \sigma_+(x)$) and $\sigma_-(x)$ is an odd one (i.e., $\sigma_-(-x) = -\sigma_-(x)$). By plugging $\sigma(x)$ into Eq. (2.22) and keeping into account that $\text{sech}^2(x/2)$ is an even function, we realize that the electrical conductivity (see Eq. (2.23)) is completely determined by the even component of the spectral conductivity, whereas the Seebeck coefficient (see Eq. (2.24)) depends on both the even (denominator) and the odd (numerator) components. Thus, if $\sigma(x)$ is an odd function (i.e., $\sigma_+(x) = 0$) we have $J_0 = 0$, thereby $\sigma(T) = 0$, and $ZT = 0$. On the other hand, if the spectral conductivity is an even function (i.e., $\sigma_-(x) = 0$), we have $J_1 = 0$ (with $J_0 \neq 0$), thereby $S(T) = 0$ and the FOM vanishes as well. Therefore, we conclude that the spectral conductivity function of promising TEMs must lack any definite parity. Thus, we realize that materials exhibiting *asymmetrically shaped* electronic structures must be the preference choice in the search for novel TEMs [98–100]. This fundamental conclusion nicely illustrates the useful guide provided by the use of the mathematically inspired spectral conductivity function. We also see that the Mott's spectral model can be regarded as a suitable instance of a spectral conductivity function having no definite parity and exhibiting good TE properties.

As another illustrative example, we can consider the model describing the metal–insulator transition referred to as *Anderson transition*. This transition is related to the wave functions localization due to the presence of disorder in the sample and typically leads to significantly large Seebeck coefficient values once the transition takes place. On this basis, the idea of exploiting this effect for TE cooling has been discussed in the recent literature [97]. In analogy with the previously considered Mott model, the Anderson transition is also described in terms of a spectral conductivity function given by the step function

$$\sigma(E) = \begin{cases} 0, & \text{if } E < E_C \\ \sigma_A \left|E - E_C\right|^{\nu}, & \text{if } E \geq E_C \end{cases} \quad (2.27)$$

where σ_A is a constant, E_C gives the mobility edge location, and $\upsilon > 0$ is a universal critical exponent [101]. We see that for $\nu \neq 0$ this step function is now continuous at $E = E_C$, whereas for $\upsilon = 0$ Eq. (2.27) reduces to the Mott's spectral conductivity model given by Eq. (2.15). Expressing Eq. (2.27) in terms of the scaled variable x, and plugging it into Eq. (2.22) we obtain

$$J_n(\beta) = \sigma_A \beta^{-\upsilon} \int_{x_C}^{\infty} x^n \left|x - x_C\right|^{\nu} \operatorname{sech}^2(x/2)dx. \quad (2.28)$$

Experimentally one can usually approach the Anderson transition from the metallic side.[a] Thus, we will consider the limit $\mu \to E_C$, with $\mu > E_C$ (i.e., $x_C \to 0^-$) in Eq. (2.28), which takes the form $J_n(\beta) = \sigma_A \beta^{-\upsilon} \tilde{J}_m$, where we have introduced the integrals

$$\tilde{J}_m \equiv \int_0^{\infty} x^m \operatorname{sech}^2(x/2)dx, \quad (2.29)$$

with $m \equiv n + \nu > 0$, $n = 0, 1, 2$. These integrals can be analytically calculated. To this end, we express them in the form

$$\tilde{J}_m \equiv \int_0^{\infty} x^m \operatorname{sech}^2(x/2)dx = 2\int_0^{\infty} x^m \frac{d}{dx}\left(\tanh\frac{x}{2}\right)dx, \quad (2.30)$$

[a]For instance, by systematically increasing the disorder amount in the sample.

and integrate by parts to obtain

$$\tilde{J}_m = 2 \lim_{x\to\infty} \left(x^m \tanh \frac{x}{2} \right) - 2m \int_0^\infty x^{m-1} \tanh \frac{x}{2} dx. \tag{2.31}$$

Making use of Eq. (2.18) in the second term of Eq. (2.31), one can express this equation as

$$\tilde{J}_m = \lim_{x\to\infty} (x^m - x^m) + 4m \int_0^\infty \frac{x^{m-1}dx}{1+e^x} = \begin{cases} 4\ln 2, \text{ if } m = 1 \\ 4mI_m, \text{ if } m \neq 1 \end{cases}, \tag{2.32}$$

where $I_m \equiv (1 - 2^{1-m})\Gamma(m)\zeta(m)$, $\Gamma(m)$ is the Gamma function and $\zeta(m)$ is the Riemann zeta function [101]. By plugging Eq. (2.32) into the expression $J_n(\beta) = \sigma_A \beta^{-\upsilon} \tilde{J}_m$, the reduced kinetic coefficients can be explicitly obtained in terms of the model parameters σ_A and υ, and substituting the obtained values in Eqs. (2.23)–(2.25) one finally derives the temperature dependence of the transport coefficients as follows,

$$\sigma(T) = \begin{cases} \sigma_A \ln 2\, \beta^{-1}, \text{ if } \upsilon = 1 \\ \upsilon\sigma_A I_\nu \beta^{-\upsilon}, \text{ if } \upsilon \neq 1 \end{cases}, \tag{2.33}$$

$$S(T) = -\frac{k_B}{|e|} \begin{cases} \dfrac{\pi^2}{6\ln 2}, \quad \text{if } \upsilon = 1 \\ \frac{\upsilon+1}{\upsilon} \dfrac{I_{\nu+1}}{I_\upsilon}, \text{ if } \upsilon \neq 1 \end{cases}, \tag{2.34}$$

$$\kappa_e(T) = \frac{\sigma_A k_B}{e^2} \begin{cases} \left(\dfrac{9}{2}\zeta(3) - \left(\dfrac{\pi^2}{3}\right)^2 \dfrac{1}{\ln 16} \right) \beta^{-2}, \text{ if } \upsilon = 1 \\ \dfrac{F(\upsilon)}{\upsilon I_\upsilon} \beta^{-(\upsilon+1)}, \qquad \text{if } \upsilon \neq 1 \end{cases}, \tag{2.35}$$

where $F(\upsilon) \equiv \upsilon(\nu + 2)I_\nu I_{\nu+2} - (\nu + 1)^2 I_{\nu+1}^2$. We note that the physical condition $\kappa_e > 0$ imposes the restriction $F(\upsilon) > 0$ which, in turn, imposes certain constrains on the possible values of the critical exponent υ. On the other hand, making use of Eqs. (2.33) and (2.35) into Eq. (2.26) we get the following expression for the Lorenz number

$$L(T) = L_0 \begin{cases} \left(\dfrac{27}{\pi^2}\zeta(3) - \dfrac{\pi^2}{3\ln 4} \right) \dfrac{1}{\ln 4}, \text{ if } \upsilon = 1 \\ \dfrac{3}{\pi^2} \dfrac{F(\upsilon)}{(\upsilon I_\upsilon)^2}, \qquad \text{if } \upsilon \neq 1. \end{cases} \tag{2.36}$$

By inspecting Eqs. (2.33)–(2.36), the following conclusions regarding the temperature dependence of the transport coefficients at the Anderson transition edge value can be drawn:

- Both the electrical and thermal conductivities obey a power law temperature dependence of the form $\sigma(T) \sim T^{\upsilon}$ and $\kappa_e(T) \sim T^{\upsilon+1}$, respectively. Therefore, both transport coefficients vanish in the $T \to 0$ limit.
- The Seebeck coefficient is temperature independent and it takes on remarkably large values. For instance, in the case of a linear spectral conductivity function (i.e., $\upsilon = 1$) one has $S(T) = -\frac{k_B}{|e|}\frac{\pi^2}{6\ln 2} \simeq -206\ \mu\text{VK}^{-1}$.
- The value of the Lorenz number is temperature independent as well, that is, the WFL holds in systems undergoing the Anderson transition, and its value depends on the adopted critical exponent value, though it is generally smaller than the Sommerfeld's reference value $L_0 = 2.44 \times 10^{-8}\ \text{V}^2\text{K}^{-2}$. For example, for $\upsilon = 1$ we have $L = 1.61 \times 10^{-8}\ \text{V}^2\text{K}^{-2}$, whereas for $\upsilon = 2$ (parabolic spectral function) one has $L = 2.24 \times 10^{-8}$ V^2K^{-2}.

The relatively large Seebeck coefficient values, along with the increase of the electrical conductivity with the temperature suggest that large power factors may be obtained for materials undergoing an Anderson transition, as far as the chemical potential is properly located close to the mobility edge. Nevertheless, the power law increase of the electronic contribution to the thermal conductivity will reduce the expected FOM as well. In order to estimate the potential of these systems as promising TEMs, we will determine the FOM expression by plugging Eqs. (2.33)–(2.36) into Eq. (2.4) to obtain

$$ZT(T) = \begin{cases} \left[\left(\frac{3}{\pi}\right)^4 \zeta(3)\ln 4 - 1 + L_0 \frac{6\ln 4}{\pi^2}\frac{\kappa_l}{\sigma_A k_B T^2}\right]^{-1}, & \text{if } \upsilon = 1 \\ \left[\frac{\upsilon I_\nu}{(\nu+1)^2 I_{\nu+1}^2}\left((\nu+2)I_{\nu+2} + \left(\frac{e}{k_B}\right)^2 \frac{\kappa_l}{\sigma_A T}\beta^{\upsilon}\right) - 1\right]^{-1}, & \text{if } \upsilon \neq 1 \end{cases} \quad (2.37)$$

As we see the temperature dependence is related to the lattice contribution only. Since $\kappa_l(T \to 0) \sim T^3$ for most materials (see Section 1.2.2.4), we conclude that the FOM value becomes temperature independent in the low temperature regime, and its value is mainly determined by the adopted critical exponent value as far as $\nu \leq 2$. For instance, in the case $\nu = 1$ after Eq. (2.37) we have $ZT(T \to 0) \simeq 2.6$, which is a large value indeed.

This theoretically derived high FOM value naturally reminds us the question regarding the possible existence of a physical upper bond for ZT, an issue that was previously discussed in Section 2.1 without any explicit consideration to the sample's electronic structure. In order to further ascertain this appealing issue, we will make use of the general expressions given by Eqs. (2.23)–(2.25) into Eq. (2.4). After some rearrangement, one obtains (we will explicitly assume $J_0 \neq 0$ and $J_2 \neq 0$, henceforth),

$$ZT = \frac{\Xi}{1 - \Xi + \Psi}, \tag{2.38}$$

where

$$\Xi \equiv \frac{J_1^2}{J_0 J_2} \geq 0, \tag{2.39}$$

is completely determined by the electronic structure, whereas

$$\Psi \equiv \left(\frac{2e}{k_B}\right)^2 \frac{\kappa_l(T)}{T} J_2^{-1} > 0 \tag{2.40}$$

depends on both the lattice dynamics (κ_l) and the electronic structure (J_2). Several important conclusions regarding FOM optimization can be drawn by inspecting Eqs. (2.38)–(2.40):

(1) The case $\Xi = 0$ (i.e., $J_1 = 0$ and $S(T) = 0$) corresponds to the *worst* TEM with $ZT = 0$ at any temperature

(2) The physical requirement $ZT \geq 0$, leads to $\Xi \leq 1 + \Psi$. The upper bond $\Xi = 1 + \Psi$, leading to $ZT \to \infty$ (thereby $\eta \to \eta_C$ after Eq. (1.73)), will correspond to the best possible TEM. Nevertheless, since $\Psi > 0$, this will imply $\Xi > 1$, thereby $J_1^2 > J_0 J_2$. According to Eq. (2.25), this condition leads to *negative* values for the charge carrier contribution to the thermal conductivity, against any physical evidence. Therefore, the condition $\Xi \leq 1$ must be fulfilled instead and ZT will always remain bounded above.

(3) The upper limit $\Xi = 1$ defines the FOM upper bond value $ZT = \Psi^{-1}(T)$. However, in this case we have $J_1^2 = J_0 J_2$, implying $\kappa_e(T) = 0$ and $L(T) = 0$ at any temperature (see Eqs. (2.25) and (2.26)). In that case, electrical conductivity decouples from the (vanishing) charge carrier contribution to the thermal conductivity, and the WFL is no longer obeyed.

As we will see in the following section, the condition $\Xi = 1$ can only be attained when *singular* electronic structures, described in terms of δ-Dirac distribution functions, are considered. Accordingly, we conclude that real materials will always satisfy the condition $\Xi < 1$. Notwithstanding this, the search for materials approaching the $\Xi \to 1$ limit provides a simple and powerful guide in TEMs theoretical research, as will be further discussed in the next section (**Exercise 2.6**).

(4) Making use of Eqs. (2.39) and (2.40), one can express Eq. (2.25) in the form $\kappa_e(T) = \kappa_l(T)(1 - \Xi)\Psi^{-1}$. Since $\kappa_e(T) \neq 0$ for realistic systems, we can replace Ψ in Eq. (2.38) to obtain the factorized expression [98, 102, 103],

$$ZT = \frac{\Xi}{1 - \Xi}\left(1 + \frac{\kappa_l}{\kappa_e}\right)^{-1}, \tag{2.41}$$

where $0 < \Xi < 1$. Since the second factor in Eq. (2.41) is always less than unity, one realizes that any significant breakthrough in the quest for new TEMs must come from suitable electronic structure based approaches, aimed at attaining the $\Xi \to 1$ optimal condition, which ideally defines the best TEM (**Exercise 2.7**).

From the general results presented in this section, three main criteria orienting the fundamental research in TEMs can be outlined:

- In order to optimize the ZT expression given by Eq. (2.41), one must consider materials exhibiting a very low lattice contribution to the thermal conductivity (i.e., $\kappa_l(T) \ll \kappa_e(T)$) over a wide temperature range.
- According to Eq. (2.4), the charge carriers contribution to the thermal conductivity $\kappa_e(T)$ must be also small. However, in order to guarantee that the small value of $\kappa_e(T)$ does not lead to a very small $\sigma(T)$ value, the materials of interest should *not obey* the WFL within the previously determined temperature range.
- For those materials satisfying the two previous requirements, one should be able to properly control their electronic structure close to the Fermi level in order to approach the

optimal limit $\Xi \to 1$. To this end, a suitable electronic structure engineering of the sample, based on both chemical structure refinement and material processing should be undertaken. This topic we will consider in the next section.

2.4 Electronic Structure Engineering

We are now ready to introduce the notion of TE *electronic structure engineering*, which refers to a tailoring process of the electronic structure in the vicinity of the Fermi level, aimed at optimizing the TE performance of a given material. Indeed, during recent years it has been progressively realized that a deeper insight into the TE properties of different materials can be obtained by closely inspecting their electronic structures. Accordingly, important aspects of TE theory deal with notions and mathematical tools borrowed from solid-state physics and condensed matter theory. Most of these concepts will be properly reminded in Section 4.1, as they address key issues related to the role that electronic structure plays in determining the transport properties of solids.

2.4.1 *Regular Electronic Structures*

Standard treatments in solid-state physics usually start by assuming that no singular topological features are present in the electronic structure close to the Fermi level.[a] In that case, one can expand the spectral conductivity function in a Taylor series about the Fermi level in order to derive approximate analytical expressions for the transport coefficients. In fact, by plugging the Taylor expansion

$$\sigma(E) = \sigma(\mu) + (E-\mu)\left(\frac{d\sigma}{dE}\right)_\mu + \frac{1}{2}(E-\mu)^2\left(\frac{d^2\sigma}{dE^2}\right)_\mu + \dots \quad (2.42)$$

[a]We note that the spectral conductivity functions corresponding to the Mott and Anderson models have no derivative at the mobility edge. Therefore, the standard treatment is not applicable in the study of these models, as we saw in the previous section.

into Eq. (2.10), and taking into account Eq. (2.21), one gets

$$\mathcal{L}_{ij} = \frac{(-1)^n}{4}\beta^{-n}\sum_{l=0}^{\infty}\frac{\beta^{-l}}{l!}\left(\frac{d^l\sigma}{dE^l}\right)_{\mu}K_l(n), \tag{2.43}$$

where $n \equiv i + j - 2$, and we have introduced the auxiliary integrals

$$K_l(n) \equiv \int_{-\infty}^{\infty} x^{n+l}\,\mathrm{sech}^2(x/2)dx, \tag{2.44}$$

which are expressed in terms of the scaled variable x. Then, keeping terms up to $l = 2$ in Eq. (2.43), Eqs. (2.12)–(2.14) can be expressed as

$$\sigma(T) \simeq \sigma(\mu) + \frac{b}{2}\sigma''(\mu)T^2, \tag{2.45}$$

$$S(T) \simeq -\frac{|e|L_0 T\sigma'(\mu)}{\sigma(\mu) + \frac{b}{2}\sigma''(\mu)T^2}, \tag{2.46}$$

$$\kappa_e(T) \simeq \frac{L_0 T}{\sigma(T)}\left[\sigma^2(\mu) + \left(\frac{13}{5}\sigma(\mu)\sigma''(\mu) - \sigma'(\mu)^2\right)bT^2\right], \tag{2.47}$$

where the prime stands for the energy derivative, $b \equiv e^2 L_0 = 2.44 \times 10^{-8}$ $(\mathrm{eV})^2\mathrm{K}^{-2}$, and we have made use of the integrals

$$\int_{-\infty}^{\infty} \mathrm{sech}^2(x/2)dx = 4, \quad \int_{-\infty}^{\infty} x^2\,\mathrm{sech}^2(x/2)dx = \frac{4\pi^2}{3},$$
$$\int_{-\infty}^{\infty} x^4\,\mathrm{sech}^2(x/2)dx = \frac{28\pi^4}{15}, \quad \int_{-\infty}^{\infty} x^l\,\mathrm{sech}^2(x/2)dx = 0, \;\; (l \text{ odd}) \tag{2.48}$$

As we can see, the temperature dependence of the transport coefficients is described by polynomial and rational functions whose coefficients depend on the slope $\sigma'(\mu)$ and the curvature $\sigma''(\mu)$ of the spectral conductivity function evaluated at the Fermi-level position. When the spectral conductivity can be approximated in terms of a simple linear function close to the Fermi level, the curvature term $\sigma''(\mu)$ becomes negligible and Eqs. (2.46) and (2.47) adopt the simpler form

$$S(T) \simeq -|e|L_0\xi T, \tag{2.49}$$

$$\kappa_e(T) \simeq L_0 T\sigma(\mu)\left[1 - b\xi^2 T^2\right], \tag{2.50}$$

where $\xi \equiv \sigma'(\mu)/\sigma(\mu)$. For materials whose electronic states exhibit a nearly constant diffusivity D_0, one has $\sigma(E) = e^2 D_0 N(E)/V$, and Eq. (2.49) reduces to the celebrated Mott's expression [8–10],

$$S(T) = -|e|L_0 \left(\frac{1}{N(E)}\frac{dN(E)}{dE}\right)_\mu T = -\frac{\pi^2 k_B^2}{3|e|}\left(\frac{d\ln N(E)}{dE}\right)_\mu T. \tag{2.51}$$

By inspecting this expression one realizes that, in order to exhibit large S values, two main features should be present in the considered material DOS, namely, (a) a sharp peak close to the Fermi level, resulting in a large DOS slope, thereby leading to large values of the DOS derivative, and (b) a substantial depletion of states (i.e., a pseudogap), thereby leading to a small value of the DOS close to the Fermi level. In this way, from very general considerations, one concludes that the *simultaneous* presence of a small DOS background and a narrow peak near the Fermi level will generally contribute to enhance the TE performance of the considered material.

Now, the presence of a pseudogap leads to two counterbalancing effects: on the one hand, a small $N(E)$ value leads to an electrical conductivity degradation (lowering the ZT value); on the other hand, according to the WFL, this lower electrical conductivity value reduces the charge carrier contribution to the thermal conductivity, hence increasing ZT. Indeed, by comparing Eqs. (2.49) and (2.50) we realize that the condition $|\xi| \gg 1$, favoring a large Seebeck coefficient value (in absolute value), also implies a substantial reduction of the $\kappa_e(T)$ value, hence further increasing the resulting ZT value. Thus, in order to properly optimize the TE performance of a promising TEM a very delicate tuning is required in its electronic structure.

2.4.2 *Singular Electronic Structures*

In Section 2.3, we learnt that promising TEMs are characterized by having electronic structures approaching the limit $\Xi \to 1$. Thus, the question concerning what is the most appropriate $\sigma(x)$ functional form in order to attain the ideal $\Xi = 1$ condition naturally appears. A relevant hint to answer this question was obtained in Section 2.4.1, where it was concluded that, in order to obtain a large S value, the electronic structure must exhibit a relatively sharp narrow feature.

Following this line of thought, Mahan and Sofo proposed in 1996 that the best TEM should be found among solids exhibiting a pronounced narrow peak in their electronic structure close to the Fermi level [104]. Indeed, it can be shown that the *ideal* electronic structure is given by a spectral conductivity function described in terms of a δ-Dirac distribution of the form $\sigma(E) = g(E)\,\delta(E - E_\delta)$, where $g(E) > 0$ is an arbitrary analytical function and E_δ indicates the location of the δ peak. In fact, taking into account the property $\delta(\lambda x) = |\lambda|^{-1}\delta(x)$, where $\lambda \in \mathbb{R}$, the spectral conductivity function reads $\sigma(x) = \beta g(x)\delta(x - x_\delta)$. Then, making use of the well-known property

$$\int_{-\infty}^{+\infty} F(x)\delta(x - x_\delta)dx = F(x_\delta), \tag{2.52}$$

in Eq. (2.22) one obtains,

$$J_0 = \beta\, g(x_\delta)\,\mathrm{sech}^2(x_\delta/2), \quad J_1 = x_\delta J_0, \quad J_2 = x_\delta^2 J_0, \tag{2.53}$$

and substituting the obtained reduced coefficients into Eq. (2.39), one realizes that the ideal value $\Xi = 1$ is satisfied by this singular electronic structure *at any temperature*. This is a very remarkable result, since Eq. (2.30) reduces to

$$ZT(x_\delta, T) = \Psi^{-1}(x_\delta) = \frac{k_B}{e^2}\frac{g(x_\delta)}{\kappa_l(T)}\theta^2(x_\delta) \equiv ZT_\delta, \tag{2.54}$$

in this case, where $\theta(x) \equiv (x/2)\,\mathrm{sech}\,(x/2)$. Thus, at any given temperature, the resulting FOM only depends on the δ-peak position through the parameter x_δ. Eq. (2.54) also indicates that the δ-Dirac function *must not* be located at the Fermi level (i.e., $x_\delta = 0$), since one gets $ZT_\delta = 0$ in that case. The most appropriate location of the δ-peak can be determined from the extreme condition $d(ZT_\delta)/dx_\delta = 0$, leading to the equation[a]

$$\frac{d\ln g}{dx_\delta} = \tanh\left(\frac{x_\delta}{2}\right) - \frac{2}{x_\delta}, \quad (x_\delta \neq 0). \tag{2.55}$$

For the particular case of a constant g, Eq. (2.55) reduces to $\tanh(x_\delta/2) = 2/x_\delta$, whose solutions, $x_\delta^\pm = \pm 2.399\ldots$, lead to the

[a]The solution $\theta(x_\delta) = 0 \Leftrightarrow x_\delta = 0$ leads to the minimum $ZT_\delta = 0$.

optimal FOM value given by [104],

$$ZT_\delta(x_\delta^\pm, T) = \frac{k_B}{e^2}\left[\left(\frac{x_\delta^\pm}{2}\right)^2 - 1\right]\frac{g}{\kappa_l(T)}. \tag{2.56}$$

In summary, the singular electronic structures based on a Dirac delta function exactly satisfy the ideal $\Xi = 1$ condition for FOM optimization. Note, however, that the precise location of this peak is critical to obtain an excellent TE performance: if it is located too close to the Fermi level then one would obtain the worst, rather than the best possible electronic structure. On the other hand, one must keep in mind that the mathematically inspired δ-Dirac electronic structure leads to unphysical behaviors for several transport coefficients. In fact, as we previously indicated in Section 2.3, the condition $\Xi = 1$ renders $\kappa_e(T) = 0$ and $L(T) = 0$ at any temperature. In addition, one obtains a divergent Seebeck coefficient of the form $S(T) \propto T^{-1}$ at low temperatures, as can be deduced from Eqs. (2.24) and (2.53).

In order to avoid these unphysical results, still preserving a Ξ value close enough to unity, some alternatives have been considered in the literature. For instance, one may include a constant background contribution in the electronic structure, namely, $\sigma(x) = \beta g\delta(x - x_\delta) + \sigma_B$ (**Exercise 2.8**). In that case, the FOM value becomes inversely proportional (in a marked nonlinear way) to the σ_B value near the singularity. This confirms that TEMs must exhibit electronic structures characterized by a severe depletion of the DOS close to the Fermi level [104].

Another possibility is to consider the presence of two narrow features in the electronic structure in the form [99, 100, 105, 106],

$$\sigma(x) = \beta\left[g_1(x)\delta(x - x_1) + g_2(x)\delta(x - x_2)\right], \tag{2.57}$$

where $x_i = (E_i - E_F)\beta$ indicates the location of each spectral feature in the energy spectrum. In the case $x_1 = x_2$ (i.e., $E_1 = E_2$) Eq. (2.57) reduces to the previously discussed single δ-Dirac electronic structure with $g \equiv g_1 + g_2$, so that we will assume $E_1 < E_2$ henceforth. Making use of Eq. (2.57) into Eqs. (2.22) and (2.39), one obtains,

$$\Xi = \frac{(x_1u_1 + x_2u_2)^2}{(u_1 + u_2)(x_1^2u_1 + x_2^2u_2)} = \left[1 + \left(\frac{x_1 - x_2}{x_1u_1 + x_2u_2}\right)^2 u_1u_2\right]^{-1} < 1, \tag{2.58}$$

where $u_i \equiv g_i(x_i)\,\mathrm{sech}^2\,(x_i/2) > 0$. Thus, the ideal value $\Xi = 1$ cannot be strictly attained in the double-δ case. Let us consider that $x_1 \neq 0$, so that we can take the spectral peak located at E_1 as a reference energy scale. Then, Eq. (2.58) can be rewritten in the form

$$\Xi(\chi, \gamma, x_1) = \frac{(1 + \chi\gamma h)^2}{(1 + \gamma h)(1 + \chi^2\gamma h)}, \tag{2.59}$$

where $\chi \equiv x_2/x_1$ can take on positive or negative values depending on the relative location of the peaks with respect to the Fermi level, $\gamma \equiv g_2(\chi x_1)/g_1(x_1) > 0$, and $h(\chi, x_1) \equiv \cosh^2(x_1/2)\,\mathrm{sech}^2\,(\chi x_1/2) > 0$ (note that at room temperature and above the $\lim_{T\to\infty} h = 1$ is confidently satisfied). As expected, in the particular cases $g_2 = 0$ (i.e., $\gamma = 0$) or, $|E_2| = |E_1|$ (i.e., $\chi = 1$, $h = 1$), Eq. (2.59) reduces to the single-δ value $\Xi = 1$. Making use of Eqs. (2.59), (2.38), (2.54), and (2.40), we get

$$ZT_{2\delta}(\chi, \gamma, x_1) = \frac{(1 + \chi\gamma h)^2}{(\chi - 1)^2\gamma h + (1 + \gamma h)ZT_\delta^{-1}}, \tag{2.60}$$

which properly reduces to ZT_δ when $g_2 \to 0$. On the contrary, if one increases the relative contribution of the second spectral feature (say, $g_1 \ll g_2$) then the FOM progressively increases, eventually outperforming any arbitrarily chosen ZT_δ value for large enough γ values. Therefore, the larger the difference between the peak heights, the *better* the resulting TE performance, confirming the convenience of considering asymmetric spectral structures in the search for good TEMs. In fact, by inspecting Eq. (2.60) we see that a necessary condition to get $ZT = 0$ is allowing for $\chi < 0$, hence indicating that it is not convenient to locate the Fermi level in between the spectral peaks.

For the sake of illustration, let us consider that each spectral feature is symmetrically located around the Fermi level ($\chi = -1$), as it is sketched in Fig. 2.4a. Then, Eq. (2.59) reads $\Xi = (1-\gamma)^2/(1+\gamma)^2$, so that, whereas we get $\Xi = 1$ in either $\gamma \to 0$ or $\gamma \to \infty$ limits, one has $\Xi = 0$ (i.e., $ZT = 0$) if both peaks have the same height and the electronic structure exhibits *mirror* symmetry ($\gamma = 1$, see Fig. 2.4a).

At any given temperature, the best possible location of the peaks is derived from the extreme condition $d(ZT_{2\delta})/dx_1 = 0$. For constant g_i values, the extreme condition is satisfied for $\gamma = 1$ (leading to the

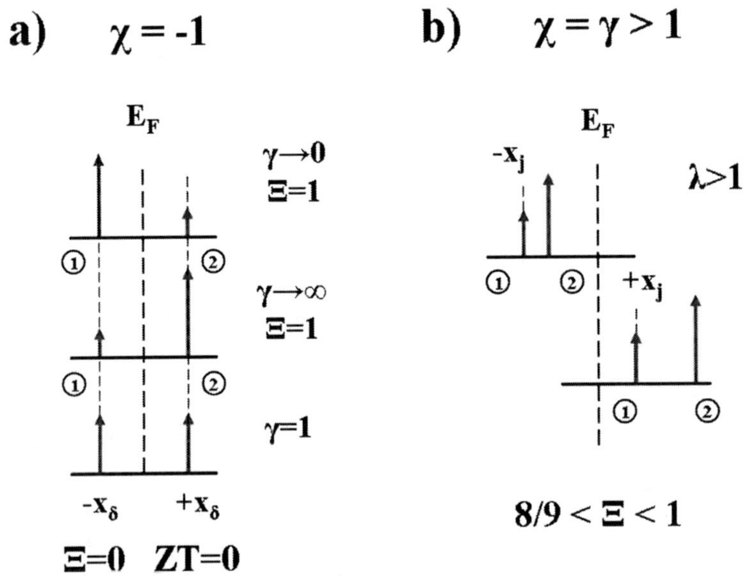

Figure 2.4 Double δ-Dirac electronic structure models of interest adopting a symmetrical (a) and a self-similar (b) configuration with respect to the Fermi energy E_F (dashed vertical line).

$ZT_{2\delta} = 0$ minimum, irrespective of the peaks location) and $x_{2\delta}^{\pm} = x_{\delta}^{\pm}$, that is, the positions maximizing the FOM in the double-δ case are just the same as those obtained in the single-δ case (see Fig. 2.4a).

An interesting particular case satisfying the $\sigma(x)$ asymmetric condition is given by the choice $\chi = \gamma \equiv \lambda > 0$, describing a *self-similar* electronic structure, where the relative position of the spectral features is related to their respective peak heights by a common ratio given by the scale factor λ (see Fig. 2.4b). In this case, Eq. (2.59) reads

$$\Xi(\lambda) = \frac{(1+\lambda^2)^2}{(1+\lambda)(1+\lambda^3)}, \tag{2.61}$$

at room temperature ($h = 1$). The $\Xi(\lambda)$ curve exhibits two minima, $\Xi(2 \pm \sqrt{3}) = 8/9$, and a maximum $\Xi(1) = 1$, which corresponds to the single-δ case ($\lambda = 1$). In Fig. 2.5 we show the $ZT_{2\delta}(\gamma)$ curve at room temperature for the two kinds of electronic structures sketched in Fig. 2.4 As we can see, in the symmetrical $\chi = -1$

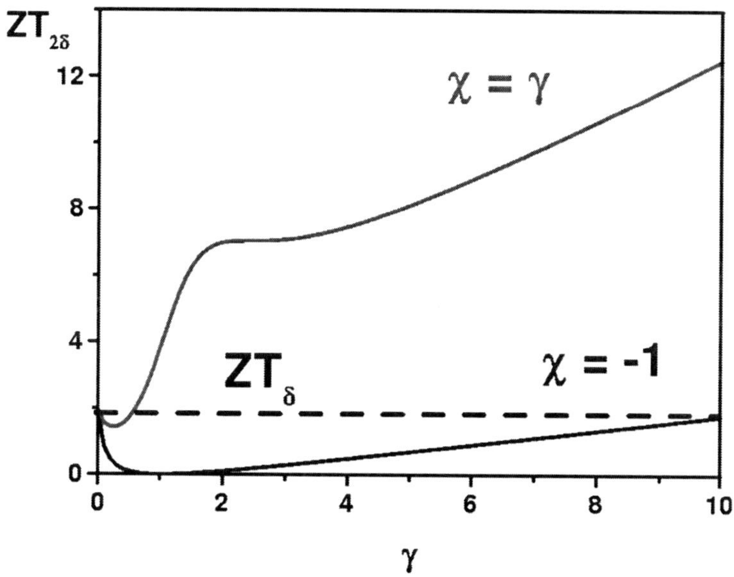

Figure 2.5 Room temperature $ZT_{2\delta}$ curves as a function of the parameter γ for double-δ electronic structure models describing (a) the symmetric configuration ($\chi = -1$) shown in Fig. 2.4a, and (b) the self-similar configuration ($\chi = \gamma$) shown in Fig. 2.4b. The dashed line indicates the $ZT_\delta(x_\delta^\pm) = 1.9$ value corresponding to the optimal single-δ case. We have adopted $g = 500$ $(\Omega\ \text{cm})^{-1}$ eV.

configuration, $ZT_{2\delta}$ monotonically decreases in the range $0 < \gamma < 1$, vanishing when $\gamma = 1$, and then progressively increases as γ is increased, reaching the optimal single-δ case $ZT_\delta(x_\delta^\pm)$ value for $\gamma \simeq 10$. Quite remarkably, significantly larger ZT values, well above the optimal single-δ one, are obtained in the self-similar electronic configuration over a wide range of γ values.

In Fig. 2.6, we compare the room temperature ($h = 1$) Ξ curves for the choices $\chi = 0$, $\chi = -1$, and $\chi = \gamma$. Note that Ξ identically vanishes in the case $\chi = -\gamma$. While the Ξ function values are significantly smaller than unity over most of the γ parameter range for both the $\chi = 0$ and $\chi = -1$ cases, we see that $8/9 \leq \Xi \leq 1$ over the entire γ range in the self-similar electronic structure case. As we can see, the self-similar electronic structure gives larger ZT values than the symmetrical one for any selection of the electronic

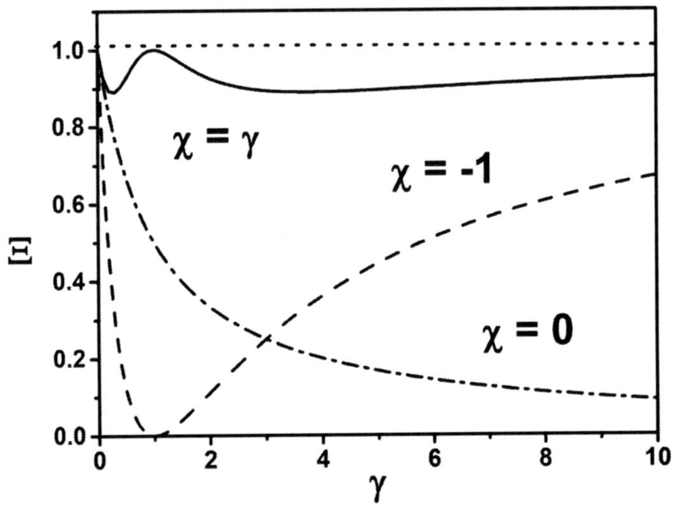

Figure 2.6 Room temperature Ξ curves as a function of the parameter γ for the double-δ electronic structure model in the cases $\chi = \gamma$ (solid line), $\chi = -1$ (dashed line) and $\chi = 0$ (dot-dashed line). We have adopted $g = 500\ (\Omega\text{cm})^{-1}\text{eV}$.

structure model parameters. Since quasicrystals and related phases are characterized by scale-invariant lattice geometries, this result significantly spurs the interest in considering them as prospective TEMs, as we will fully discuss in Chapter 5.

In summary, large $ZT > 1$ values can be obtained over a wide range of model parameters in electronic structures containing sharp features close to the Fermi level, with the proviso that: (i) they have different heights, the more different they are the better; and (ii) none of the features is exactly located at the Fermi level. These results properly illustrate the high sensitivity of TE properties to minor changes in the electronic structure of the sample, as well as the importance of using electronic structure engineering concepts in order to guide the experimental research.

2.4.3 *Spectral Conductivity Shape Effect*

A δ-Dirac function is just a mathematically convenient way of describing a very narrow, sharp peak in the electronic structure.

In practice, δ functions can be approximated in terms of suitable analytical expressions such as Gaussian or Lorentzian functions satisfying appropriate limits, and some recent works have studied in detail the role played by the shape of the spectral conductivity function in the resulting TE performance in that case. Two broad categories of spectral functions have been considered, namely, normalized (i.e., $\int_{-\infty}^{+\infty} \sigma(x)dx = C_1$) and bounded (i.e., $\sigma(x) \leq C_2$, $\forall x$) ones, where C_1 and C_2 are arbitrary constants. For instance, the spectral conductivity function $\sigma(x) = \sigma_c(1 + \tanh(x/2))$, studied in the **Exercise 2.7**, is bounded, since $\sigma(x) \leq 2\sigma_c\ \forall x$,[a] but it is not normalizable, since ($\sigma_c = 1$)

$$\int_{-\infty}^{+\infty} \sigma(x)dx = \int_{-\infty}^{+\infty} \left(1 + \tanh\frac{x}{2}\right) dx = 2\int_{-\infty}^{+\infty} \frac{e^x dx}{1+e^x}$$
$$= 2\lim_{x\to+\infty} \ln(1+e^x) - 2\lim_{x\to-\infty} \ln(1+e^x) \to +\infty$$

In the case of normalized distributions, it has been confirmed that the Dirac delta function gives the highest possible FOM, though similarly large values can be obtained for both Gaussian and rectangular $\sigma(x)$ functions, centered at $x_0 = x_\delta^{\pm}$, in the vanishing width limit $\Delta x \to 0$. In addition, in the case of the rectangularly shaped spectral functions, ZT decreases very slowly as Δx is increased [100]. When the normalization condition is relaxed, one excludes the very possibility of obtaining a Dirac delta function as a (limit) solution to the optimization problem. In that case, numerical simulations showed that the optimized x_0 value shifts to $x_0 \simeq 3.3$ for bounded Gaussian and to $x_0 \simeq 3.5$ for bounded rectangular functions, respectively. In addition, rectangularly shaped $\sigma(x)$ functions exhibit higher ZT values than Gaussian profiles for any considered width value [100].

Bounded step functions of the form

$$\sigma(E) = \begin{cases} \sigma_a, & \text{if } x < x_0 \\ \sigma_b, & \text{if } x \geq x_0 \end{cases} \tag{2.62}$$

where x_0 measures the location of the step discontinuity with respect to the Fermi level, have been considered as well [98]. Note that this spectral function reduces to the previously considered

[a] Recall the limits $\lim_{x\to\pm\infty} \tanh(x/2) = \pm 1$.

Mott's model when $\sigma_a = 0$. Depending on the adopted x_0 value relatively large values of the $\Xi(1 - \Xi)^{-1}$ factor in Eq. (2.41) can be obtained (**Exercise 2.9**).

In summary, the search for materials exhibiting narrow features in the electronic structure close to the Fermi level is a promising approach in current TE research. These narrow features may be related to impurity levels, be introduced via a DOS modulation due to stoichiometric changes (e.g., modifying the number of atoms contributing to core f levels in the total DOS of a rare-earth bearing compound) or may stem from hybridization effects between partially filled d-bands and the (s, p)-bands. We will consider these topics in more detail in Chapters 4 and 5.

2.5 Exercises

(1) The room temperature resistivity and Seebeck coefficient values of the clathrate compounds (see Section 3.5.3) $Sr_8Ga_{16}Ge_{30}$ and $Eu_8Ga_{16}Ge_{30}$ are $\rho = 12.8$ and $\rho = 2.52$ mΩ cm, and $S = -313$ and $S = -152$ μVK^{-1}, respectively [108]. Obtain and compare their respective power factors.

(2) The $CoGe_{1.5}Se_{1.5}$ skutterudite (see Section 3.5.2) has a very high Seebeck coefficient value at room temperature ($S = -570$ μVK^{-1}). Determine its room temperature ZT value making use of the transport data $\kappa_l = 6.0$ Wm^{-1}K^{-1} and $\rho = 3.37$ Ωcm [107]. Discuss the obtained results.

(3) The longitudinal and transversal sound velocities (obtained from room temperature ultrasonic measurements) of the skutterudite compound $IrSb_3$ (with a cubic lattice parameter $a = 9.250$ Å and containing $N = 16$ atoms in the primitive unit cell, see Section 3.5.2) are $v_l = 4675$ ms^{-1} and $v_t = 2717$ ms^{-1}, respectively [109]. Determine the minimal thermal conductivity of this compound.

(4) The longitudinal and transversal sound velocities of the clathrate compound $Ba_8Ga_{16}Ge_{30}$ (with a cubic lattice parameter $a = 10.78$ Å and containing $N = 23$ atoms in the

primitive unit cell, see Section 3.5.3) are $v_l = 4096$ ms^{-1} and $v_t = 2795$ ms^{-1}, respectively [110]. Determine the minimal thermal conductivity of this compound.

(5) Taking the appropriate limits in Eq. (2.20), obtain the asymptotic temperature dependence of the Seebeck coefficient for the Mott's spectral conductivity model in the cases $\mu > E_C$, and $\mu < E_C$.

(6) Obtain the analytical expression for the Ξ function for (a) the Mott model, and (b) the Anderson model (with $\nu = 1$) in the case $\mu = E_C$.

(7) Let us consider a spectral conductivity model given by the function $\sigma(E) = \sigma_c \left[1 + \tanh\left(\frac{E-\mu}{2k_B T}\right)\right]$. Show that this spectral conductivity lacks a definite parity. Obtain the analytical expressions for (a) the electronic Ξ function, (b) the electronic contribution to the thermal conductivity, and (c) the FOM of this system.

(8) Demonstrate that the condition $\Xi = 1$ is no longer satisfied when a constant background contribution of the form $\sigma(x, \beta) = \beta g \delta(x - x_\delta) + \sigma_*$ is added to the δ-Dirac based spectral conductivity model.

(9) Let us consider the spectral conductivity model given by the step function ($\sigma_a < \sigma_b$)

$$\sigma(E) = \begin{cases} \sigma_a, \text{ if } E < E_C \\ \sigma_b, \text{ if } E \geq E_C \end{cases} \tag{2.63}$$

which reduces to the Mott's model in the case $\sigma_a = 0$. Obtain the analytical expressions for (a) the electronic Ξ function, (b) the electronic contribution to the thermal conductivity, and (c) the FOM of this system when the Fermi level is located at the mobility edge E_C. Discuss the obtained results.

2.6 Solutions

(1) The power factor is given by $P = \sigma S^2$. Thus, for the $Sr_8Ga_{16}Ge_{30}$ compound we have

$$P = \frac{1000}{12.8}\Omega^{-1}\text{cm}^{-1}(-313\,)^2 \times 10^{-12}\,\text{V}^2\text{K}^{-2}$$
$$= 7.65\ \mu\text{Wcm}^{-1}\text{K}^{-2},$$

whereas for the $Eu_8Ga_{16}Ge_{30}$ clathrate we get

$$P = \frac{1000}{2.52}\,\Omega^{-1}\text{cm}^{-1}(-152\,)^2 \times 10^{-12}\,\text{V}^2\text{K}^{-2}$$
$$= 9.17\ \mu\text{Wcm}^{-1}\text{K}^{-2}.$$

Therefore, the most convenient material for TE applications is $Eu_8Ga_{16}Ge_{30}$, albeit it exhibits a significant lower Seebeck coefficient. This result illustrates that a large Seebeck coefficient value does not guarantee by its own that the considered sample is a good TEM, since extremely small electrical conductivity values can counterbalance the resulting power factor

(2) Assuming the WFL is valid for this sample, after Eq. (1.49), we have

$$\kappa = \kappa_l + L_0\sigma T$$
$$= 6.0\ \text{Wm}^{-1}\text{K}^{-1} + 2.44 \times 10^{-8}\ \text{V}^2\text{K}^{-2}\frac{10^2}{3.37}\ \Omega^{-1}\text{m}^{-1}\ 300\ \text{K}$$
$$= 6.0002\ \text{Wm}^{-1}\text{K}^{-1}.$$

Thus, we realize that, due to its high resistivity, the electron contribution to the thermal conductivity is negligible at room temperature. Making use of Eq. (2.4), we have

$$ZT = 29.67\ \Omega^{-1}\ \text{m}^{-1}\frac{(-570)^2 \times 10^{-12}\ \text{V}^2\text{K}^{-2} \times 300\ \text{K}}{6\ \text{Wm}^{-1}\text{K}^{-1}}$$
$$= 4.8 \times 10^{-4}.$$

This result further illustrates that a large Seebeck coefficient value does not guarantee by its own that the considered sample is a good TEM, since small electrical conductivity values (which are beneficial in order to reduce the charge carrier contribution to the thermal conductivity) result in a low power factor, leading to a quite small ZT value (comparable to those of typical metallic systems for this sample).

(3) Substituting the reported data into Eq. (2.5), we have

$$\kappa_m = 0.4 \times 1.38 \times 10^{-23}\,\text{JK}^{-1} \left(\frac{16}{9.25^3}\right)^{2/3} 10^{20}\,\text{m}^{-2}$$
$$\times(4675 + 2 \times 2717)\,\text{ms}^{-1} = 0.414\ \text{Wm}^{-1}\text{K}^{-1}.$$

(4) Substituting the reported data into Eq. (2.5), we have

$$\kappa_m = 0.4 \times 1.38 \times 10^{-23}\,\text{JK}^{-1} \left(\frac{23}{10.78^3}\right)^{2/3} 10^{20}\,\text{m}^{-2}$$
$$\times(4096 + 2 \times 2795)\,\text{ms}^{-1} = 0.372\ \text{Wm}^{-1}\text{K}^{-1}.$$

(5) In order to obtain the low ($T \to 0$) and high ($T \to \infty$) temperature limits, it is convenient to write Eq. (2.20) in the form

$$S(T) = -\frac{k_B}{|e|}\left[x_C + 2\frac{\ln\cosh\frac{x_C}{2} - \frac{x_C}{2} + \ln 2}{1 - \tanh\frac{x_C}{2}}\right], \tag{2.64}$$

where we have used Eq. (2.18). In this way, in the high temperature limit we get

$$S(T \to \infty) = -\frac{2k_B}{|e|}\lim_{x_C \to 0^{\pm}} \frac{\ln\cosh\frac{x_C}{2} - \frac{x_C}{2} + \ln 2}{1 - \tanh\frac{x_C}{2}} = -\frac{2k_B}{|e|}\ln 2$$
$$= -\frac{k_B}{|e|}\ln 4 \equiv S_M,$$

where the limit $x_C \to 0^+$ ($x_C \to 0^-$) applies to the case $E_C > \mu$ ($E_C < \mu$), respectively. On the other hand, in the case $E_C > \mu$ the low temperature limit ($x_C \to +\infty$) is given by

$$S(T \to 0) = -\frac{k_B}{|e|}\left[\lim_{x_C \to +\infty} x_C + 2\lim_{x_C \to +\infty}\frac{\ln\cosh\frac{x_C}{2} - \frac{x_C}{2} + \ln 2}{1 - \tanh\frac{x_C}{2}}\right]. \tag{2.65}$$

Recalling that $\lim_{x_C \to +\infty}\tanh\frac{x_C}{2} = 1$, and taking into account the asymptotic limit $\lim_{x_C \to +\infty}\ln\cosh\frac{x_C}{2} \simeq \frac{x_C}{2} - \ln 2$, the second term in Eq. (2.65) leads to an indefinite limit of the form 0/0, which can be solved by using the L'Hôpital rule twice to obtain

$$\lim_{x_C \to +\infty}\frac{\ln\cosh\frac{x_C}{2} - \frac{x_C}{2} + \ln 2}{1 - \tanh\frac{x_C}{2}} = \lim_{x_C \to +\infty}\frac{1 - \tanh\frac{x_C}{2}}{\text{sech}^2\frac{x_C}{2}}$$
$$= +\lim_{x_C \to +\infty}\frac{1}{2\tanh\frac{x_C}{2}} = 1/2.$$

Thus, Eq. (2.65) reads

$$S(T \to 0) = -\frac{k_B}{|e|} \lim_{x_C \to +\infty} x_C = -\infty$$

Finally, in the case $E_C < \mu$ the low temperature limit ($x_C \to -\infty$) is given by

$$S(T \to 0) = -\frac{k_B}{|e|} \left[\lim_{x_C \to -\infty} x_C + 2 \lim_{x_C \to -\infty} \frac{\ln \cosh \frac{x_C}{2} - \frac{x_C}{2} + \ln 2}{1 - \tanh \frac{x_C}{2}} \right]. \tag{2.66}$$

Recalling that $\lim_{x_C \to -\infty} \tanh \frac{x_C}{2} = -1$, and taking into account the asymptotic limit $\lim_{x_C \to -\infty} \ln \cosh \frac{x_C}{2} \simeq \frac{|x_C|}{2} - \ln 2$, Eq. (2.66) can be written in the form

$$S(T \to 0) = -\frac{k_B}{|e|} \lim_{x_C \to -\infty} \left(x_C + 2 \frac{\frac{|x_C|}{2} + \frac{|x_C|}{2}}{2} \right) = 0,$$

so that $S(T \to 0) = 0$ in this case.

(6) (a) In terms of the dimensionless x variable, the Mott spectral function reads

$$\sigma(x) = \begin{cases} 0 & \text{if } x < 0 \\ \sigma_M & \text{if } x \geq 0 \end{cases} \tag{2.67}$$

when the Fermi level is located at the mobility edge. Plugging this expression in Eq. (2.22) and making use of the integrals

$$\int_0^\infty \operatorname{sech}^2(x/2)dx = 2, \quad \int_0^\infty x \operatorname{sech}^2(x/2)dx = 4\ln 2,$$
$$\int_0^\infty x^2 \operatorname{sech}^2(x/2)dx = \frac{2\pi^2}{3} \tag{2.68}$$

we get $J_0 = 2\sigma_M$, $J_1 = 4\sigma_M \ln 2$, and $J_2 = 2\pi^2\sigma_M/3$, and substituting these reduced kinetic coefficient values in Eq. (2.39), we finally obtain $\Xi = 12\left(\frac{\ln 2}{\pi}\right)^2 \simeq 0.584\ldots < 1$.

(b) Making use of expression $J_n(\beta) = \sigma_A \beta^{-\upsilon} \tilde{J}_m$, along with Eq. (2.32) for $\upsilon = 1$ we have

$$J_0 = 4\sigma_A \beta^{-1} \ln 2,$$
$$J_1 = 4\sigma_A \beta^{-1}(1 - 2^{-1})\Gamma(2)\zeta(2) = \frac{2\pi^2}{3}\sigma_A \beta^{-1},$$
$$J_2 = 4\sigma_A \beta^{-1} 3(1 - 2^{-2})\Gamma(3)\zeta(3) = 18\zeta(3)\sigma_A \beta^{-1}.$$

Substituting these reduced kinetic coefficient values in Eq. (2.39), we finally obtain $\Xi = \frac{\pi^4}{162\zeta(3)\ln 2} = 0.721\ldots < 1$.
By comparing with the result obtained in (a) we realize that the electronic Ξ function for the Anderson model is closer to the ideal condition $\Xi = 1$.

(7) In terms of the scaled variable x the spectral conductivity function reads $\sigma(x) = \sigma_c(1 + \tanh(x/2))$, so that $\sigma(-x) = \sigma_c(1 - \tanh(x/2))$, an expression which differs from both $\sigma(x)$ and $-\sigma(x)$. Therefore, $\sigma(x)$ has no definite parity.

(a) Plugging $\sigma(x)$ into Eq. (2.22), keeping into account that $\tanh(x/2)\,\mathrm{sech}^2(x/2)$ is an odd function, and making use of the integrals given by Eq. (2.48), we obtain $J_0 = 4\sigma_c$ and $J_2 = 4\pi^2\sigma_c/3$. On the other hand, we have

$$J_1 = \sigma_c \int_{-\infty}^{+\infty} x \tanh\frac{x}{2}\,\mathrm{sech}^2\frac{x}{2}\,dx, \qquad (2.69)$$

which can be integrated by parts using the integral

$$\int \tanh\frac{x}{2}\,\mathrm{sech}^2\frac{x}{2}\,dx = \int \frac{d}{dx}\left(\tanh^2\frac{x}{2}\right)dx$$
$$= \tanh^2\frac{x}{2} = 1 - \mathrm{sech}^2\frac{x}{2},$$

to obtain

$$\int x \tanh\frac{x}{2}\,\mathrm{sech}^2\frac{x}{2}\,dx = x\left(1 - \mathrm{sech}^2\frac{x}{2}\right) - x + \int \mathrm{sech}^2\frac{x}{2}\,dx$$
$$= -x\,\mathrm{sech}^2\frac{x}{2} + \int \mathrm{sech}^2\frac{x}{2}\,dx. \qquad (2.70)$$

Since

$$\lim_{x\to\pm\infty}\left(x\,\mathrm{sech}^2\frac{x}{2}\right) = 0, \qquad (2.71)$$

making use of Eqs. (2.70) and (2.71) into Eq. (2.69) and taking into account Eq. (2.48), we get

$$J_1 = \sigma_c \int_{-\infty}^{+\infty} \mathrm{sech}^2\frac{x}{2}\,dx = 4\sigma_c = J_0.$$

Making use of the obtained reduced kinetic coefficients into Eq. (2.39), we finally obtain $\Xi = 3/\pi^2 \simeq 0.304 < 1$. Thus the proposed spectral conductivity does not improves the Ξ values previously obtained for both the Mott and Anderson models in **Exercise 2.6**.

(b) Making use of the reduced kinetic coefficients obtained in (a) into Eq. (2.25) after some arrangement we have

$$\kappa_e(T) = \frac{\pi^2 - 3}{\pi^2} L_0 \sigma_c T.$$

Thus, the considered system obeys the Wiedemann–Franz law, with an effective Lorenz number which is smaller than the Sommerfeld value (i.e., $\simeq 0.7 L_0$).

(c) Making use of the results obtained in (a) and (b) into Eq. (2.41) we have

$$ZT = a^{-1}\left(1 + \frac{\pi^2}{3}\frac{k_l(T)}{a L_0 \sigma_c T}\right)^{-1},$$

where $a \equiv (\pi^2 - 3)/3$.

(8) Plugging the proposed spectral conductivity function into Eq. (2.22) and making use of the integrals given by Eq. (2.48), we have

$$J_0 = x_\delta^{-1} J_1 + 4\sigma_B, \quad J_2 = x_\delta J_1 + \frac{4\pi^2}{3}\sigma_B, \quad J_1 = \beta x_\delta g \operatorname{sech}^2(x_\delta/2),$$

and substituting into Eq. (2.39), we get

$$\Xi = \left[1 + \left(\frac{\pi^2}{3} + x_\delta^2\right) h + \frac{\pi^2}{3} x_\delta^2 h^2\right]^{-1} < 1$$

where, at any given temperature, $h \equiv \frac{4\sigma_B}{x_\delta J_1} > 0$ is proportional to the spectral background contribution. This result illustrates the fact that the ideal optimal condition $\Xi = 1$ is very hard to be obtained in practice in real systems.

(9) When the Fermi level is located at the mobility edge the scaled variable $x_C = (E_C - \mu)\beta = 0$ and the spectral conductivity reads

$$\sigma(x) = \begin{cases} \sigma_a, & \text{if } x < 0 \\ \sigma_b, & \text{if } x \geq 0. \end{cases} \tag{2.72}$$

Plugging it into Eq. (2.22) and making use of the integrals

$$\int_{-\infty}^{0} \operatorname{sech}^2(x/2)dx = 2, \quad \int_{-\infty}^{0} x \operatorname{sech}^2(x/2)dx = -4\ln 2,$$

$$\int_{-\infty}^{0} x^2 \operatorname{sech}^2(x/2)dx = \frac{2\pi^2}{3}$$

along with those given in Eq. (2.68), we obtain $J_0 = 2(\sigma_a + \sigma_b)$, $J_1 = 4(\sigma_b - \sigma_a)\ln 2$, and $J_2 = 2\pi^2(\sigma_a + \sigma_b)/3$.

(a) Substituting into Eq. (2.39), we get

$$\Xi = 12\left(\frac{\ln 2}{\pi}\right)^2\left(\frac{q-1}{q+1}\right)^2, \tag{2.73}$$

where $q \equiv \sigma_b/\sigma_a > 1$. This function has its minimum ($\Xi = 0$) for $q = 1$ (when the step function reduces to a constant background value) and its maximum value $\Xi^* = 12\left(\frac{\ln 2}{\pi}\right)^2 = 0.584\ldots < 1$ is attained in the limits $q \to 0$ and $q \to \infty$ (which correspond to the Mott model studied in the **Exercise 2.6**). The obtained results nicely illustrate that the more asymmetrical the spectral conductivity function the better the TE performance of the considered system.

(b) Making use of the reduced kinetic coefficients obtained in (a) into Eq. (2.25), after some arrangement we have

$$\kappa_e(T) = \frac{1}{2}(1+q)(1-\Xi)\, L_0\sigma_a T. \tag{2.74}$$

where we have used Eq. (2.73). Taking into account that the electrical conductivity is given by

$$\sigma(T) = \frac{J_0}{4} = \frac{\sigma_b + \sigma_a}{2} = \frac{1+q}{2}\sigma_a,$$

Eq. (2.74) can be expressed in the form

$$\kappa_e(T) = (1-\Xi)\, L_0\sigma(T)T.$$

Thus, the considered system obeys the Wiedemann–Franz law, with an effective Lorenz number whose value depends on the ratio σ_b/σ_a.

(c) Making use of the results obtained in (a) and (b) into Eq. (2.41) we have

$$ZT = \frac{\Xi}{1-\Xi}\left(1 + \frac{2k_l(T)}{(1+q)(1-\Xi)\, L_0\sigma_a T}\right)^{-1},$$

which, in the optimal TE performance limit $q \to 0$, reduces to

$$ZT^* = \left[\frac{1}{\Xi^*} - 1\right]^{-1}\left(1 + \frac{2k_l(T)}{(1-\Xi^*)\, L_0\sigma_a T}\right)^{-1}.$$

In the high temperature limit, the lattice contribution to the thermal conductivity vanishes due to the onset of Umklapp phonon–phonon interactions and we can estimate the following upper bond for the FOM in this temperature regime

$$(ZT^*)_{\max} \simeq \left(\frac{1}{12}\left(\frac{\ln 2}{\pi}\right)^{-2} - 1\right)^{-1} \simeq 1.405.$$

Quite interestingly, this figure is larger than that obtained in Section 2.1 for semiconductor materials obeying Eq. (2.1), namely, $(ZT)_{\max} = 12/\pi^2 \simeq 1.2$.

Chapter 3

The Structural Complexity Approach

In the previous chapters we have described the main TE effects and their related transport coefficients, by considering both macroscopic and microscopic points of view. The efficiency of TE devices for both power generation and cooling applications has then been expressed in terms of a basic parameter, namely, the dimensionless FOM, which depends on the transport coefficients of the considered material and it is, therefore, temperature dependent. In order to optimize the ZT value, two main approaches has been also introduced. The first one focuses on the lattice contribution to the thermal conductivity, while the second one mainly deals with the close relationship among the electrical conductivity, the Seebeck coefficient and the charge carriers contribution to the thermal conductivity.

The next two chapters will be devoted to gain a deeper understanding on these optimization routes by considering the role played by the lattice structure and the electronic structure in the resulting TE performance of different materials of current interest. In this chapter, firstly we will review the most important bulk TEMs considered to date by following a structural viewpoint based on the *structural complexity* degree of crystalline solids. Indeed, as we will see throughout this Chapter, current research in novel TEMs is guided by the very notion of atomic lattice complexity and the role that this complexity plays in the electrical and thermal properties of a given material at different temperatures.

Thermoelectric Materials: Advances and Applications
Enrique Maciá-Barber

ISBN 978-981-4463-52-2 (Hardcover), 978-981-4463-53-9 (eBook)
www.panstanford.com

The structural complexity can be measured in terms of a series of parameters which we will introduce in Section 3.1. In Section 3.2 we will describe the main physical and transport properties of several elemental solids of interest in TE research. These include representatives of group 16 chalcogens (selenium and tellurium), group 15 pnicogens (antimony and bismuth) and group 14 elements (silicon, germanium, and lead). Afterward, in Section 3.3 we will present the so-called *traditional* TEMs. These include BiSb binary alloys for low temperature (liquid nitrogen range) applications, Bi_2Te_3-Bi_2Se_3-Sb_2Se_3 pseudoternary alloys for room temperature applications, ZnSb binary alloys for applications within the 400–600 K range, lead chalcogenide binary alloys for moderate temperature (600–800 K) applications, and SiGe binary alloys for high temperature ($>$1000 K) TE generation.

As we see, most of these traditional materials are binary compounds, so that search for improved TEMs has naturally proceeded by considering materials with a higher chemical diversity. Thus, in Section 3.4 we will consider the so-called *complex chalcogenides* family. These materials are based on the ternary compound $AgSbTe_2$ and their solid solutions with GeTe and PbTe alloys (Sections 3.4.1 and 3.4.2). We will also describe some thallium bearing compounds, characterized by extremely low thermal conductivity values (Section 3.4.3) and alkali metal bismuth chalcogenides of interest for low temperature ($\sim$200–250 K) applications (Section 3.4.4). To conclude, in Section 3.5 we will present a detailed study of the so-called *inclusion compounds*, where certain concepts originally inspired in the phonon glass-electron crystal notion discussed in Section 2.2.1 will be thoroughly analyzed. The four main classes of open structured compounds will be studied in this section, namely, half-Heusler, skutterudites, clathrates, and Chevrel phases, which will be respectively described through Sections 3.5.1–3.5.4.

3.1 Structural Complexity and Physical Properties

Most materials can be synthesized in either glassy or crystalline form. Amorphous materials are characterized only by the presence of short-range atomic order, whereas crystals are characterized by the presence of long-range atomic order (either periodic or quasiperiodic). Though glasses typically show very small thermal conductivity values, they are very poor electrical conductors as well, which leads to very small ZT values in turn. This intrinsic drawback of glasses cannot be alleviated by doping them, as it is usually done when dealing with semiconducting materials, or even for certain types of insulators, such as the so-called *conducting polymers* (see Section 6.3.1). This feature then precludes the possible use of amorphous materials in competitive TE devices.[a] Accordingly hereafter, we will focus our attention on the study of materials grown in crystalline form (**Exercise 3.1**). Broadly speaking, the *structural complexity* of a crystalline solid can be measured in terms of several parameters, including:

- the number of different chemical species, n_S, forming the compound. Thus, $n_S = 1$ for elemental solids, $n_S = 2$ for binary compounds, and so on,
- the number of lattice parameters, n_L, required to fully describe the unit cell geometry. For the different crystallographic systems we have: $n_L = 1$ (cubic), $n_L = 2$ (rhombohedral, tetragonal, and hexagonal), $n_L = 3$ (orthorhombic), $n_L = 4$ (monoclinic), and $n_L = 6$ (triclinic),
- the unit cell volume, V_U,
- the number of atoms present in the primitive unit cell, n_U,
- the number, n_E, and nature of stable *chemical entities* which can be present as inclusions (e.g., interstitial atoms, guest atoms in voids, and small solvent molecules) or building blocks (e.g., chains of bonded atoms, small molecules, or molecular cluster frameworks) in the unit cell,
- the coordination index of different atoms within the unit cell, n_C
- the number of stable isotopes, n_I, for a given element as well as the atomic mass ratio between the atoms composing binary and ternary compounds, respectively accounting for isotopic and alloying dynamical effects.

[a]Some years ago, Nolas and Goldsmid proposed a possible scenario under which the FOM may be higher in the amorphous state than in the crystalline state for a given solid. Essentially, this may occur when the mean free path of phonons is greater than that of charge carriers [71].

In order to illustrate the role of atomic structure on the transport properties of TE interest, let us consider the effect of (a) increasing the unit cell size and (b) the inclusion of a second alloying element in a simple elemental solid having a cubic lattice with lattice constant a. In the former case $V_U = a^3$, so that to increase the a value leads to a smaller reciprocal lattice volume $(2\pi/a)^3$, resulting in a smaller Brillouin zone. This Brillouin zone reduction, in turn, gives rise to the presence of a large number of phonon bands in the frequency spectrum of the solid, due to an enhanced band-folding effect.

A similar effect is obtained when an alloy is formed from a solid solution process, keeping the original unit cell size almost constant,[a] but increasing the number of atoms n_U present in it. In this case, a larger number of optical phonons, $3(n_U - 1)$ in number, must be accommodated in the frequency spectrum when n_U is increased (see Fig. 3.1).

In both cases, one gets quite fragmented frequency spectra, characterized by the presence of many low dispersion, flat bands, for energies, say, higher than $\sim$10 meV. Since the group velocity of the related phonons is determined by the slope of their dispersion relations, according to the expression $v_{ph} \sim d\omega/dq$ (where ω is the phonon frequency and q is its momentum), one realizes that the presence of flat bands results in a significant reduction of the phonons' group velocities, thereby degrading the thermal conductivity. In addition, in the case of the lattice size increase, the Brillouin size reduction in reciprocal space also enhances the role of phonon–phonon Umklapp processes in the solid (see Section 1.2.2.3), further reducing the thermal conductivity. Thus, from general principles, we conclude that solids characterized by large unit cells containing many atoms therein must exhibit remarkably low *lattice* thermal conductivities. We will further discus on this important result throughout this chapter.

---

[a] According to the empirical Vegard's law, a linear relation exists between the crystal lattice parameter of an alloy and the concentration of its constituent elements at constant temperature. Therefore, the lattice volume usually changes upon alloying (see Tables 3.5 and 3.6). Notwithstanding this, for the sake of clarity we are now ignoring this fact in order to conceptually distinguish the effects arising from an increase in the *number* of atoms in the unit cell from those related to its *volume* change.

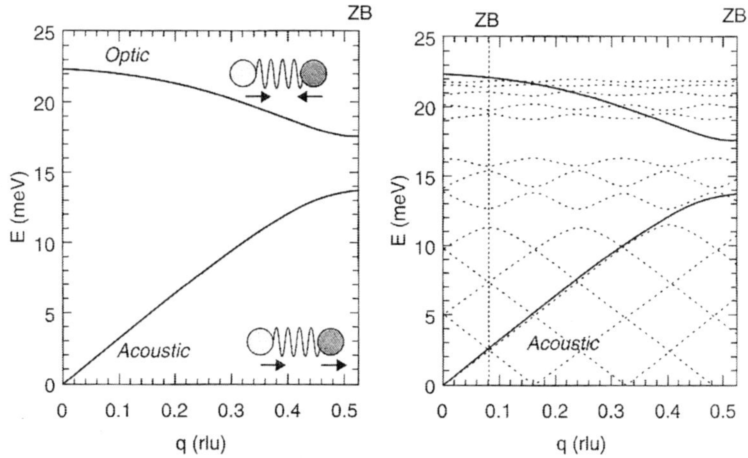

Figure 3.1 Extended zone scheme dispersion relation for two binary chains ($n_S = 2$) of atoms A and B whose one-dimensional unit cells are given by AB ($n_U = 2$, left) and $ABAABABAABAAB$ ($n_U = 13$, right), respectively. For the sake of comparison, the dispersion relation of the simple binary chain with unit cell AB is shown by black solid curves on the right panel [23]. Adapted from T. Janssen, G. Chapuis, and M. de Boissieu, *Aperiodic Crystals: From Modulated Phases to Quasicrystals*, Oxford (2009). By permission of Oxford University Press. The significant shortening of the Brillouin zone boundaries (ZB, shown by a vertical dashed line) can be clearly appreciated. The phonon wave vector q is measured in relative lattice units.

3.2 Elemental Solids of TE Interest

In the previous two chapters we have learned that the quest for good TEMs involves the search for peculiar arrangements of matter, able to simultaneously exhibit a very low thermal conductivity along with both high electrical conductivity and Seebeck coefficient values. In order to properly fulfill these requirements one should discard metallic substances, which are characterized by large thermal conductivity values arising from the presence of a high concentration of charge carriers in these materials. In a similar way, insulating materials must be also discarded, since their electrical conductivity is generally too low.

On the other hand, in bulk materials the lattice contribution to the thermal conductivity is proportional to the squared sound velocity (see Eq. (1.28)). Now, the speed of sound depends on both the bonding properties of the solid and its density, according to the expression $v = \sqrt{C/\rho_a}$, where, C is the elastic stiffness constant and ρ_a is the atomic density of the material. In this way, phonon transport is closely related to the bonding strength and the atomic mass, so that the thermal conductivity gets smaller as the atomic bond strength is weaker,[a] and solid density is increased. Accordingly, one should start by considering loosely bonded solids (small C, hence low κ_l), exhibiting semiconducting or semimetallic transport properties (large S values and reasonably high σ, yet keeping a moderate enough κ_e value) made of heavy atoms tightly packed in the unit cell (large ρ_a, hence low κ_l).

Indeed, most broadly used TEMs to date are based on binary and ternary alloys involving just a few chemical elements located on the lower right corner of the periodic table, namely, lead, antimony, bismuth, selenium, and tellurium. To these elements, which are mainly used for low and intermediate temperature applications, we must add silicon and germanium for their use in high temperature TEGs.

In Fig. 3.2 we summarize some basic data for the aforementioned elemental solids. In fact, by attending to their physical properties, these elemental solids can be classified into two groups (separated by a diagonal dashed line in Fig. 3.2). On the one hand, we have low density silicon and germanium elemental solids, which crystallize into the cubic diamond structure, forming strong covalent bonds leading to both high melting points and Debye temperature values. On the other hand, we have high density elemental solids, which crystallize into more complex lattice structures and exhibit significantly low melting and Debye temperatures.

Two questions then naturally appear. First, what are the physical and chemical reasons explaining the success of these selected elements in the TE industry? Second, what its the physical reason (if any) accounting for the need of relying on binary and ternary

---

[a]Thus, strongly bonded materials like diamond ($\kappa \simeq 2200$ Wm^{-1}K^{-1}) or graphene ($\kappa \simeq 5000$ Wm^{-1}K^{-1}) exhibit very high thermal conductivities at room temperature.

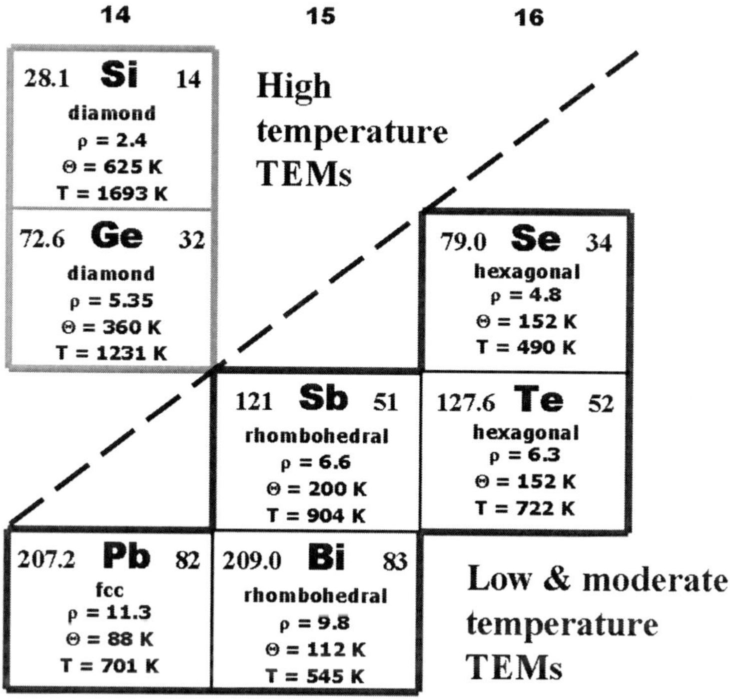

Figure 3.2 Chart showing the atomic number (Z, right upper corner), atomic mass M_a (in uma, left upper corner), crystalline structure, solid density (ρ_a in gcm^{-3}), Debye temperature, and melting temperature of elemental solids of interest in thermoelectric research.

compounds rather than using simple elemental solids as useful TEMs?

To properly address these questions we will start by considering the data presented in Table 3.1, where we list the room temperature transport coefficients values of the elemental solids depicted in Fig. 3.2, along with their resulting ZT values. In the first place, we note that ZT varies over nine orders of magnitude, ranging from $ZT = 0.36$ for bismuth to $ZT = 2 \times 10^{-10}$ for silicon. This wide FOM window mainly stems from the electrical conductivity contribution, which spans over about ten orders of magnitude, ranging from $\sigma = 5.2 \times 10^4\ \Omega^{-1}\text{cm}^{-1}$ for metallic lead, down to $\sigma = 5 \times 10^{-6}\ \Omega^{-1}\text{cm}^{-1}$ for semiconducting silicon. By contrast, the Seebeck coefficient *only*

Table 3.1 Room temperature transport properties of elemental solids of thermoelectric interest arranged according to their FOM values. SM stands for semimetals, M for metals, and SC for semiconducting materials. Most single crystals exhibit quite anisotropic transport properties. The values reported correspond to those measured along the trigonal direction for bismuth and antimony, and along the *c* direction for selenium and tellurium

Element	Type	σ (Ω^{-1}cm^{-1})	S (μVK^{-1})	κ (Wm^{-1}K^{-1})	ZT
Bi	SM	7400	-100	6.1	0.36
Sb	SM	41000	$+40$	24	0.08
Te	SC	2	$+290$	2.3	0.002
Pb	M	51700	-1.25	36	7×10^{-5}
Ge	SC	0.02	$+400$	60	2×10^{-6}
Se	SC	2×10^{-6}	$+1250$	2.0	5×10^{-8}
Si	SC	5×10^{-6}	$+450$	148	2×10^{-10}

varies over three orders of magnitude in absolute value, ranging from $S = -1.25$ μVK^{-1} for metallic lead, to $S = +1250$ μVK^{-1} for semiconducting selenium, whereas the thermal conductivity varies by less than two orders of magnitude, between $\kappa = 148$ Wm^{-1}K^{-1} for silicon and $\kappa = 2$ Wm^{-1}K^{-1} for selenium.

By inspecting Table 3.1 several additional conclusions can be drawn:

- Of the considered elements only bismuth and antimony semimetallic elemental solids are suitable TEMs on their own. Bismuth displays a moderate Seebeck coefficient and a relatively small thermal conductivity, attaining a $ZT \simeq 0.36$ value at room temperature in high quality samples. Indeed, elemental bismuth was one the first materials to be studied for TE applications and it was used in earlier thermopiles for many years [111]. Since it has a negative Seebeck coefficient, it was common practice to complete the TE module with another group 15 element, antimony, which shares a similar power factor (see Table 3.2). However, its four-fold higher thermal conductivity value results in a four-fold lower ZT value, as compared with that of bismuth.
- Elemental tellurium and selenium are semiconductors exhibiting small electrical and thermal conductivity values and

high Seebeck coefficients. Remarkably enough, selenium has the largest Seebeck coefficient and the smallest thermal conductivity values of all the listed solids, but these favorable contributions to the FOM are overwhelmed by an extremely small electrical conductivity, ultimately leading to a very small ZT value. On the contrary, tellurium has a significantly higher σ value, ranking the third position in Table 3.1.[a]

- Elemental lead is a typical metal characterized by large electrical and thermal conductivities (**Exercise 3.2**), and a very small Seebeck coefficient value. Accordingly, it has a small FOM value. Notwithstanding this, this element played a significant role in earlier TE research, serving as a metallic contact in TE devices (see Section 1.3) and providing a reference standard for thermocouples calibration (see Section 1.1).
- Elemental silicon and germanium are typical semiconductors (see Section 4.2.3) characterized by low intrinsic electrical conductivity values and both high Seebeck coefficient and thermal conductivity values. The combined effect of their small electrical conductivity and large thermal conductivity values cannot be compensated by their relatively large S figures, so that these materials have very small FOM values, even smaller than that of metallic lead. Fortunately, their intrinsically low electrical conductivity can be significantly improved upon appropriate doping, and their high thermal conductivity properly lowered by alloying these isoelectronic elements to each other.

A deeper understanding on the structural complexity role in the TE performance of the materials listed in Table 3.1 can be gained from the data presented in Table 3.2, where we list the thermal conductivity and the power factor of these elemental solids but now arranged according to the structural complexity parameters introduced in Section 3.1. Silicon and germanium crystallize in the diamond type cubic system ($n_L = 1$) and have two atoms in the unit cell ($n_U = 2$). Lead crystallizes in the face centered cubic system ($n_L = 1$) with $n_U = 4$. In standard conditions bismuth and antimony

[a]As we will see in Section 4.2.2, the large difference between the electrical conductivities of elemental selenium and tellurium arises from significant differences in their respective electronic structures.

Table 3.2 Room temperature thermal conductivity, and thermoelectric power factor of the elemental solids listed in Table 3.1 arranged according to their structural complexity

Element	Structure	n_L	n_U	n_E	V_U (Å^3)	κ (Wm^{-1}K^{-1})	P (μWcm^{-1}K^{-2})
Si	diamond	1	2	0	160.2	148	1×10^{-6}
Ge	diamond	1	2	0	181.1	60	3×10^{-3}
Pb	fcc	1	4	0	121.3	36	0.08
Sb	rhombohedral	2	2	0	181.3	24	66
Bi	rhombohedral	2	2	0	212.5	6.1	74
Se	hexagonal	2	3	1	245.9	2.0	3×10^{-6}
Te	hexagonal	2	3	1	306.2	2.3	0.2

crystallize in the rhombohedral structure (space group $R\bar{3}m$, $n_L = 2$, Fig. 3.3) with $n_U = 2$. Selenium and tellurium crystallize into a trigonal structure (space group $P3_121$, $n_L = 2$) with $n_U = 3$. Their atomic arrangement can be visualized as an array of parallel helical chains of Te (Se) atoms ($n_E = 1$) located at the center and six corners of a hexagon (Fig. 3.4). Three trends can be clearly appreciated from the data listed in Table 3.2:

- the unit cell volume progressively increases as the structural complexity increases,
- the larger the structural complexity the smaller the thermal conductivity for all the considered elemental solids,
- the power factor systematically increases for the elemental solids silicon, germanium, lead, antimony, and bismuth.

Since the power factor appears in the numerator of the FOM expression, and the thermal conductivity appears in its denominator, we realize that the resulting ZT value directly correlates with the structural complexity for all the elements belonging to the cubic and rhombohedral systems.[a] This trend, however, is broken by the structurally more complex selenium and tellurium crystals. As we will discuss in more detail in Section 4.2.1 the electronic structure of

[a]The crystal structure of antimony and bismuth can be regarded as a distorted fcc lattice along the [111] direction. The close relationship between electrical transport properties and structural aspects of antimony and bismuth elements is discussed in detail in Section 4.2.1.

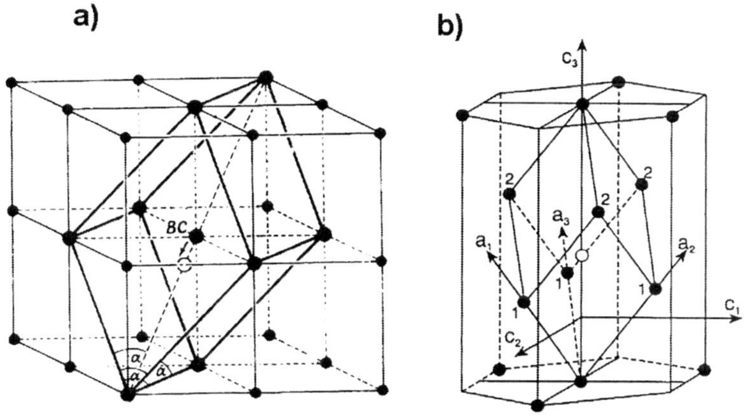

Figure 3.3 (a) Crystal structure of elemental bismuth and antimony. The rhombohedral cell can be formed by starting from a simple cubic lattice and then sharpening the rhombohedral angles from 60° to $\alpha = 57^\circ\ 14.2'$ (for bismuth) or $\alpha = 57^\circ\ 6.5'$ (for antimony) and shifting the body centered atom (labeled BC) down the diagonal of the rhombohedron, which defines its trigonal axis (see Section 4.2.1). (b) The crystal structure of these elements is commonly referred to a hexagonal basis with lattice parameters $a =$ 4.546 Å (a = 4.308 Å), c = 11.862 Å (c = 11.274 Å), and the internal displacement u = 0.237 (u = 0.233) measured in units of the c axis, for Bi (Sb), respectively. First and second neighbors to the central atom (open circle) are labeled 1 and 2, respectively. The bisectrix (C_1), binary (C_2) and trigonal (C_3) axes, and primitive vectors ($\mathbf{a}_1, \mathbf{a}_2, \mathbf{a}_3$) are also shown. Adapted from Ref. [10], U. Mizutani, *Introduction to the Electron Theory of Metals*, Copyright 2001, by permission of Cambridge University Press, and reprinted from Ref. [112], with permission from Liu Y. and Allen E., 1995, *Phys. Rev. B* **52**, 1566, Copyright 1995, American Physical Society.

selenium and tellurium crystals is determined by their characteristic molecular helical chains, ultimately determining the low electrical conductivity and power factor values. In addition, viscosity data suggest that these chains are, in some form, retained upon melting, hence indicating strong interactions among the involved atoms along the chain. In this regard, it was pointed out by Slack some time ago that the thermal conductivity of a solid may critically depend on the possible existence of molecular groupings among its constituent atoms, due to differences in thermal energy transport between the (low frequency) atomic lattice phonon modes and

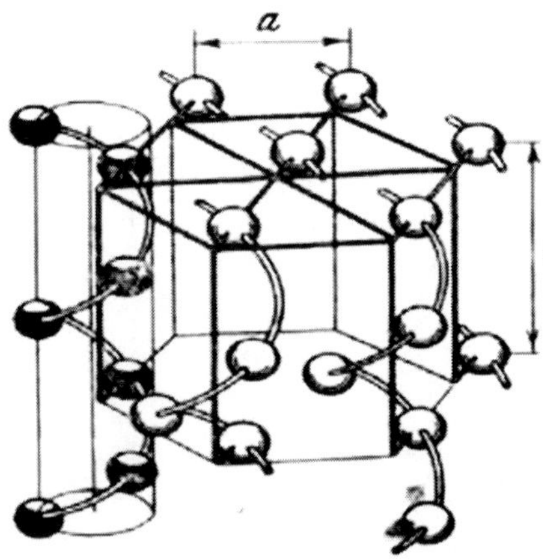

Figure 3.4 Crystal structure of the elemental selenium and tellurium. The direction of the chains is the c crystallographical axis and every third atom is directly above another atom along the chain, so that in a plane perpendicular to the chain the atoms project into an equilateral triangular pattern. Each atom has a two-plus-four coordination: two interchain nearest neighbors at a distance of $d_1 = 2.835$ Å (2.373 Å) and four interchain second-nearest neighbors at a distance $d_2 = 3.495$ Å (3.426 Å), for Te and Se, respectively. The bond angles along the chain are 103.1° and 103.2° for selenium and tellurium, respectively. As we see, d_2 is considerably larger than the distance between two neighboring atoms along the chain, but still shorter than the van der Waals contact of 4.12 Å, indicating the presence of some interchain covalent bonding. As a consequence, these compounds exhibit very anisotropic transport properties.

the (higher frequency) modes related to such molecular groupings [113, 114]. Thus, the possible coupling between these two classes of normal modes will result in a substantial decrease of the resulting thermal conductivity. In fact, from Table 3.1 we see that tellurium and selenium elemental solids, where such groupings exist in the form of helical chains, show very low thermal conductivity values.

3.3 Traditional Thermoelectric Materials

The quest for materials exhibiting very low thermal conductivities is the main topic of this chapter. In the previous section we have seen that the best elemental solids for TE applications are bismuth, antimony, and tellurium. The atoms of these elemental solids are next neighbors in the periodic table, located in the right bottom corner of the chart shown in Fig. 3.2, so that all of them have relatively large atomic mass values, as required. Nevertheless, their thermal conductivity values are still too large for their direct use in TE devices.

It is well known that the lattice thermal conductivity is generally reduced after alloying with a second element forming a solid solution. This is due to three main mechanisms:

(1) the chemical complexity of the resulting compound increases, since we now have two kinds of atoms which can be more or less randomly distributed in the nodes of an otherwise perfectly regular crystal lattice (chemical disorder). Each atom has a different mass value, thereby introducing very effective point defect scattering centers for heat carrying phonons, which reduces the relaxation time value in Eq. (1.39). This, in turn, results in a decrease of the thermal conductivity at low temperatures, as prescribed by Eq. (1.34).
(2) the presence of two different atoms in the unit cell produces a splitting of the phonon bands in the frequency spectrum. This leads to flat bands, resulting in slow group velocities phonons, further reducing κ_l, as we explained in Section 3.1.
(3) the lattice volume expansion due to alloying effects also contributes to the formation of low dispersion phonon band in the frequency spectrum, further contributing to reduce the phonons group velocity.

The concerted action of all these effects leads to a substantial thermal conductivity reduction in the solid solution as compared to its value for the elemental solids. It is then natural to consider the TE properties of binary compounds based on bismuth, antimony, and tellurium, in order to assess their TE performance. To this end, in Table 3.3 we list relevant transport properties for: a representative of the BiSb alloy, two properly doped Sb_2Te_3 and Bi_2Te_3 compounds, and two representative ternary alloys obtained from the latter

Table 3.3 Room temperature transport properties of binary and ternary compounds based on antimony, bismuth, and tellurium arranged according to their FOM values. The data listed in the first row were measured at $T = 80$ K. These compounds crystallize in the rhombohedral system and exhibit quite anisotropic transport properties. The values reported correspond to those measured along the trigonal direction

Compound	σ (Ω^{-1}cm^{-1})	S (μVK^{-1})	P (μWcm^{-1}K^{-2})	κ (Wm^{-1}K^{-1})	ZT	Ref.
$Bi_{0.88}Sb_{0.22}$	2300	−95	20.8	2.6	0.40	[115]
$Bi_{0.88}Sb_{0.22}$	4000	−95	36.1	4.3	0.25	[116]
Sb_2Te_3	1200	+133	21.2	2.4	0.26	Pb doped
Sb_2Te_3	3000	+100	30.0	2.4	0.38	In doped
Bi_2Te_3	1000	+185	34.2	1.9	0.54	p-type
Bi_2Te_3	1000	−205	42.0	1.9	0.66	n-type
$Bi_{0.5}Sb_{1.5}Te_3$	305	+265	21.4	0.9	0.70	[117]
$Bi_2Te_{2.7}Se_{0.3}$	1000	−210	44.1	1.4	0.95	[117]

binary compounds. By comparing the data listed in Tables 3.1, 3.2, and 3.3 the following conclusions can be drawn:

- First, alloying effects in the isoelectronic BiSb compound gives rise to a significantly lower room temperature thermal conductivity value ($\kappa = 4.3$ Wm^{-1}K^{-1}) than that of both antimony ($\kappa = 24$ Wm^{-1}K^{-1}) and bismuth ($\kappa = 6.1$ Wm^{-1}K^{-1}) elemental solids, still preserving a Seebeck coefficient value comparable to that of elemental bismuth ($S \simeq -100$ μVK^{-1}). Unfortunately, the electrical conductivity is substantially reduced in the BiSb alloy ($\sigma = 4000$ Ω^{-1}cm^{-1}), as compared to the values of the parent elements ($\sigma = 7400$ Ω^{-1}cm^{-1} for bismuth and $\sigma = 41000$ Ω^{-1}cm^{-1} for antimony). As a consequence, the resulting ZT is reduced by about 30% with respect to the bismuth value in the BiSb compound. We note, however, the FOM value of this alloy slightly improves that of the elemental bismuth at nitrogen temperatures, hence suggesting its possible use for low temperature cooling applications.
- Second, the FOM values of the binary Sb_2Te_3 compound ($ZT = 0.3 - 0.4$), significantly improve those of the corresponding elemental solids ($ZT = 0.002 - 0.08$). Three main features contribute to this improvement: (1) the thermal conductivity

of the alloy essentially coincides with the low value of the elemental tellurium ($\kappa \simeq 2.3$ Wm^{-1}K^{-1}), (2) the alloy Seebeck coefficient ($S \simeq +100-130$ μVK^{-1}) is about three times higher than that of the elemental antimony ($S \simeq +40$ μVK^{-1}), and (3) the electrical conductivity of the alloy is three orders of magnitude higher than that of elemental tellurium.

- Third, the bismuth telluride compounds further improve the room temperature FOM values attained by the antimony tellurides, achieving $ZT = 0.5-0.7$ values. This is mainly due to an additional reduction (increment) of the thermal conductivity (Seebeck coefficient) values of Bi_2Te_3 alloys as compared to those of Sb_2Te_3 ones, yet keeping electrical conductivity values around $\sigma = 1000\ \Omega^{-1}$cm^{-1}. Thus, albeit elemental tellurium crystals are not good TEMs by themselves, they form good TEMs when properly alloyed with bismuth or antimony (see Section 3.3.2).
- Fourth, by alloying the Bi_2Te_3 binary compounds with selenium as a third element the thermal conductivity is further reduced, still preserving both the electrical conductivity and Seebeck coefficient values. This yields close to the unity FOM values. On the contrary, we note that alloying with antimony as a third element does not result in a substantial ZT improvement due to a remarkable reduction of the electrical conductivity of the ternary alloy ($\sigma \simeq 300\ \Omega^{-1}$cm^{-1}), even though its thermal conductivity is considerably reduced as compared to that of Bi_2Te_3.
- Finally, we note that none of the compounds listed in Table 3.3 exhibits a power factor comparable to that of bismuth ($P = 74$ μWcm^{-1}K^{-2}) or antimony ($P = 66$ μWcm^{-1}K^{-2}) elemental solids, whereas most of them have lower thermal conductivities than elemental chalcogenides. Accordingly, although the electronic properties determining P are generally degraded when going from elemental solids to their related binary or ternary compounds, this trend is properly compensated by a substantial reduction of the thermal conductivity due to alloying effects.

This κ reduction can be nicely correlated to a systematic unit cell volume increase in these compounds, all of them sharing the

Table 3.4 Room temperature crystal lattice parameters, thermal conductivity, and thermoelectric power factor of the compounds listed in table 3.3 arranged according to their unit cell volume

Compound	a (Å)	c (Å)	V_U (Å^3)	κ (Wm^{-1}K^{-1})	P (μWcm^{-1}K^{-2})
BiSb	4.536	11.841	633.4	4.3	36.1
Bi_2Se_3	4.143	28.676	1297.7	2.7	19.5
Sb_2Te_3	4.264	30.458	1439.8	2.4	20 – 30
Bi_2Te_3	4.383	30.487	1522.8	1.9	34 – 42
$Bi_{0.5}Sb_{1.5}Te_3$				0.9	21.4

same crystal structure (rhombohedral, $n_L = 2$ with $n_U = 5$). To this end, in Table 3.4 we list the room temperature crystal lattice parameters, the thermal conductivity and power factor values of the compounds listed in Table 3.3 arranged according to their atomic volume value. It is clearly seen that the thermal conductivity progressively decreases as V_U increases.

Up to now, we have focused on room temperature transport properties only. As we saw in Section 1.4, commercially available TEMs are generally considered for use in different possible applications, covering a temperature range between 200 K and 1300 K. In what follows we will review the main features of benchmark materials for different temperature windows within the above temperature range. These include BiSb alloys ($ZT \sim 0.4$ at $T = 80$ K) for low temperature (liquid nitrogen range) applications, Bi_2Te_3-Bi_2Se_3 alloys ($ZT \sim 1$ at $T = 350$ K) for room temperature applications, ZnSb alloys ($ZT \sim 0.6 - 0.8$ at $T = 570$ K) within 400–600 K temperature range, PbTe alloys ($ZT \sim 0.8$ at $T = 750$ K) for moderate temperature (600–800 K) applications, and SiGe alloys ($ZT \sim 1$ at $T = 1200$ K) for high temperature (>1000 K) TE generation.

3.3.1 *BiSb Alloys*

$Bi_{1-x}Sb_x$ alloys form a solid solution over the entire composition range, $0 \leq x \leq 1$, and crystallize in the rhombohedral structure $R\bar{3}m$ (as their parent elements), with a rhombohedral angle deviation

Table 3.5 Room temperature crystal lattice parameters, unit cell volume, and thermal conductivity of bismuth, antimony, and its solid solution

Compound	a (Å)	c (Å)	V_U (Å^3)	κ (Wm^{-1}K^{-1})
Sb	4.308	11.274	544.0	24.0
BiSb	4.536	11.841	633.4	4.3
Bi	4.546	11.862	637.4	6.1

from the cubic $\alpha = 60^\circ$ value which progressively increases as x increases. Since Bi and Sb atoms have similar sizes the bond characteristics (i.e., bond length and bond angle) are not greatly modified and the alloy unit cell volume does not change very much as compared to those of the parent elemental compounds (see Table 3.5).

At the same time, due to the large mass difference between Bi ($M = 209$ uma) and Sb ($M = 121$ uma) atoms the presence of chemical disorder due to the substitution of Bi by Sb atoms in the $Bi_{1-x}Sb_x$ solid solution leads to a substantial thermal conductivity reduction in the alloy, as compared to its value for the elemental solids (see Table 3.5). Indeed, it was recognized long ago that $Bi_{1-x}Sb_x$ alloys could be among the best available TEMs in the liquid nitrogen temperature range, when properly doped and oriented. Thus, the Seebeck coefficient is highest at low temperatures (~80 K for $x = 0.1$–0.2) along the trigonal direction ($ZT = 0.41$ at 80 K and $ZT = 0.33$ at 300 K). Along the plane perpendicular to this direction the FOM is lower ($ZT = 0.24$ at 80 K).

Since BiSb alloys are n-type materials (typically $S = -100\ \mu VK^{-1}$ at room temperature along the trigonal direction, see Table 3.3) low temperature thermocouples use BiSb alloys for n-type branches (and Bi_2Te_3 for the p-type ones, see below). However, single crystals are not ideal for large scale manufacturing due to their difficulty in growing and cleaving, as well as their slow growth rate. More importantly, single crystals are mechanically weak, limiting their use in commercial applications. In an effort to increase their mechanical strength, polycrystalline alloys have been produced, but all of them have a lower ZT than single crystals.

Table 3.6 Room temperature crystal lattice parameters, unit cell volume, and thermal conductivity of bismuth, antimony, selenium, and tellurium elemental solids and their non-isoelectronic solid solutions

Compound	a (Å)	c (Å)	V_U (Å^3)	κ (Wm^{-1}K^{-1})
Se	4.368	4.958	245.9	2.0
Te	4.457	5.929	306.2	2.3
Bi_2Se_3	4.143	28.676	1279.7	2.7
Bi_2Te_3	4.383	30.487	1522.8	1.9
Sb_2Te_3	4.264	30.458	1439.8	2.4
Bi	4.546	11.862	637.4	6.1
Sb	4.308	11.274	544.0	24.0

3.3.2 *Bi_2Te_3-Sb_2Te_3-Bi_2Se_3 Alloys*

The transport properties of elemental bismuth were definitively overshadowed during the 1950s by the tellurium-based binary alloys Bi_2Te_3 and PbTe (see Section 3.3.4). Indeed, bismuth telluride was first identified as a material for TE refrigeration in 1954 (see Section 1.3). Its properties were later improved by the addition of Bi_2Se_3 and Sb_2Te_3,[a] to form the pseudoternary solid solutions that form the basis of today's thriving TE industry [118]. Thus, $Bi_{2-x}Sb_xTe_{3-y}Se_y$ alloys have been used for decades, and to date these materials continue to dominate the field of TEs for both TE generation and refrigeration at room (and lower) temperatures, because their TE performance has reached optimal values through composition optimization, doping, and device design [33]. Moreover, their potential application in solid state solar TEGs has further spurred these materials interest [45].

The improved TE performance of $Bi_{2-x}Sb_xTe_{3-y}Se_y$ alloys, as compared to that of their parent elemental solids, is due to the significant role played by alloying effects in decreasing the thermal conductivity value of these non-isoelectronic compounds. In fact, in addition to chemical substitutional disorder related effects, when compounds involving elements from different groups of the periodic

[a]The compounds Bi_2Te_3, Bi_2Se_3 and Sb_2Te_3 belong to the so-called tetradymite group, named after the mineral Bi_2Te_2S, which has essentially the same crystalline structure.

table are formed, the bond characteristics of the resulting compound generally change as compared to those of the parent elemental solids, and its unit cell volume usually increases.[a] As it was described in the introduction of this chapter, this volume expansion naturally leads to band folding in the frequency spectrum so that, the phonons' group velocity reduces and the thermal conductivity gets lower. This feature can be readily appreciated from the data listed in Table 3.6, where we see that bismuth and antimony chalcogenides unit cell volumes are three-fold to four-fold increased as compared to those corresponding to the parent elemental solids. Accordingly, room temperature thermal conductivity values of binary alloys are significantly lower than that of elemental bismuth, and the κ value of bismuth telluride is even lower than that of both tellurium and selenium elemental solids.

The crystal structure of Bi_2Te_3 single crystals ($T_m = 858$ K) can be given in a hexagonal unit cell in terms of successive groupings of layers of atoms (plates), which are stacked along the c axis (see Fig. 3.5a). Each plate consists in five atomic monolayers in the sequence Te(1)-Bi-Te(2)-Bi-Te(1), where Te(1) and Te(2) denote two different types of tellurium atoms in the crystal, according to their respective atomic environments and bonding conditions. Te(2) is coordinated nearly octahedrally by pnictide atoms, whereas Te(1) has three pnictide and three Te(1) as nearest neighbors and the coordination has not exactly octahedral symmetry. Thus, the bonding within each plate unit is considered to be ionic covalent, and no bonding electrons remain to connect the adjacent plates, so that the bonding between their border layers (involving Te(1)-Te(1) atoms) is of weak van der Waals type (see Fig. 3.5b). This layered structure can be easily cleaved perpendicular to the c-axis along the Te-Bi-Te-Bi-Te layers. Accordingly, Bi_2Te_3 crystals are rather soft and they can be easily deformed mechanically. To overcome this disadvantage many synthetic methods were employed to obtain alloys with better mechanical properties. These methods include hot pressing, hot extrusion, shear extrusion, physical vapor deposition, and spark plasma sintering. This material is therefore, only being used in sintered form. Indeed, conventional polycrystalline Bi_2Te_3

[a]At variance with alloys involving isoelectronic elements, such as $Bi_{1-x}Sb_x$ (see Table 3.5).

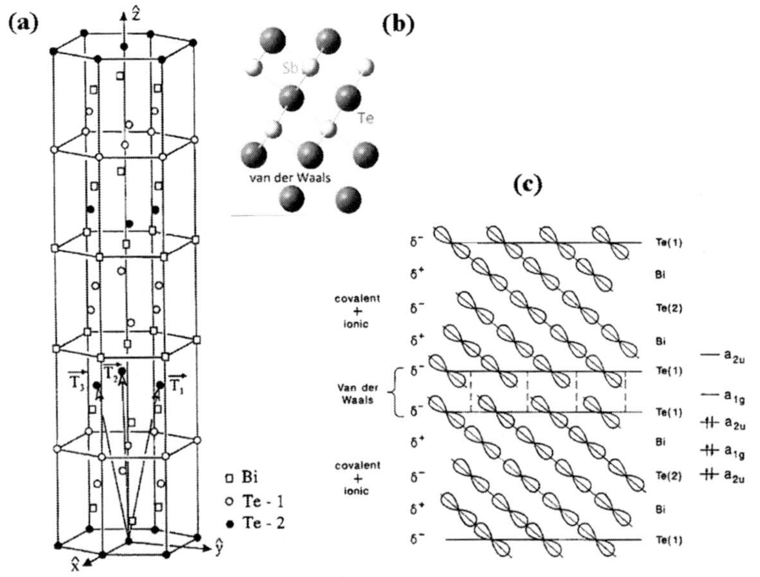

Figure 3.5 (a) Crystal structure of bismuth telluride. The open and dark circles denote the Te(1) and Te(2) atoms, respectively, with the Bi atoms shown as squares. (b) Schematic atomic structure of the isostructural Sb_2Te_3 crystal depicting a mixed ionic-covalent bonded plate unit held by van der Waals bonding to the next plates along the c direction. The Sb (Bi) atoms are coordinated with six Te atoms in octahedral geometry. (c) A sketch of the chemical bonding network in the Bi_2Te_3 crystal. The molecular orbitals produced by the $\pi\pi$p interaction on the five-atom linear chain connecting consecutive layers are shown on the right. Adapted from Ref. [119], with permission from IOP Publishing Ltd.

based materials have much better mechanical properties than their single crystalline counterpart, showing a slightly lower ZT value only, hence being widely used for industry applications. Furthermore, Bi_2Te_3 alloys can be tailored as either n-type or p-type (see Table 3.3) by varying the composition with slight deviations from its stoichiometric value, so that both legs can be made of the same materials in a TE device.

Although neither antimony telluride Sb_2Te_3 ($T_m = 895$ K) nor bismuth selenide Bi_2Se_3 ($T_m = 979$ K) are particularly good TEMs, the addition of one or both to Bi_2Te_3 improves the ZT value of the resulting alloy, mainly by reducing the lattice thermal conductivity, not affecting the favorable Bi_2Te_3 electronic structure very much (see Section 4.3.2). Within the Bi_2Te_3-Sb_2Te_3-Bi_2Se_3 pseudoternary system the optimal compositions for TE cooling use are the solid solutions of nominal composition $Bi_2Te_{2.7}Se_{0.3}$ and $Bi_{0.5}Sb_{1.5}Te_3$ for the n-type and p-type materials, respectively. In actual devices the p-type legs are generally hot-pressed and annealed-pellets, which have good mechanical properties. The n-type counterpart is typically an ingot form grown by zone melting techniques. A maximum $ZT = 1.05$ at $T = 420$ K was reported for the n-type Se substituted ternary solid solutions $Bi_2(Te_{1-x}Se_x)_3$ with $x = 0.0 - 0.1$. The sample also shows a high average $ZT = 0.97$ between 300 and 500 K [120]. Recently, large FOM values, within the range $ZT = 1.3$–1.4 have been reported for nanostructured $Bi_xSe_{x-2}Te_3$ compounds [121].

3.3.3 *ZnSb Alloys*

Zinc antimonide ($T_m = 819$ K) has been known for decades as a TEM, being used in some of the earliest TEGs (see Section 1.3) [73]. Indeed, along with its sibling Zn_4Sb_3, ZnSb is one of the best materials in the important temperature range between 400 and 600 K for which only a limited number of efficient TEMs are known.[a] The best reported FOM value is around $ZT = 0.6$ for the p-type doped material. Some studies of n-doped compounds with impurities such as In, Te, or Se have been reported, but the obtained ZT values are too low.

ZnSb crystallizes into the orthorhombic structure (space group Pbca) with lattice parameters $a = 6.2016$ Å, $b = 7.7416$ Å, and $c = 8.0995$ Å and it contains $n_U = 16$ atoms in the unit cell. Its structure can be described in terms of an arrangement of interconnected planar Zn_2Sb_2 rhomboid rings, as shown in Fig. 3.6 The small thermal conductivity of this compound ($\kappa = 1.4$ Wm^{-1}K^{-1}) can then be understood in terms of its relatively complex structure, containing small atomic groupings ($n_E = 1$) within a relatively large unit cell ($V_U = 388.9$ Å^3). Single phase ZnSb alloys prepared using

[a]Other possible candidates are the so-called lead, antimony, silver, and telluride (LAST) and tellurium, antimony, germanium, and silver (TAGS) tellurides, which will be described in Section 3.4.2.

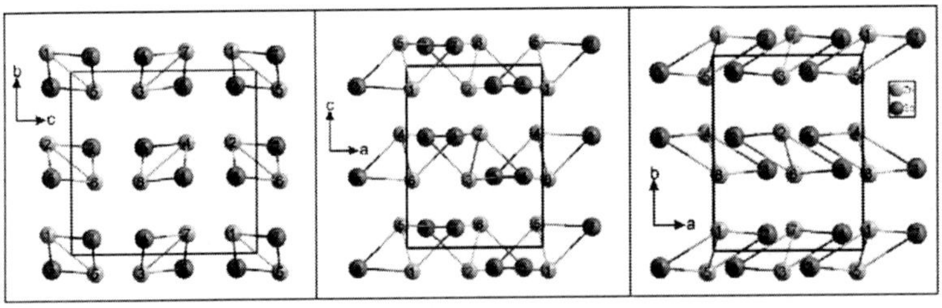

Figure 3.6 Crystal structure of the ZnSb alloy showing the Zb_2Sb_2 rhomboid rings projected along the three directions of the orthorhombic cell [122]. Reprinted with permission from P. Jund, R. Viennois, X. Tao, K. Niedziolka, and J. C. Tédenac, 2012, *Phys. Rev. B* **85**, 224105, Copyright 2012, American Physical Society.

a mechanical grinding process have been reported to exhibit $ZT = 0.6$–0.8 values at $T = 573$ K [123].

The TE properties of Zn_4Sb_3 compounds ($T_m = 836$ K) were first investigated as part of a broad research for more efficient TEMs undertaken at the Jet Propulsion Laboratory during the late 1990s. Three modifications, labeled α-, β-, and γ-Zn_4Sb_3 are known for this compound, which are stable below 263 K, between 263 and 765 K, and above 765 K, respectively. β-Zn_4Sb_3 has a rhombohedral crystal structure (space group $R\bar{3}c$, Fig. 3.7), with $a = 12.228$ Å and $c = 12.407$ Å hexagonal lattice parameters ($V_U = 4823.4$ Å^3). The power factor of this compound is only moderate ($P = 13$ μWcm^{-2}K^{-2}) as compared to other TEMs, like half-Heusler compounds ($P = 70$ μWcm^{-2}K^{-2}), for instance, see Section 3.5.1. Possible doping concentrations are very limited, since heavy doping conditions drastically rise the κ value, and no n-doped materials have been found so far (May 2014). Notwithstanding this, due to its very low thermal conductivity ($\kappa = 0.9$ Wm^{-1}K^{-1} at room temperature, which decreases to $\kappa = 0.7$ Wm^{-1}K^{-1} at 650 K) the FOM of β-Zn_4Sb_3 reaches $ZT = 1.3$ at 670 K. Its low κ value has been explained in terms of migration energies of Zn atoms in the β-Zn_4Sb_3 lattice. Indeed, according to detailed numerical studies, Zn atoms can flow around this material like highly viscous liquid in the lattice frame of Sb atoms [124]. Since it has a large ZT value within the 500–650 K temperature range this material could be of interest for automobile

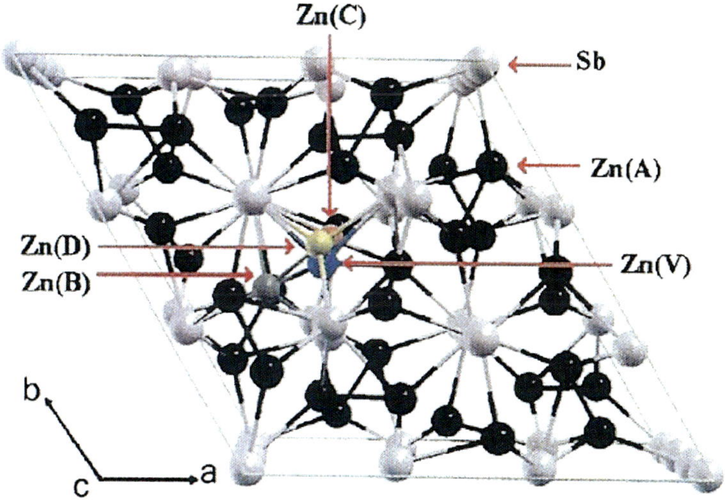

Figure 3.7 Crystal structure of β-Zn_4Sb_3 [124]. Reprinted with permission from W. Chen and J. Li, *Appl. Phys. Lett.* **98**, 241901 (2011), Copyright 2011, American Institute of Physics.

industry in order to use the waste heat generated by the vehicle exhaust to replace or supplement the alternator. In this regard, one should take into account that β-Zn_4Sb_3 is relatively inexpensive as compared to state-of-the-art TEMs and more environmentally friendly.

3.3.4 *Lead Chalcogenides*

Bi_2Te_3-based alloys are not suitable for use at temperatures much higher than about 400 K since they become chemically unstable.[a] This shortcoming spurred the search for alternative TEMs for their possible use in TEGs. To this end, lead chalcogenides, bearing the Pb heavy atom, whose elemental solid ranges as the fourth best TE performing solid in Table 3.1, appear as a natural choice. Indeed, lead telluride has a long history as TEM, since it was one of the first TEMs studied by Ioffe and his colleagues for practical applications (see Section 1.3).

[a] Even though this temperature value is not close to the alloy melting point.

Table 3.7 Room temperature transport coefficients values, power factor, and FOM, along with the optimal FOM value at $T^* = 723$ K of undoped binary lead chalcogenides arranged according to their ZT value [125]. These compounds crystallize in the cubic system ($n_L = 1$) with four atoms in the unit cell ($n_U = 4$)

	σ	S	P	κ	ZT	ZT^*
Compound	(Ω^{-1}cm^{-1})	(μVK^{-1})	(μWcm^{-1}K^{-2})	(Wm^{-1}K^{-1})		
PbS	144	−207	6.2	2.8	0.07	0.30
PbSe	373	−154	8.9	2.1	0.13	0.45
PbTe	207	−189	7.4	1.8	0.13	0.48

The FOM of the PbTe compound is only just over $ZT \simeq 0.1$ at room temperature (see Table 3.7), but at higher temperatures it shows a peak ZT of about 0.8, eventually becoming superior to that of most TEMs ($ZT \simeq 0.8$–1.0, see Fig. 1.15). Accordingly, PbTe based alloys have been the premiere TE material for moderate temperature (600–800 K) regime in TE power generation applications for more than 50 years, specially in deep space exploring programs. Thus, the first TE powered spacecraft launched by NASA used a radioisotope TEG containing both n- and p-type alloys of PbTe, and related materials continue to be used for remote power today, as in the latest NASA Mars rover mission.

Lead chalcogenides crystallize in a cubic (NaCl-type) rock-salt structure with $Fm\bar{3}m$ space group symmetry, with lead atoms occupying the cation sites and chalcogenide atoms forming the anionic lattice. The cubic symmetry implies that the TE properties are the same in all directions in randomly oriented polycrystalline materials (easier to grow than single crystals). PbTe single crystals can be prepared by almost any standard method, including the Bridgmann, Czochralski, and vapor growth techniques, though obtaining large single crystals with a high perfection degree and homogeneous composition is difficult. PbTe easily cleaves because of its layered structure and is a rather soft material, so that it is mainly used in sintered form.

By inspecting Table 3.7, we see that lead chalcogenides exhibit low thermal conductivities and moderate Seebeck coefficient values (n-type). However, their electrical conductivity is about one order of magnitude smaller than that of the binary compounds considered

Table 3.8 Crystal lattice size, density, melting temperature, and Debye temperature for the compounds listed in Table 3.7 arranged according to their unit cell volume [125]

Compound	a (Å)	V_U (Å^3)	ρ (g cm^{-3})	T_m (K)	Θ (K)
PbS	5.936	209.2	7.61	1384	145
PbSe	6.121	229.3	8.27	1355	141
PbTe	6.454	268.8	8.24	1253	136

in the previous sections. In order to improve its low electrical conductivity properly doped samples are generally used instead of using pristine PbTe. Lead chalcogenides are classified as polar materials with a mixed ionic-covalent chemical bond. Thus, these alloys can be doped by introducing alkali metals (acceptors) or halogen atoms (donors), so that both n-type and p-type samples can be produced, either by doping or through slight stoichiometric changes. For instance, the combination of precise control of the sodium doping level resulted in a large $ZT = 1.4$ value for both p-type (n-type) PbTe alloys (see Fig. 4.11). In a similar way, a $ZT = 1.5$ value at 625 K has been reported for $PbSe_{0.5}Te_{0.5}$ alloys doped with iodine (n-type) [126]. Thallium-doped samples, exhibiting $ZT = 1.5$ values at 773 K, have been considered within the framework of electronic engineering approach (see Sections 2.4 and 4.4) [127]. In fact, band structure engineering can help to improve the power factor, thereby the FOM, up to $ZT = 1.8 \pm 0.1$ at $T = 850$ K in heavily Na-doped $Na_{0.02}Pb_{0.98}Te_{0.85}Se_{0.15}$ p-samples [76]. High average FOM values of $ZT \simeq 1$ between 300 and 800 K have been reported as well [38].

Since tellurium is very rare in the Earth's crust (0.001 ppm),[a] it is desirable to develop some alternative material with less tellurium content. PbSe and PbS alloys (sharing the same rock-salt crystalline structure) have received some attention recently, and certain promising FOM values have been reported: $ZT = 1.2$ at 850 K for Na-doped p-type PbSe, and $ZT = 1.3$ at 850 K in Al-doped

[a]Even less than Pt (0.005 ppm) and Au (0.004 ppm). Data relative to the market concentration, scarcity and supply risk of several elements of interest for the TE industry can be found in the following Web site: http://www.mrl.ucsb.edu:8080/datamine/thermoelectrics.jsp [305].

n-type PbSe. The PbSe analog[a] has a higher melting point than PbTe (see Table 3.8), so that it can sustain higher operation temperatures up to 1000 K. Therefore, very high temperature applications, including solar thermal energy conversion, could benefit from the development of PbSe-based TEGs [128]. PbSe is also attractive for high temperature TE because it exhibits a monotonically increase in Seebeck coefficient up to at least 1000 K and has lower thermal conductivity than PbTe in this temperature range.

3.3.5 *SiGe Alloys*

The study of SiGe alloys as possible materials for TE generation was undertaken as early as 1954 by Ioffe's group. This alloy is the preferred TEM for TE power generation at elevated temperatures, since bulk SiGe devices ($ZT = 0.7$–1.0 at about 1200 K) can operate at temperatures up to 1300 K without significant degradation. This material was first used in space probes nuclear reactors and has been the preferred choice for radioisotope TEGs launched by the U.S. space program since 1976 [30].

$Si_{1-x}Ge_x$ alloy is a solid solution semiconductor with a cubic diamond-type structure. As compared with the TEMs considered up to now, made of heavy atoms arranged into relatively complex crystal structures, this TEM belongs to an entirely different class. In these compounds chemical substitutional disorder, whereby atoms with two different masses randomly occupy the nodes of the diamond type crystal structure plays a crucial role. In fact, while silicon and germanium elemental solids exhibit very large thermal conductivity values (see Tables 3.1 and 3.2), their thermal conductivity is significantly lowered upon alloying (e.g., $\kappa = 4$ $Wm^{-1}K^{-1}$ at room temperature for $Si_{80}Ge_{20}$ alloy which has an average FOM of approximately $ZT = 0.5$ from 570 to 1370 K). These alloys can be tailored in either n-type or p-type.

To conclude this section, in Table 3.9 we list the structural complexity parameters of the traditional TEMs considered up to now, along with their thermal conductivity, and optimal FOM values arranged according to their optimal temperature of performance T^*.

---

[a]Selenium is less expensive and about 50 times more abundant than tellurium in Earth's crust.

Table 3.9 Room temperature thermal conductivity and FOM of the binary compounds considered in this section arranged according to their temperature of use

Compound	Structure	n_L	n_U	n_E	κ ($Wm^{-1}K^{-1}$)	ZT	T^* (K)
BiSb	R3m	2	2	0	4.3	0.40	80
Sb_2Te_3	R3m	2	5	0	2.4	0.26	350
Bi_2Te_3	R3m	2	5	0	1.9	0.66	300
ZnSb	Pbca	3	16	1	1.4	0.70	570
Zn_4Sb_3	R3c	2	66	0	0.9	1.30	670
PbTe	Fm3m	1	4	0	1.8 – 2.4	0.8 – 1.0	600–800
SiGe	diamond	1	8	0	4–10	0.7–1.0	1200

By inspecting this table we see that TEMs for low and intermediate temperature applications are more structurally complex than those used in high temperature applications. This trend can be easily understood, since materials to be used in TEGs must have high enough melting points. This generally requires strong bonds, so that stable compounds at high temperatures usually crystallize into relatively simple, mainly cubic lattice structures. As we commented in Section 3.2 the presence of strong bonds leads to large sound velocities and, consequently, large thermal conductivity values as well, which restrain good TE performance. The goal, therefore, would be to find out materials able to keep complex structures up to high enough working temperatures. To this end, one may start by considering solids composed of more than two different chemical elements, with very large unit cells containing many atoms within them. This approach will be our main guideline through the next two sections.

3.4 Complex Chalcogenides

In the previous section we have considered in detail the TE properties of binary tellurides and their pseudo-ternary solid solutions. In this section we will consider the ternary $AgSbTe_2$ compound and its solid solutions with GeTe and PbTe alloys. The resulting quaternary compounds generally contain nano-inclusions which favor the thermal conductivity degradation due to phonon–phonon anharmonicity driven effects. These materials are predominantly semiconductors, with energy gaps appropriate for TE (e.g., 0.1–0.8 eV) over a wide temperature range, they are stable, melt congruently, and have high melting points, which favor their use in high temperature TE applications.

3.4.1 *$AgSbTe_2$ Compound*

The $AgSbTe_2$ compound was synthesized in 1957 and it is known as one of the bulk p-type TEM with the highest FOM value: $ZT = 1.6$ at 673 K [129]. Like PbTe, it crystallizes in the rocksalt cubic structure (space group $Fm\bar{3}m$), where Ag and Sb atoms occupy the cation sites. In fact, the $AgSbTe_2$ compound is isoelectronic with PbTe in which the lead atom (with a 2+ valence) is replaced by one Ag^+ and one Sb^{+3} in $AgSbTe_2$. Its lattice constant, $a = 12.152$ Å, is almost twice that of PbTe ($a = 6.462$ Å). As we discussed at the beginning of this chapter, due to both its larger unit cell size (1794.5 Å^3) and its higher chemical complexity ($n_S = 3$), one expects a significant reduction of the thermal conductivity coefficient of this ternary compound, as compared to the value reported for the PbTe alloy. Indeed, it was noted early on that the room temperature thermal conductivity of this material is extremely small, on the order of $\kappa = 0.3$–0.4 Wm^{-1}K^{-1} within the range 300–700 K. This extremely low κ value is related to the random distribution of Ag and Sb atoms occupying analogous lattice sites (chemical disorder) as well as to a great anharmonicity of the lattice vibrational dynamics, that gives rise to strong phonon–phonon interactions.

Experiments suggest that this material is a semiconductor with a very narrow gap of 7 meV, with highly mobile electrons that dominate the electrical conductivity, and holes in a heavy band that dominates the Seebeck coefficient ($S \simeq +260$ μVK^{-1} at room temperature). Unfortunately, the $AgSbTe_2$ compound has a complicated phase diagram stability indicating that it is metastable at low temperatures and becomes prone to decomposition to Ag_2Te and Sb_2Te_3 binary alloys. In addition, it is difficult to control its doping degree. These shortcomings prevent a direct use of this material in actual TE devices. A possible way to circumvent this limitation, still fully exploiting the extraordinary thermal properties of this material, is to alloy $AgSbTe_2$ with other structurally similar tellurium based compounds.

3.4.2 *TAGS and LAST Materials*

In fact, large FOMs ($ZT > 1$) have been reported for $AgSbTe_2$ alloyed with PbTe, SnTe, and GeTe, mainly resulting from their low thermal conductivities. In particular, p-type alloys of the form $(AgSbTe_2)_x(GeTe)_{1-x}$, have been studied for several decades and they are commonly referred to as TAGS-x, where x represents the mole percent GeTe.[a] The report of high ZT values, of approximately $ZT = 1.4$ and 1.7 at $T = 750$ K and $T = 700$ for $x = 80$ and $x = 85$, respectively, motivated several studies during the last decades.[b] Thus, TAGS-85 has been used in NASA missions since the early 1970s.

The $AgSbTe_2$-GeTe alloy is structurally related to the PbTe alloy (both share the same cubic NaCl-type crystal structure over a wide solid solution range), though the $AgSbTe_2$-GeTe system undergoes a polymorphic transformation from a low temperature rhombohedral (R$\bar{3}$m) structure to the high temperature NaCl-type cubic one. Quite remarkably, the highest FOM is displayed close to this phase transformation, presumably because the lattice strain associated with this transformation may be effective in reducing the thermal conductivity, thereby increasing the ZT value. The transition temperature depends primarily on the ratio of GeTe to $AgSbTe_2$, so that TAGS compounds are not suitable for TE applications when the concentration of GeTe becomes less than 80%. As the composition is varied from $AgSbTe_2$ to GeTe in the solid solution, the transport properties vary smoothly, except for a double minimum observed in the thermal conductivity ($\kappa = 0.3$ and 1.0 $Wm^{-1}K^{-1}$ at 80% and 85% GeTe, respectively).

---

[a] This semiconducting material ($E_g \sim 1$ eV) melts at $T_m = 1197$ K and exhibits a moderate $ZT \simeq 0.7$ at 800 K, as well as relatively low thermal conductivity values $\kappa = 1$–3 $Wm^{-1}K^{-1}$ at room temperature.

[b] This material was one the first "second generation" traditional TEMs to exhibit $ZT > 1$ at high temperatures [130], hence indicating that the $ZT \simeq 1$ practical barrier was not necessarily a physical barrier.

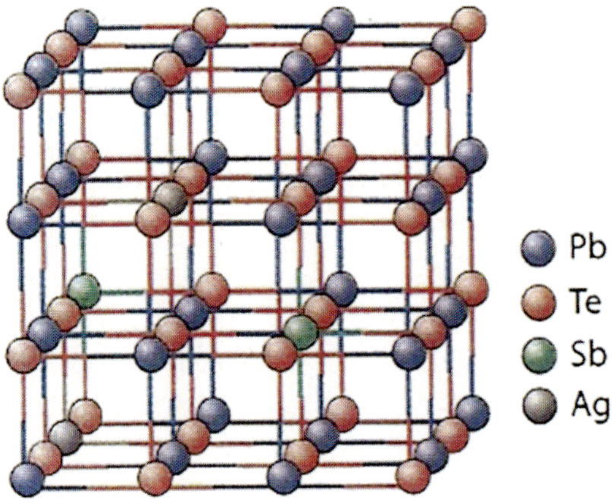

Figure 3.8 Crystal structure of the LAST materials.

Notwithstanding their large FOM values, TAGS compounds have found little use due to their sublimation rate, high cost, and the presence of the low temperature phase transition. In addition, only p-type TAGS materials, without equivalent n-type counterparts, have been found to date. These limitations spurred the search for alternative $AgSbTe_2$-based alloys.

In 2004 it was reported that bulk materials belonging to the $Ag_nPb_xSb_nTe_{2+x}$ system, fabricated from $AgSbTe_2$ and PbTe by melting and slow cooling process, reached the very high $ZT = 2.1$ value at 800 K for $x = 0.05$ [131]. This materials class has become known as LAST compounds and they have a typical NaCl structure (Fig. 3.8), where the Ag, Sb, and Pb atoms are substitutionally disordered in the structure on the cation sites, whereas, the chalcogen atoms occupy the anion sites. Thus, this compound can be considered as an antimony and silver co-doped PbTe alloy with p-type properties and a low lattice thermal conductivity. The LAST materials are stable up to their melting point (>1200 K), and doping is generally controlled by non-stoichiometry of the Ag, Pb, or Sb fractions in the form of $Ag_{1-x}Pb_{m+y}Sb_{1+z}Te_{2+m}$ compounds, with $m = 10$ and $m = 18$. Subsequent studies on $(Ag_{1-y}SbTe_2)_{0.05}(PbTe)_{0.95}$ samples with $0.7 \leq 1-y \leq 1$ gave significantly lower maximum FOM values, namely, $ZT = 1.07$ and $ZT = 0.33$ at $T = 673$ K for $Ag_{0.6}Pb_{18}SbTe_{20}$

and $Ag_{0.7}Pb_{18}SbTe_{20}$ compounds, respectively [132, 133]. These conflicting results were explained as stemming from the fact that $(Ag_{1-y}SbTe_2)_{0.05}(PbTe)_{0.95}$ materials are multiphase on the scale of millimeters despite appearing homogeneous by X-ray diffraction and routine electron microscopy techniques. Indeed, using a scanning Seebeck microprobe, a significant variation of Seebeck coefficient (including both n-type and p-type behavior in the same sample) was found [134]. Thus, although from a metallurgical perspective these compounds can be viewed as a solid solution of the form $(AgSbTe_2)_x(PbTe)_{1-x}$, this picture may be an oversimplification. In fact, microstructural analysis showed the presence of nano-sized regions rich in AgSb composition, whereas the surrounding structure is AgSb poor in composition with a unit cell parameter (6.44 Å) close to that of PbTe (6.46 Å). The presence of these nanocrystals will produce local distortions due to the presence of strain at the interfaces, affecting both structural (resulting in enhanced phonon scattering) and electronic properties of this quaternary alloy. However, one must keep in mind that the nanostructures may grow or dissolve during prolonged high-temperature operation.

On the other hand, both n- and p-type LAST materials can be obtained by adjusting the chemical composition. For instance, $ZT = 1.45$ at 630 K was achieved in p-type $Ag(Pb_{1-x}Sn_x)_mSbTe_{2+m}$ by partially substituting the lead in the LAST system by tin (the so-called LASTT compounds) [135]. Subsequent research showed that simply replacing silver in the LAST system with sodium or potassium atoms (referred to as SALT compounds) leads to a higher $ZT = 1.7$ at 650 K and $ZT = 1.6$ at 750 K in p-type $Na_{0.95}Pb_{20}SbTe_{22}$ and $K_{0.95}Pb_{20}SbTe_{22}$, respectively [136]. Anyway, these high ZT values are mainly attributed to the very low thermal conductivity of the materials, which is as low as 0.85 $Wm^{-1}K^{-1}$, of which approximately 0.5 $Wm^{-1}K^{-1}$ corresponds to the lattice contribution.

3.4.3 *Thallium Bearing Compounds*

The route toward progressively lower thermal conductivity values naturally led to the consideration of the possible replacement of thallium by lead in binary chalcogenides. Indeed, thallium is a close

Table 3.10 Room temperature structural parameters, thermal conductivity value, and optimal FOM at T^* of thallium bearing tellurides arranged according to their atomic volume [137–140]. All compounds crystallize in the tetragonal system ($n_U = 32$) except the last one which crystallizes into the rhombohedral one ($n_U = 12$)

Compound	a (Å)	c (Å)	V_U (Å^3)	κ (Wm^{-1}K^{-1})	ZT^*	T^* (K)
Tl_5Te_3	8.929	12.607	1005.1	3.5	0.07	560
Tl_2GeTe_5	8.243	14.918	1013.6	0.5	0.60	300
Tl_2SnTe_5	8.306	15.161	1046.0	0.5	0.60	300
Tl_9BiTe_6	8.860	13.050	1024.4	0.5	0.9–1.2	500–600
Ag_9TlTe_5	11.431	41.945	4750.1	0.2	1.20	700

neighbor of lead element in the periodic table, sharing a similar atomic mass value ($M = 204.4$ uma), density ($\rho_a = 11.85$ gcm^{-3}), and melting temperature ($T_m = 576$ K). In its elemental form thallium crystallizes into the hexagonal form, which is closely related to the elemental tellurium crystal structure. Notwithstanding this, the Tl_5Te_3 compound crystallizes in the tetragonal structure (space group $I4/mcm$) and has relatively large unit cell parameters (see Table 3.10). Albeit its relatively complex structure, the thermal conductivity of this binary compound is not particularly small as compared to that of the Bi_2Te_3 ($\kappa = 1.9$ Wm^{-1}K^{-1} at room temperature). In addition, Tl_5Te_3 is reported to be semimetallic and its Seebeck coefficient ($S = 25$ μVK^{-1} at room temperature) is about eight times smaller than that reported for the narrow band semiconductor Bi_2Te_3,[a] leading to small power factor ($P = 1.8$ μWcm^{-1}K^{-2}) and FOM ($ZT \simeq 0.01$) values at room temperature. Although this FOM slightly improves at higher temperatures ($ZT = 0.07$ at 560 K), it is clear that Tl_5Te_3 cannot be regarded as a promising enough TEM. Therefore, other Tl bearing compounds should be considered.

Tl_9BiTe_6 ($T_m = 813$ K) belongs to a large group of ternary compounds which can be derived from the isostructural Tl_5Te_3 (see Fig. 3.9a). Indeed, assuming that the three tellurium atoms are in the valence state Te^{-2}, the five thallium atoms have to provide six electrons. This is achieved by expanding the formula unit to

[a]As expected from Eq. (4.4), see Section 4.1.

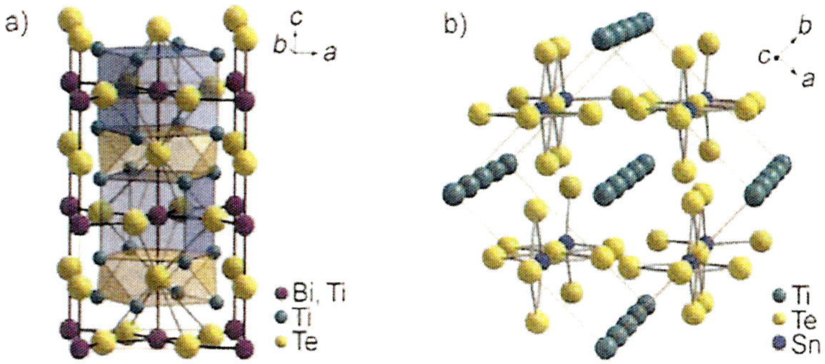

Figure 3.9 (a) The crystal structure of Tl_9BiTe_6. There are two alternating cages of Tl atoms (yellow and blue). The heavy-atom positions are randomly occupied by Bi and Tl. (b) The crystal structure of Tl_2SnTe_5 is composed of tetrahedral $SnTe_4$ units bridged by Te atoms in square-planar geometry. There are not Tl-Tl bonds [141]. Reprinted with permission from M. G. Kanatzidis et al., Thermoelectric materials, *Angew. Chem. Int. Ed.* **48**, 8616 (2009), Copyright 2009, Wiley.

$Tl_9^+Tl^{+3}Te_6^{-2}$, where the 4c site, which accommodates two Tl atoms per expanded formula unit, is equally occupied by Tl atoms in each possible valence state. The Tl^{+3} atoms can then be substituted with the trivalent elements Bi or Sb, resulting in ternary compounds with 9-1-6 stoichiometry.

The unit cell parameters of Tl_9BiTe_6 are somewhat larger than those of the isostructural binary Tl_5Te_3 (see Table 3.10), which combined with the increased chemical complexity inherent to any ternary compound ($n_S = 3$), suggests a substantial thermal conductivity degradation. In fact, not only atomic disorder (typically introduced by alloying) but also the disorder arising from the different valence values at certain atomic sites (the so-called *valence fluctuation*) can strongly reduce the thermal conductivity. In Tl_9BiTe_6, the Bi^{+3} and Tl^+ atoms are distributed randomly at the 4c site, introducing in this way a significant valence disorder through the lattice. Thus, the thermal conductivity of Tl_9BiTe_6 is extremely small ($\kappa = 0.5$ $Wm^{-1}K^{-1}$ at room temperature). The Seebeck coefficient of Tl_9BiTe_6 increases monotonically with the temperature reaching a value of $S = 400$ μVK^{-1} at 500 K and the power factor value is of approximately $P = 10$ $\mu Wcm^{-1}K^{-2}$ at this temperature. While this is a significant improvement with respect to the Tl_5Te_3 power factor value, it is still low compared to state-of-the-art TEMs ($P = 30 - 40\ \mu Wcm^{-1}K^{-2}$), mainly due to the low values of its electrical conductivity ($\sigma = 150–200\ \Omega^{-1}cm^{-1}$), which is an order of magnitude smaller than those of traditional TEMs. Anyway, the relatively large S value, combined to its very low κ, lead to FOM values comparable to that of Bi_2Te_3 at room temperature ($ZT = 0.6$) and to remarkably high FOM values in the 500–600 K temperature range ($ZT = 1.2$ at $T = 500$ K [137], $ZT = 0.9$ at $T = 590$ K [138]).

The TE properties of Tl_2SnTe_5 and Tl_2GeTe_5 compounds, reaching a FOM comparable to that of Bi_2Te_3 at room temperature, have been also explored [139, 140]. Both compounds crystallize in the tetragonal structure (space group $I4/mcm$) and have relatively large unit cell parameters containing $n_U = 32$ atoms within (see Fig. 3.9b). These compounds are characterized by the presence of columns of Tl ions along the crystallographic c axis with large interatomic distances, indicative of loose Te–Te bonds for the eight-fold coordinated Tl atoms. Indeed, neutron diffraction measurements showed large atomic displacements parameters for these atoms and the measured low value for the thermal conductivity ($\kappa = 0.5\ Wm^{-1}K^{-1}$ at room temperature) support the notion of weakly bounded thallium atoms producing very low frequency phonons.

Even smaller thermal conductivity values have been reported for some compounds obtained in the Tl_2Te-Ag_2Te pseudo-binary system. Among them the compounds $Tl_2Te + Ag_2Te \rightarrow TlAgTe$ and $Tl_2Te + 9Ag_2Te \rightarrow Ag_9TlTe_5$ are of special interest. Ag_9TlTe_5 ($T_m = 772$ K) exhibits a rhombohedral crystal structure (space group $R\bar{3}c$) with huge lattice parameters (see Table 3.10). Note that the unit cell volume of this compound is about four times larger than those corresponding to the previously discussed thallium bearing ternary alloys. Ag_9TlTe_5 combines extremely low thermal conductivity of $0.22\ Wm^{-1}K^{-1}$ and a relatively low electrical resistivity to give $ZT = 1.2$ at 700 K (see Fig. 3.10). A similarly low thermal conductivity value ($\kappa = 0.22\ Wm^{-1}K^{-1}$ at room temperature) has been reported for the related compound AgTlTe, although its FOM is remarkably lower ($ZT = 0.6$ at 550 K) [143].

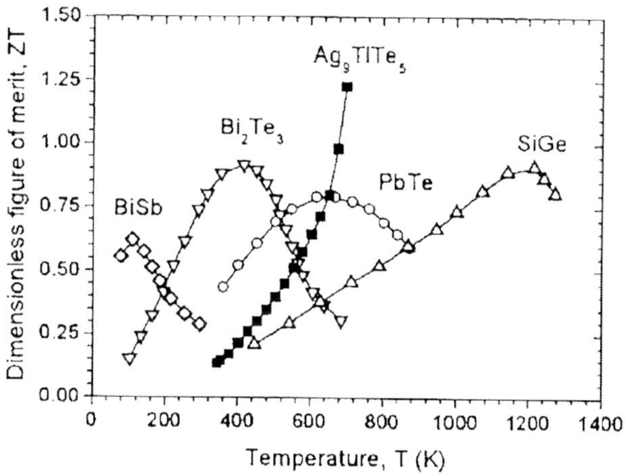

Figure 3.10 Comparison between the ZT temperature dependence curves of Ag_9TlTe_5 and those corresponding to several traditional TEMs [142]. Reprinted with permission from K. Kurosaki, A. Kosuga, H. Muta, M. Uno, and S. Yamanaka, *Appl. Phys. Lett.* **87**, 061919 (2005), Copyright 2005, American Institute of Physics.

Because of significant toxicity and environmental issues, thallium bearing compounds are unlikely to be accepted for practical use. Despite this, studies on this class of compounds are important from a scientific point of view because they could provide some insight on how to substantially reduce the lattice contribution to the thermal conductivity.

3.4.4 *Alkali-Metal Bismuth Chalcogenides*

To conclude this section on complex chalcogenides we will consider bismuth chalcogenides containing alkali elements. This type of compounds show low thermal conductivity, high Seebeck coefficient, and relatively high electrical conductivity reaching, in the case of $CsBi_4Te_6$, a $ZT = 0.8$ at 225 K, which makes it the best bulk TEM below room temperature [144]. The $CsBi_4Te_6$ compound ($T_m = 818$ K) was synthesized by reacting Cs_2Te and Bi_2Te_3 at 973 K. This compound crystallizes in the monoclinic system (space group $C2/m$) with lattice parameters $a = 51.9205$ Å, $b = 4.4025$ Å, $c = 14.5118$ Å, and $\beta = 101.48^\circ$ ($V = 3250.7$ Å^3). It has a layered

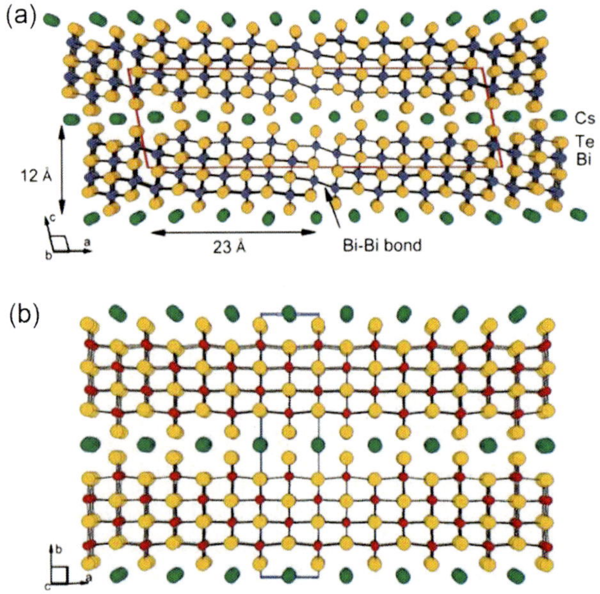

Figure 3.11 Perspective view of the structures of (a) $CsBi_4Te_6$ along the *b*-axis. Bi atoms are blue; Te atoms are yellow. (b) $CsPbBi_3Te_6$ composed of NaCl-type layers. The red atoms are Bi or Pb atoms (mixed occupancy). In $CsBi_4Te_6$, the $[Bi_4Te_6]$ slabs have a finite width, with 12×23 Å^2 cross-section area, are interconnected side by side, and are linked by Bi–Bi bonds at 3.238(1) Å. In $CsPbBi_3Te_6$, the slabs have infinite width and extend continuously along the crystallographic *a*-axis. Reprinted with permission from Ref. [147]. Copyright (2004) American Chemical Society.

structure composed of anionic Bi_4Te_6 slabs alternating with layers of Cs^+ ions which act as electron donors to the framework (Fig. 3.11). The added electrons localize on the Bi atoms, sticking the Bi_4Te_6 slabs to form Bi–Bi bonds that are 3.238 Å long.[a] Such confinement restricts the motions of Cs^+ ions in the direction perpendicular to the crystal *b* axis. Therefore, these ions are able to act as rattlers and their resulting localized vibrations are strongly coupled with the lattice framework phonons. This results in resonant scattering of phonons and reduction in the thermal conductivity measured value ($\kappa = 1.48$ Wm^{-1}K^{-1} at 250 K), which is appreciably smaller than that of Bi_2Te_3 ($\kappa = 1.85$ Wm^{-1}K^{-1}) and comparable to that

[a]The Bi–Bi bond is unusual as it involves a reduction of Bi^{+3} to Bi^{+2} seldom seen in bismuth chalcogenide systems.

of $Bi_{2-x}Sb_xTe_{3-y}Se_y$ optimized alloys ($\kappa = 1.56$ $Wm^{-1}K^{-1}$) at the same temperature.

The common feature of these materials is their high structural anisotropy. The crystals grow with long needle-like morphology, and the direction of rapid growth along the needle axis is also the direction of maximum TE performance. Measurements performed in the perpendicular direction to the growth b axis yield sharply lower values ($\kappa = 0.6$ $Wm^{-1}K^{-1}$), which reflect the highly anisotropic nature of $CsBi_4Te_6$. The measured Seebeck coefficient rises monotonically up to approximately $S = 175$ μVK^{-1} at 275 K, and starts to decrease at higher temperatures. This is in good agreement with the experimentally found narrow bandgaps, ranging from 0.04 to 0.1 eV, and it means that $CsBi_4Te_6$ has only favorable TE properties at low temperatures. Indeed, this compound exhibits a high $\sigma = 1450$ $\Omega^{-1}cm^{-1}$, leading to a high power factor value $P = 44.4$ $\mu Wcm^{-2}K^{-2}$, a figure which compares well with that of Bi_2Te_3.

3.5 Large Unit Cell Inclusion Compounds

During the past decade we have witnessed a growing interest in "open structure" semiconducting compounds, which exhibit remarkably low thermal conductivity values. Among these compounds we find half-Heusler phases (Section 3.5.1), skutterudites (Section 3.5.2), and clathrates (Section 3.5.3) representatives. These materials (see Fig. 3.12) belong to the general group of substances known as *inclusion compounds*, in which atoms, or comparatively simple molecular species, are trapped in cage-like voids of the crystal structure.

By inspecting Fig. 3.12 we see that the voids in half-Heusler compounds are much smaller than the ones present in the family of skutterudite compounds which, in turn, are somewhat smaller than those present in clathrates. The relative structural complexity of these ternary compounds (all of them sharing a cubic lattice structure) can be roughly estimated in terms of the increasing number of atoms in their unit cells, n_U, along with the presence of certain characteristic building blocks ($n_E = 1$), like Sb_4 molecular rings in skutterudites or polyhedral cages framework in clathrates. For the sake of illustration, in Table 3.11 we list some typical

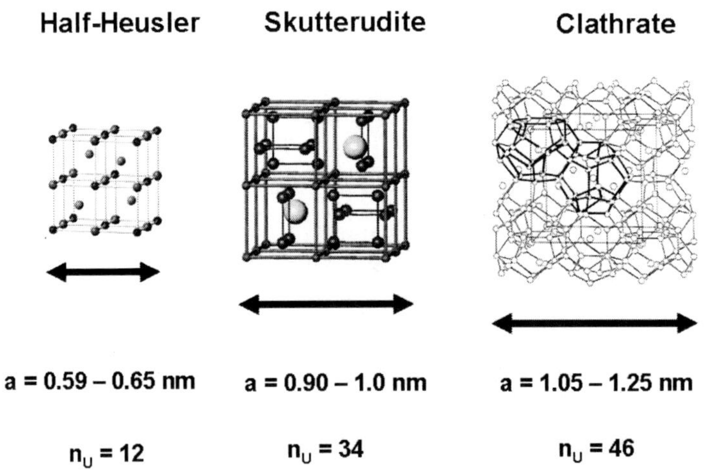

Figure 3.12 Comparison between the structural complexity of the half-Heusler, skutterudite, and clathrate families of open-structured cubic unit cell ($n_L = 2$), ternary ($n_S = 3$) compounds in terms of the average lattice parameter value, a, and the number of atoms in the unit cell, n_U.

structural parameters of half-Heusler, skutterudite, and clathrate compounds, along with two relevant thermal magnitudes. By inspecting this table, one sees that the average thermal conductivity of the half-Heusler, skutterudite, and clathrate compounds systematically decreases when both their a and n_U parameter values are progressively increased, as expected from the discussion presented in Section 3.1. We also see that the averaged κ decreases when the unit cell atom number density $\nu_a \equiv n_U / V_U$ (properly combining the n_U-related and a-related effects) decreases.

3.5.1 *Half-Heusler Phases*

Heusler phases are ternary compounds with the formula XY_2Z, where X is a transition metal of the left-hand side of the periodic table (Ti or V group elements), Y is a transition metal of the right-hand side of the periodic table (Fe, Co, or Ni group elements), and Z is one of the main-group elements (Ga, Sn, or Sb). Lanthanides can also occupy the X and Y sites. The Heusler phases crystallize in the so-called MgAgAs-type cubic structure, with space group

Table 3.11 Values of some parameters measuring the structural complexity of three classes of compounds of interest in TE research (see Fig. 3.12), along with two relevant thermal magnitudes: the observed Debye temperature (Θ_D) range, and their mean thermal conductivity value, κ (averaged from data listed in Tables 3.12, 3.15, and 3.16, respectively)

Property	half-Heusler	Skutterudites	Clathrates
crystal structure	F43m	Im3	Pm3n
a (nm)	0.62	0.95	1.15
V_U (Å^3)	238	857	1521
n_U	12	34	46
ν_a (nm^{-3})	50.4	39.7	30.2
chemical units	–	Sb rings	Ge,Ga polyhedra
Θ_D (K)	250–470	270–320	275–360
κ (Wm^{-1}K^{-1})	6.2	4.4	1.4

$F\bar{4}3m$ (Fig. 3.13). This structure is built from three independent interpenetrating face-centered-cubic (fcc) sublattices of equal unit cell size. Two of the fcc sublattices combine to form a rocksalt (NaCl) substructure (X, Z). The third fcc sublattice (Y) is displaced along the body diagonal of the rocksalt XZ substructure by one fourth of the unit cell. Generally, the smaller of the two metal atoms occupies the eight-coordinated Y site positions within the unit cell. In the half-Heusler alloys, only *half* of those interstitial positions are filled with transition metals. Thus, the structural formula now reads XYZ. Some representative compounds are TiNiSn, ZrNiSn, and HfNiSn, with lattice parameters of $a = 5.93$ Å, $a = 6.11$ Å, and $a = 6.08$ Å respectively.

The half-Heusler structure is quite open, thereby prone to site substitutional disorder. For instance, if the two transition metal atoms are of similar size, a substantial amount of sublattice intermixing can occur. In that case, the degree of site exchanging between the atoms in different sublattices is very sensitive to the sample preparation and annealing conditions, and can have a profound effect on the physical properties.

Two main half-Heusler families, respectively based on Sn or Sb atoms in the Z sites, have been considered in TE research. In Tables 3.12 and 3.13 we list the room temperature transport

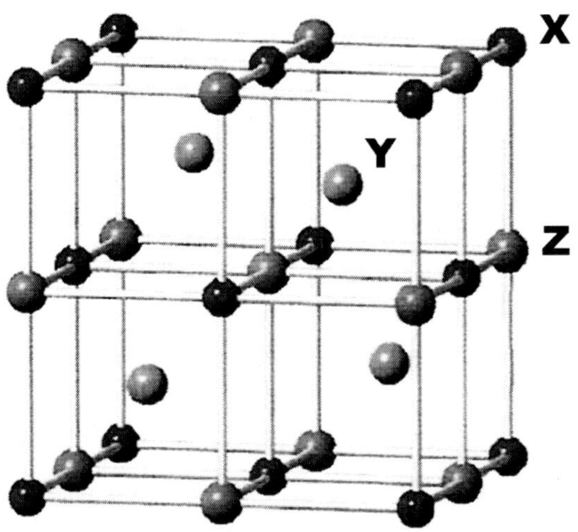

Figure 3.13 Ideal structure of the XYZ half-Heusler phase.

properties, power factors, and FOM values for a series of Sn-based and Sb-based representatives, respectively. Let us start by considering the tin-based compounds. These materials can be easily prepared from their constitutive elements by arc melting, and the resulting buttons are mechanically sturdy. Reliable contacts to the

Table 3.12 Room temperature transport coefficients for tin based half-Heusler compounds

Sample	σ (Ω^{-1}cm^{-1})	S (μVK^{-2})	P (μWcm^{-1}K^{-1})	κ (Wm^{-1}K^{-1})	ZT
TiNiSn	100	−142	2.0	9.3	0.007
ZrNiSn	91	−176	2.8	8.8	0.010
HfNiSn	77	−124	1.2	6.7	0.005
HfPtSn	52	+189	1.9	14.9	0.004
$Ti_{0.5}Hf_{0.5}NiSn$	24	−281	1.9	3.6	0.016
$Zr_{0.5}Hf_{0.5}NiSn$	121	−163	3.2	4.4	0.022
$Ti_{0.5}Zr_{0.5}NiSn$	91	−289	7.3	4.9	0.045
$Zr_{0.5}Hf_{0.5}NiSn_{0.98}Bi_{0.02}$	633	−139	12.2	7.4	0.050
$(ZrHf)_{0.99}Ta_{0.01}NiSn$	1000	−147	21.6	5.4	0.120
$(ZrHf)_{0.99}Nb_{0.01}NiSn$	667	−181	21.9	5.2	0.126
$(Zr_{0.5}Hf_{0.5})_{0.5}Ti_{0.5}NiSn_{0.998}Sb_{0.002}$	400	−250	25.0	2.9	0.254

Source: Refs. [145–148]

Table 3.13 Room temperature transport coefficients for antimony based half-Heusler compounds

Sample	σ (Ω^{-1}cm^{-1})	S (μVK^{-2})	P (μWcm^{-1}K^{-1})	κ (Wm^{-1}K^{-1})	ZT
TiCoSb	13	−230	0.7	24	9×10^{-4}
ZrCoSb	11	−110	0.1	21	2×10^{-4}
HfCoSb	50	−20	0.02	12	3×10^{-5}
VFeSb	1327	−70	6.5	13	0.015
HoPdSb	126	+175	3.9	5.5	0.021
$Eu_{0.25}Dy_{0.75}PdSb$	63	+260	4.2	3.1	0.040
$Eu_{0.25}Dy_{0.75}Pd_{1.02}Sb$	1666	+60	6.0	3.6	0.050
$Ho_{0.5}Er_{0.5}PdSb_{1.05}$	137	+205	5.7	3.1	0.055
$Zr_{0.65}Hf_{0.35}Co_{0.65}Pt_{0.35}Sb_{0.65}Sn_{0.35}$	59	+60	0.2	3.0	0.002

Source: Refs. [149–152]

samples are obtained with silver epoxy or by soldering with tin. In addition, XNiSn alloys have high melting points, within the range 1400–1600 K, and excellent chemical stability with negligible sublimation near their melting points. By inspecting Table 3.12 we see that ternary Sn-based half-Heusler alloys with the formula XNiSn phases generally exhibit n-type electronic transport,[a] with relatively small electrical conductivities, and Seebeck coefficients of the order of a few hundred μVK^{-1} in magnitude, resulting in generally low power factors. In addition, their thermal conductivities are too large for TE applications, leading to ZT values within the range 0.004–0.010.[b] Accordingly, one should manage to properly enhance the electrical conductivity and to reduce the thermal conductivity values of these ternary compounds.

As we know, a common way to reduce the thermal conductivity of solids is by alloying. Thus, in solid solutions of composition $X_xX'_{1-x}NiSn$, (where X and X' are isoelectronic atoms) the mass disorder in the X site sublattice is expected to cause additional

[a] It is interesting to note that HfPtSn compound shows a p-type behavior, while HfNiSn exhibits a n-type one. Although Ni and Pt are elements of the same group, the difference in overlapping of valence electrons, 3d for Ni and 5d for Pt, strongly affects atomic bonding properties, hence the resulting electronic band structure (see Section 4.1).

[b] For the sake of comparison we note that these figures are similar to that of elemental selenium, see Table 3.1.

phonon scattering, thereby reducing the lattice thermal conductivity. According to the mass-fluctuation model the strongest effect should occur near $x = 0.5$. Indeed, the quaternary $Zr_{0.5}Hf_{0.5}NiSn$ compound has a thermal conductivity reduced to one-half of the value of the ternary ZrNiSn compound, and the corresponding ZT value is doubled. The lowest thermal conductivity among the quaternary phases listed in Table 3.12 is found in $Ti_{0.5}Hf_{0.5}NiSn$. This result is not unexpected because of the very different atomic weights of Ti and Hf (ratio 1:3.7), whereas the mass ratios of Zr and Hf (ratio 1:2) and of Ti and Zr (ratio 1:1.9) are almost the same. The fact that $Zr_{0.5}Hf_{0.5}NiSn$ shows a lower thermal conductivity than $Ti_{0.5}Zr_{0.5}NiSn$ is, in turn, explained by its larger mean atomic weight. Notwithstanding this, by inspecting Table 3.12 we see that the ZT values of the listed quaternary compounds run in reversed order to their thermal conductivity values, while their FOM values are directly related to their power factor values. This result clearly indicates that the resulting ZT enhancement mainly comes from the power factor improvement, hence highlighting the important role that chemical effects play in determining the electronic transport properties of these materials.

Following this line of thought and keeping in mind that atoms occupying the Z sites are the primary providers of carriers for charge transport in half-Heusler compounds, one may try to improve the electrical properties of the previously optimized thermal conductivity quaternary compounds, by doping the Sn atoms with Sb or Bi in the Z sites. In so doing, it is observed that the resulting electrical conductivity increases by an order of magnitude, yet the Seebeck coefficient remains relatively high (e.g., ZrHfNiSnBi compound in Table 3.12). Thus, the originally semiconducting XNiSn system is transitioned into semimetallic material with a resulting net gain in power factor. Note that doping also increases the thermal conductivity (**Exercise 3.3**), which is however, properly compensated by a collateral increase in σ, hence leading to a remarkable increase of the ZT value in doped *plus* alloyed compounds with respect to the values obtained by alloying only. Quite interestingly, since each sublattice can be tuned independently through chemical substitution in the XYZ structure, we may dope on the X sublattice with Nb or Ta instead of doping the Z sublattice. This leads to a remarkable improvement of the power factor value, which is accompanied by a quite moderate thermal conductivity increase in this case, resulting in a significant ZT enhancement.

Let us now consider the antimony-based half-Heusler compounds listed in Table 3.13. In the first place, we note that the (Ti,Zr,Hf)CoSb compounds exhibit significantly smaller ZT values than those observed in their Sn-based analogs (Ti,Zr,Hf)NiSn listed in Table 3.12, since they exhibit both lower power factors and substantially larger thermal conductivity values. This shortcoming naturally led to the consideration of alternative compounds. For instance, VFeSb samples exhibited a good electrical conductivity value, but their κ value was still too large and their Seebeck coefficient too small. More promising candidates were found in the (RE)PdSb system, where early period transition metal atoms are replaced by rare-earth elements. These atoms are characterized by the presence of $4f$ electrons, which may hybridize with the conduction band of the parent compound, giving rise to fine spectral features close to the Fermi level and thus to higher Seebeck coefficients (see Section 2.4.1). In this case, the half-Heusler compounds become mainly p-type and similar trends to those reported for Sn-based compounds (regarding favorable alloying and doping effects) are observed for the (RE)PdSb compounds listed in Table 3.13. Nevertheless, we also see that the progressive inclusion of more and more alloying elements does not guarantee a subsequent ZT improvement by itself. Thus, for instance, the six-component compound $Zr_{0.65}Hf_{0.35}Co_{0.65}Pt_{0.35}Sb_{0.65}Sn_{0.35}$, which exhibits one of the smaller thermal conductivity values reported for half-Heusler compounds, also yields a substantially low ZT value. Thus, we realize that by simultaneously disordering the three sublattice systems both thermal *and* electrical currents are degraded. In summary, by comparing the data listed in Tables 3.12 and 3.13 we conclude that Sn-based half-Heusler compounds are better TEMs than antimony based ones.

At this point, we must highlight that the reported room temperature best ZT values are relatively small for both Sn-based and antimony-based half-Heusler compounds. Certainly, they are smaller than the $ZT = 0.36$ value listed for elemental bismuth in Table 3.1. In fact, the potential of half-Heusler compounds as TEMs arose from the FOM values they attain at elevated temperatures. For instance, a maximum $ZT = 1.5$ was reported at 700 K for the compound $(Zr_{0.5}Hf_{0.5})_{0.5}NiTi_{0.5}Sn_{0.998}Sb_{0.002}$, whereas it only attains $ZT = 0.25$ at room temperature (see Table 3.12). The best *reproducible* $ZT \simeq 1$ value at $T = 1000$ K has been obtained for the quinary $Hf_{0.6}Zr_{0.4}NiSn_{0.98}Sb_{0.02}$ half-Heusler material (n-type) [153]. Somewhat lower FOM values have been reported for $Zr_{0.5}Hf_{0.5}Ni_{0.8}Pd_{0.2}Sn_{0.99}Sb_{0.01}$ ($ZT \simeq 0.7$ at 800 K) [154], and $Zr_{0.25}Hf_{0.75}NiSn_{0.975}Sb_{0.025}$ ($ZT \simeq 0.8$ at 1025 K) [155]. The fact that these materials are composed of five or six atoms clearly illustrates the need for both reducing the thermal conductivity (by alloying) and improving the electronic properties (by doping), as can be readily seen by comparing the transport data corresponding to the last entries in Tables 3.12 and 3.13, respectively.

Accordingly, the TE performance of half-Heusler alloys must be optimized via selectively substituting the three sublattice systems. Thus, doping on the Z sites mainly tunes carrier concentration, while substitutions on X and Y sites promotes mass fluctuation and strain fields, respectively, that help reduce the lattice thermal conductivity by suppressing the coherence of the vibrational modes. There is still hope that by attacking the integrity of the Y sublattice one may disrupt the normal modes of the structure enough to drastically diminish the lattice thermal conductivity. In many aspects the Y sublattice is pivotal to the structure, and it would be of interest to study to what extent it can be disordered without destroying the underlying half-Heusler crystal structure itself.[a]

The half-Heusler structure also offers a sublattice of voids. It is intriguing to consider what would happen if one could introduce disorder in this empty sublattice. In principle, it is reasonable to expect that such crystal structure modifications could reduce the thermal conductivity and thus improve the FOM. For instance, the room temperature thermal conductivity was lowered from $\kappa \simeq 9$ to $\kappa = 2.9$ Wm^{-1}K^{-1} in TiNiSn samples with a Ni/(Ti+Sn) ratio equal to 0.54, where excess of Ni introduces interstitial defects leading to stress in the structure [156].

[a]However, it has been reported that alloying Pd on the Ni site reduces the absolute value of the Seebeck coefficient [154].

A final consideration is in order. To produce TE modules based on half-Heusler alloys, finely matching n- and p-type materials should be designed in the same family of compounds with similar physical and chemical properties (see Section 1.3). Compared with n-type half-Heusler alloys, which are made of rather inexpensive elements, the extensive use of p-type half-Heusler compounds like HfPtSn, ZrPdSn, or HfPdSn is restricted by their high cost. Therefore, the possible conversion of n-type alloys into p-type ones resulting from the controlled introduction of acceptor dopants in order to change the major carriers from electrons to holes should be further explored [157].

3.5.2 *Skutterudites*

The word "skutterudite" is derived from a town in Norway where minerals with this crystalline structure were first discovered. Binary skutterudites compounds crystallize in a body-centered-cubic (bcc) structure (space group $Im\bar{3}$) with $n_U = 32$ atoms in the unit cell,[a] and obey the structural formula MX_3, where M is Fe or a Co group transition metal and X is a pnicogen (group 15) atom. The skutterudite structure consists of six square planar rings formed by four pnicogen atoms each, with the rings oriented along the [100], [010], and [001] crystallographic directions, while the metal atoms complete a cubic lattice (Fig. 3.14a).

Binary skutterudites are semiconductors with small band gaps ($E_g \sim$ 0.05–0.5 eV), very high carrier mobilities ($\mu =$ 200–2000 $\text{cm}^2\text{V}^{-1}\text{s}^{-1}$), and moderate to large Seebeck coefficients ($|S| =$ 50–350 μVK^{-1}). These properties motivated the early TE research in this class of compounds. Nevertheless, they also have low carrier densities ($n = 10^{17}$–10^{19} cm^3) in the native form, leading to relatively small electrical conductivities ($\sigma =$ 20–2000 $\Omega^{-1}\text{cm}^{-1}$), and have too high thermal conductivities ($\kappa =$ 10–20 $\text{Wm}^{-1}\text{K}^{-1}$) to compete with state-of-the-art TEMs. For instance, taking $\sigma = 20$ $\Omega^{-1}\text{cm}^{-1}$, $|S| = 300$ μVK^{-1}, and $\kappa = 10$ $\text{Wm}^{-1}\text{K}^{-1}$ as representative

[a]The *primitive* unit cell, which is usually used in theoretical estimates of the thermal conductivity (**see Exercise 3.4**), has $n_U = 16$ atoms.

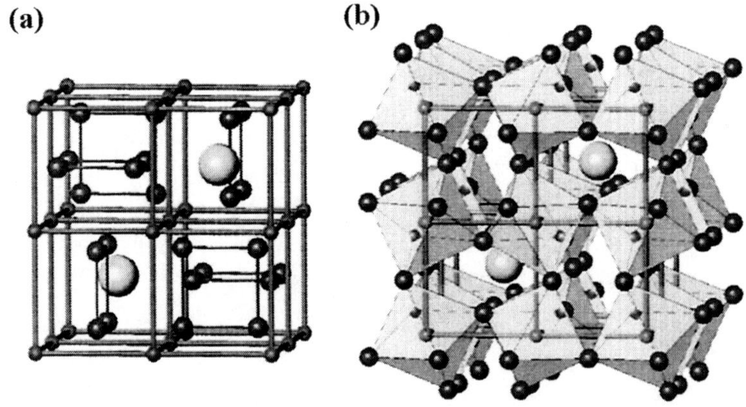

Figure 3.14 (a) Ideal structure of filled skutterudite compounds. Large spheres represent the guest cations, small and intermediate size spheres indicate the transition metal and pnicogen elements, respectively. (b) Sketch highlighting the X type atoms octahedra surrounding the guest sites [158]. Reprinted with permission from M. M. Koza, L. Capogna, A. Leithe-Jasper, H. Rosner, W. Schnelle, H. Mutka, M. R. Johnson, C. Ritter, and Y. Grin, 2010, *Phys. Rev. B* **81**, 174302, Copyright 2010, American Physical Society.

room temperature values for $CoSb_3$,[a] one gets $ZT = 0.005$. Accordingly, the rationale inspiring the use of skutterudites as potential TEMs is based on the very possibility of chemically modifying the binary compounds in order to simultaneously improve (reduce) their electrical (thermal) conductivities.

The lattice thermal conductivity can be straightforwardly reduced by alloying with a third heavy element. For example, the room temperature lattice thermal conductivity of $IrSb_3$ ($\kappa_l = 16$ $Wm^{-1}K^{-1}$) decreases by a 43% to $\kappa_l = 9.1$ $Wm^{-1}K^{-1}$ upon the addition of Rh to form the ternary-mixed crystal $Ir_{0.5}Rh_{0.5}Sb_3$. This figure, however, comes nowhere close to the ideal minimum thermal conductivity value ($\kappa_m \simeq 0.3$–0.6 $Wm^{-1}K^{-1}$) for these compounds (**Exercise 3.4**). In order to better approach this minimum value, Slack and Tsoukala suggested to study the so-called *filled* skutterudites, which were discovered by Jeitschko and Braun in 1977 (**Exercise 3.5**) [159].

[a] $CoSb_3$ skutterudite can exhibit both n-type and p-type thermopower depending on a small deviation on stoichiometry in the Sb sublattice.

Table 3.14 Occupation filling fraction limits in $G_xCo_4Sb_{12}$ skutterudites

G	Ce^{+3}	Nd^{+3}	In^{+3}	Eu^{+2}	Tl^{+}	La^{+3}	Yb^{+2}	Ba^{+2}
x	0.1	0.13	0.2	0.22	0.22	0.23	0.25	0.44
REF	[160]	[161]	[162]	[163]	[164]	[165]	[166]	[167]

The filled skutterudites are stable in an inert atmosphere up to temperatures near 1000 K and can be formed by inserting guest metal atoms interstitially into the 12-coordinated X atom large voids (two per unit cell) present in the crystal structure of the binary compounds, whose structural formula can be expressed in the form $\square M_4Sb_{12}$, where $\square$ denotes the void (**Exercise 3.6**). These cages are formed by eight corner-sharing MX_6 octahedra (Fig. 3.14b) and their radii range from 1.763 to 2.040 Å for CoP_3 and $IrSb_3$, respectively. These sites are large enough to easily accommodate a filling atom each, allowing them to rattle about (**Exercise 3.7**). In terms of phonon scattering, these rattling atoms correspond to additional phonon–phonon interaction contributions due to the presence of additional phonon modes in the filled compounds. In this way, one expects the lattice thermal conductivity of filled skutterudites to be significantly reduced, over a wide temperature range, compared to the binary counterparts (**Exercise 3.4**).

The chemical composition for filled skutterudites can be written as $G_xM_4X_{12}$, where G represents a guest (electropositive) atom and x is its filling factor. There are many possible guest atoms that can be inserted into the voids of the skutterudite crystal structure and, consequently, alter the properties of the material. In fact, we must consider that, besides affecting the lattice dynamics, the incorporation of these electropositive elements also adds electrons to the structure, which must be partially compensated by properly alloying on the M site. Accordingly, only a relatively small fraction of the voids can be filled without any charge compensation (Table 3.14). This gives an opportunity to tune the transport properties of skutterudites by altering the size, oxidation state, and concentration of the filling atoms. In the past several years, a number of filling atoms, including group 13 (Tl), group 14 (Sn), alkaline-earth (Ba), and rare-earth (La, Ce, Pr, Nd, Eu, Yb) atoms

Table 3.15 Room temperature transport coefficients, power factor, and FOM values of Co- and Fe-based skutterudites. The last two columns list the FOM values obtained at high temperatures

Sample	σ (Ω^{-1}cm^{-1})	S (μVK^{-1})	P (μWcm^{-1}K^{-2})	κ (Wm^{-1}K^{-1})	ZT	ZT^*	T^* K	Ref.
$Tl_{0.1}Co_4Sb_{12}$	235			4.0	0.175	0.8	800	[164]
$LaFe_3CoSb_{12}$	670	+80	4.3	1.6	0.080	0.9	800	[169]
$Yb_{0.19}Co_4Sb_{12}$					0.300	1.0	600	[170]
$Ca_{0.18}Co_{3.97}Ni_{0.03}Sb_{12.4}$	2000	−170	31	5.5	0.190	1.0	800	[171]
$K_{0.38}Co_4Sb_{12}$	2800	−90	32	5.5	0.180	1.0	800	[172]
$Ba_{0.24}Co_4Sb_{12}$	1050	−100	10.5	4.8	0.068	1.1	850	[167]
$Ba_{0.3}Ni_{0.05}Co_{3.95}Sb_{12}$	1940	−90	15.7	2.8	0.130	1.2	800	[167]
$Eu_{0.42}Co_{3.95}Ge_{0.5}Sb_{11.5}$	1330	−210	58.7	3.5	0.260	1.1	675	[163]
$In_{0.25}Co_4Sb_{12}$	625	−50	1.6	2.6	0.020	1.2	573	[162]
$CeFe_4P_{12}$	49	+58	0.2	14	0.0004	0.1	800	[173]
$CeFe_4As_{12}$	2040	+37	2.8	5.3	0.016	0.4	850	[173]
$Pr_{0.9}Fe_4Sb_{12}$	2400	+80	15.4	3.6	0.128	0.9	750	[168]
$CeFe_4Sb_{12}$	1333	+59	4.6	2.5	0.056	1.1	800	[173]
$CeFe_4Sb_{12}$	769	+70	3.8	0.9	0.125			[174]
$Ce_{0.9}Fe_3CoSb_{12}$						1.1	700	[175]

have been explored in an effort to optimize the TE FOM of these compounds. Indeed, the properties of the particular filler atom, such as atomic radius, mass, and valence, have considerable influence on the electrical and thermal transport. For instance, the high ZT values reported for barium bearing compounds (see Table 3.15) are explained in terms of barium being a very good filling element, achieving a higher cage filling fraction (44%) than other guest atoms (see Table 3.14). In fact, barium filling not only reduces the thermal conductivity but also increases the electrical conductivity, leading to good power factor values.

Theoretical studies demonstrated that the filling fraction in $G_xCo_4Sb_{12}$ compounds is associated with the electronegativity difference between the filling atom G and the host lattice atom Sb. Only the elements satisfying the electronegativity difference $\chi_G - \chi_{Sb} < 0.8$ could get into the cage site. This filling fraction limit can be controlled by adding some electron-charge-compensation elements. Thus, in order to allow the skutterudite structure to accommodate a higher concentration of guest atoms, one usually replaces Fe, Ru, or Os on the Co site, or Sn on the Sb site. Since Co (Sb) has one more outer electron than Fe (Sn), this substitution results in an effective p-doping, properly stabilizing the overall structure, although also leading to an appreciable carrier mobility reduction. In the limiting sublattice substitution $Co_4Sb_{12} \to Fe_4Sb_{12}$, where all Co atoms are replaced by Fe ones, the valence electrons donated by the fillers are insufficient to compensate the holes because the oxidation states of the fillers are usually around +2 and +3 (see Table 3.14).[a] Hence, GFe_4Sb_{12} phases are expected to possess high hole densities and positive S value (see Table 3.15). This behavior confirms that the divalent fillers donate one electron less than the trivalent fillers, so that the filled skutterudites with divalent fillers exhibit higher hole densities.

In Table 3.15 we list the transport coefficients, power factor, and FOM values for several representatives of the two main

[a]At the early stages of research of Yb-filled skutterudites the role of itinerant magnetism of the Fe–Sb host structure was not fully explored, resulting in an inadequate attribution of Yb-valence state. Subsequent extensive experimental as well as theoretical studies have shown that Yb is unambiguously divalent in $YbFe_4Sb_{12}$ compounds [176].

skutterudite families. We note that most room temperature FOM reported values are not specially large, ranging from $ZT = 4 \times 10^{-4}$ for $CeFe_4P_{12}$ to $ZT = 0.3$ for $Yb_{0.19}Co_4Sb_{12}$. Conversely, most of the listed skutterudites exhibit large FOM values ($ZT \sim 0.8$–1.2) in the intermediate temperature range ($T \sim 600$–800 K), which determines the working temperature range for TE devices based on these materials.

The reason for the significant increase of the FOM when going from room temperature to higher temperatures is related to the substantial reduction of the κ_l contribution as the temperature is increased. To gain a deeper understanding on the physical mechanisms controlling this transport coefficient, let us first consider the behavior of thermal conductivity as a function of the anionic atoms mass in the case of both binary and ternary compounds. Experimental data show the lattice thermal conductivity decrease in binary skutterudite compounds CoX_3 (X = P, As, and Sb), with increasing anionic mass in the X sublattice [173]. This κ_l reduction is mainly due to a more efficient scattering between heat-carrying acoustic phonons (spanning the energy range within 1–5 meV) and the vibrational modes of the progressively heavier atoms in this sublattice. A significant additional reduction (amounting to 50–85%) takes place when the voids in the GFe_4X_{12} skutterudite families are filled with Ce atoms, the largest reduction occurring for antimonides [173]. Two main conclusions can be drawn from these results:

- the inclusion of guest atoms is essential to attain low enough lattice thermal conductivity values for these materials.
- Sb atoms provide the best possible choice for the X sublattice system. In this way, antimonide-based skutterudite compounds become the material of choice in current TEM research.

This significant κ_l reduction was earlier attributed to the so-called *rattling* effect of the void filling G atoms. Indeed, the rattling of these atoms can be inferred from: (a) their large X-ray thermal parameters, (b) their effects on the lattice modes as investigated by Raman spectroscopy, and (c) by inelastic neutron scattering. A direct evidence of rattling comes from structure refinements obtained using either X-ray or neutron diffraction on single crystals.

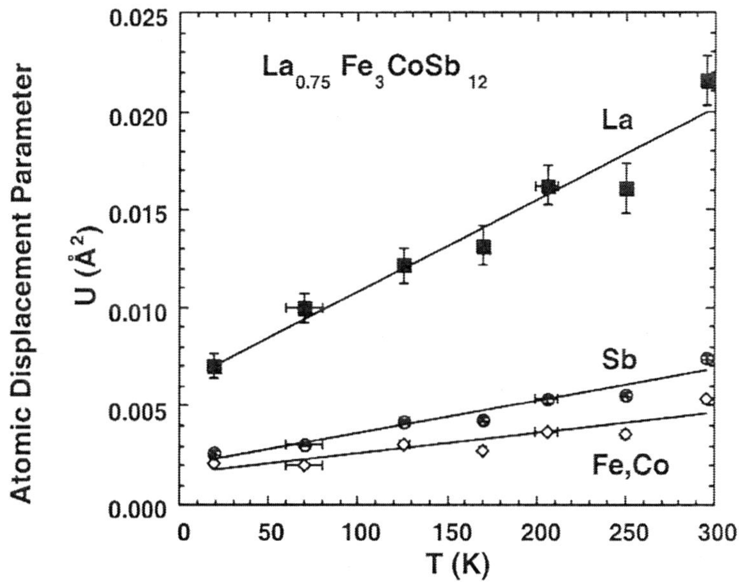

Figure 3.15 Temperature dependence of the isotropic displacement parameters measured on a single crystal of $La_{0.75}Fe_3CoSb_{12}$ skutterudite [175]. Reprinted with permission from B. C. Sales, D. Mandrus, B. C. Chakoumakos, V. Keppens, and J. R. Thomson, 1997, *Phys. Rev. B* **56**, 15081, Copyright 1997, American Physical Society.

In this way, the so-called *atomic displacement parameters* can be obtained for each distinct atom in the crystal structure. These parameters measure the mean-square displacement of the atom about its equilibrium position, hence providing a direct measure of its rattling motion. For the sake of illustration Fig. 3.15 shows the atomic displacement parameters for La, Fe, Co, and Sb atoms as obtained from a single skutterudite crystal with the composition $La_{0.75}Fe_3CoSb_{12}$. The atomic displacement parameters for Fe, Co, and Sb atoms are typical for these elements in compounds with similar coordination numbers. The values for La atoms, however, are anomalously large, particularly near room temperature. These values clearly indicate that the La is loosely bonded in the structure and moves back and forth about its equilibrium position.

Interpreting the atomic displacement data requires a microscopic model. In the simplest approximation, the guest atom dynamics can be described in terms of a localized phonon mode, usually called the rattling mode,[a] whereas the remaining atoms in the lattice are described within the usual Debye mode (see Section 1.2.2.3). For a quantized harmonic oscillator the mean-square displacement amplitude is given by (**Exercise 3.8**)

$$U \equiv \frac{1}{3}\left\langle u^2 \right\rangle = \frac{\hbar}{2M\omega}\coth\left(\frac{\hbar\omega}{2}\beta\right), \qquad (3.1)$$

(measured in $Å^2$), where M is the mass of the rattler and ω is its frequency of vibration. At high temperatures, where $\hbar\omega < 2k_BT$, Eq. (3.1) reduces to the classical expression $U = k_BT/K$, where $K = M\omega^2$ is the force constant of the oscillator.[a] Therefore, the so-called Einstein temperature of vibration, $\Theta_E \equiv \hbar\omega_E/k_B$, can be derived from the slope of the $U(T)$ curve in the high temperature regime. Compelling evidence for the existence of Einstein oscillator modes in several skutterudites has been provided by inelastic neutron time-of-fight and specific-heat measurements in combination with first-principles calculations. For instance, the presence of two Einstein modes at $\hbar\omega_{E1} = 5.0$ meV and $\hbar\omega_{E2} = 12.8$ meV has been reported in the filled skutterudite $Yb_{0.2}Co_4Sb_{12}$ [178]. Thus, typical Einstein temperature values for filled skutterudites are within the range $\Theta_E = 50$–70 K, which are significantly lower than their typical Debye temperatures, comprised in the range $\Theta_D = 270$–320 K (see Table 3.11).

Once the existence of rattler-like localized modes was confirmed, an inelastic scattering mechanism was proposed to explain the interaction between the propagating lattice phonons and these localized modes. According to the proposed mechanism, the local modes may first absorb a small amount of energy from the lattice phonons to get into an excited rattling state, and then later release the energy back to the lattice phonons system, with wave vectors incoherent with those of the originally absorbed phonons. Therefore, only the lattice phonons with similar energy to that of the local modes would interact, and such phonon–phonon mechanism was regarded as a resonant phonon scattering. The effect of these resonant scattering processes can be incorporated into the general

[a] Also known as an Einstein oscillator. The interested reader is referred to Ref. [177] for a detailed account of this topic.

[a] Recall the Taylor expansion $\coth x \sim 1/x$ for small enough x values.

lattice thermal conductivity expression given by Eq. (1.38) through the A_3 related term in Eq. (1.39). In this way, resonant scattering due to rattling guest atoms was naturally considered to be the dominant mechanism responsible for the rapid decrease in the thermal conductivity of filled skutterudites as small amounts of guest atoms are systematically placed into the voids.[a] Several factors contributed to support this view, namely:

- a rapid increase in the thermal resistivity is observed, followed by a gradual saturation of κ_l^{-1}, as higher concentrations of G atoms are added to the voids (**Exercise 3.9**).
- from the study of ytterbium-filled skutterudites where a systematic Sn substitution for Sb at different concentration levels was performed, it was observed that κ_l does not appreciably depend on the X atom content, while it is drastically reduced upon G atom filling [160]. This observation was interpreted as indicating that relatively heavy Sb atoms sublattice play a subsidiary role in determining κ_l as compared to that played by the G atoms.
- it was also observed that fillers with similar chemical characteristics demonstrate a similar resonance frequency value ω_0, while filler atoms from different groups of the periodic table show considerably different ω_0 values. Consequently, one may expect that the κ_l value of multi-filled skutterudites with fillers belonging to different groups should be lower than that of single-filled skutterudites, since one would have several resonant terms contributing to Eq. (1.39) instead of just one.

Indeed, experimental thermal conductivity measurements for the multi-filled system $(Ca/Ba)_yDD_xFe_4Sb_{12}$ (where DD stands for the so-called didymium metal: a natural mix consisting of 4.76 mass per cent Pr and 95.24 mass per cent Nd),[b] indicate that the thermal current degradation using two different fillers is much more efficient than for only one filler [69]. Figure 3.16 shows the lattice thermal conductivities of representative single-, double-, and triple-filled skutterudites at 300 K and 850 K. A remarkable reduction in the κ_l

[a] Note that τ^{-1} diverges when $\omega \to \omega_0$ in Eq. (1.39), so that $\kappa_l \to 0$ after Eq. (1.38).
[b] Cheaper than the respective pure rare-earth elements!

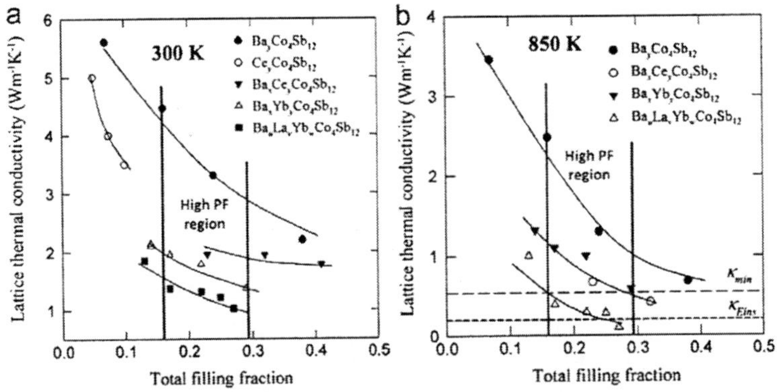

Figure 3.16 (a) Room temperature and (b) high temperature lattice thermal conductivities as a function of total filling fraction for single-, double-, and triple-filled skutterudites. The solid lines are a guide for the eyes. Vertical lines delimite the estimated higher power factor region for these compounds. Dashed lines in (b) indicate the estimated minimum thermal conductivity range [179]. Reprinted with permission from X. Shi, J. Yang, J. R. Salvador, M. F. Chi, J. Y. Cho, H. Wang, S. Q. Bai, J. H. Yang, W.Q. Zhang, and L. D. Chen, 2011, *J. Am. Chem. Soc.* **133**, 7837, Copyright 2011, American Chemical Society.

value is observed when one goes from single- to double- to triple-filled samples. In particular, one obtains $\kappa_l \simeq 1\text{–}2$ Wm^{-1}K^{-1} at 300 K and $\kappa_l \simeq 0.2$ Wm^{-1}K^{-1} at 850 K for the triple-filled skutterudites $Ba_xLa_yYb_zCo_4Sb_{12}$. Consequently, one gets the large FOM values $ZT = 1.7$ at $T = 850$ K for n-type $Ba_{0.08}La_{0.05}Yb_{0.04}Co_4Sb_{12}$, [179] and $ZT = 1.3$ at $T = 800$ K for p-type $Sr_{0.12}Ba_{0.18}DD_{0.39}Fe_3CoSb_{12}$ [180]. Quite interestingly, since most of the pure rare-earth elements are quite expensive, the above results suggests a cheaper way to obtain multi-filled skutterudites making use of natural multiple fillers, such as didymium or mischmetal (Mm, a natural mix composed of La 50.8%, Ce 28.1%, Nd 16.1%, and Pr 5.0%).

From all these facts, it was initially concluded that the observed lattice thermal conductivity decrease should be interpreted as arising from the localized incoherent motion of the guest atoms in the voids, ultimately leading to shorter mean free paths of the propagating lattice phonons. This picture, based on a decoupled dynamics of the guest atoms and lattice phonons systems, was in turn framed within the phonon glass-electron-crystal concept introduced by Slack in the 1990s, on the basis of independent, uncorrelated vibrations of fillers scattering of phonons in a random way.

Nevertheless, a number inelastic neutron scattering experiments subsequently revealed a larger than anticipated guest–host coupled dynamics, and thus, the need for a more refined microscopic picture for the phonon–phonon interaction mechanism. In fact, it has been confirmed that the vibration of the guest atoms is actually coupled to the vibration of the cage itself, in such a way that the cage vibration modes are qualitatively modified by the movement of the guests. Thus, it has been progressively realized that the overall phonon spectrum must be properly understood in the usual framework of well-ordered structures, and the use of concepts borrowed from the study of glassy materials may be quite misleading. Accordingly, the current emerging physical picture is that filled skutterudite compounds are able to achieve a low thermal conductivity mainly by lowering the effective *phonon speed* rather than by getting shorter mean free path values for the phonons. In this picture, the guest atoms modes hybridize with lattice acoustic modes, so that the presence of guest atoms does not turn the phonon spectra into that typical of glass-like materials, but they radically alter the phonon density of states in a way that results critical for the efficiency of the heat carrier phonon frequencies [178, 181].

Indeed, numerically calculated phonon dispersion relations of Ca-, Sr-, Ba-, and Yb-filled iron antimonides demonstrated that the vibrational dynamics of these filled skutterudites can be well approximated by a set of usual normal modes. These studies highlight the dominant effect of the bonding properties of the cation rattlers to the anion host matrix rather than the masses of the cation on the frequencies of collective excitations of the compounds (**Exercise 3.10**). Thus, it has been recently conjectured that the Sb ring normal modes, which are isolated from one another in the unfilled $CoSb_3$ skutterudite, become weakly coupled through the guest atoms in filled compounds and split into two separate groups of vibrations (each one characterized by a typical frequency range) referred to as cis- and trans-modes (Fig. 3.17). In addition, due to the hybridization of acoustic lattice modes and guest atoms

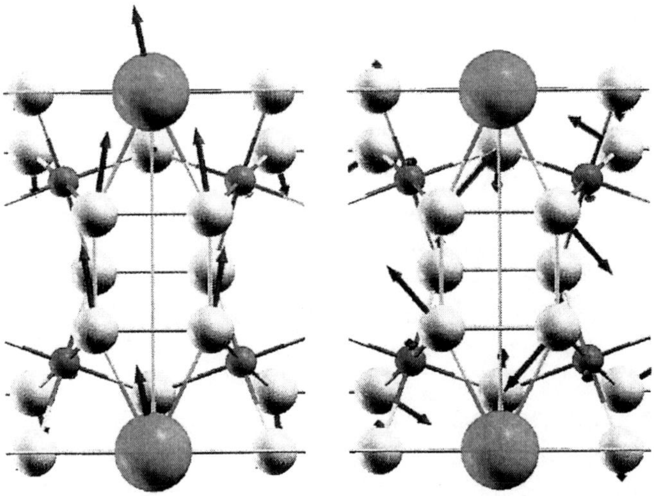

(a) (b)

Figure 3.17 Typical relevant vibration modes obtained in numerical models for $Sr_{0.5}Co_4Sb_{12}$ skutterudite: (a) cis-mode and (b) trans-mode [182]. Reprinted with permission from D. Wee, B. Kozinsky, and M. Fornari, 2013, *J. Phys. Soc. Jpn.* **82**, 014602, Copyright 2013, Physical Society of Japan.

dominated lower energy optical modes, the Debye model becomes an insufficient approximation to the low-energy dynamics of these compounds. As a consequence, some properties in their low-energy inelastic response upon heating can be captured in terms of anharmonic models only [176].

In summary, the physics of filled skutterudites is governed by the interplay of filler ions and their host structure, and the TE properties can be enhanced with the right amount and type of filler atoms. Improvements in TE efficiency mainly result from intrinsic properties of filling atoms, like the ionic radii, atomic mass, and phonon–phonon interaction details involving guest atoms and the host framework. We should remark that this interaction will generally result in a decrease of *both* thermal and electrical conductivities with respect to those observed in unfilled skutterudites.[a] In addition, one must consider that as the filling atom fraction is increased in the

[a]Indeed, the guest ions in the voids appear to scatter charge carriers as well as phonons.

skutterudite structure, the carrier concentration typically changes by several orders of magnitude. Therefore, part of the increase in thermal resistivity may be due to an enhanced electron–phonon scattering not included in Eq. (1.39) (**Exercise 3.11**). Accordingly, it was proposed that partial filling of the voids may be a suitable approach for obtaining improved materials since (i) less filler atoms reduce scattering with the carriers, increasing their mobility and (ii) a random distribution of the filler atoms would be more effective in scattering phonons than an ordered arrangement in which all of the voids are filled [165]. In this regard one must consider that disorder introduced in the metal atom sublattice as a consequence of the Fe substitution on the Co sites is another source of scattering for both phonons and charge carriers. In order to minimize the degradation of electrical conductivity due to these alloying effects, substitution at the Sb was thought to be more effective.

Finally, although there are a series of works reporting $ZT > 1.3$ values for n-type partially filled skutterudites, the ZT values in p-type filled skutterudites (fully filled GFe_4Sb_{12} being one of the most important representatives, see Table 3.15) are experimentally confined below $ZT \simeq 1$. The stagnation of this relatively low FOM value in p-type skutterudites limits the further development of TE modules and devices. Therefore, searching for p-type skutterudites with comparable ZT values to n-type ones is urgent for actual industry applications [303].

3.5.3 *Clathrates*

For more than a century many different clathrate hydrates have been discovered in nature. In *ice clathrates* water molecules form a hydrogen-bonded framework where each water molecule is tetrahedrally bonded to four H_2O neighbors, such as in normal ice, but with a more open structure which forms different types of cavities that can enclose atoms or small molecules. Quite interestingly, the group 14 tetravalent atoms silicon, germanium, and tin also form similar clathrate structures, hosting alkaline atoms inside. In this case the host lattice framework is based on strong covalent bonds with lengths comparable to those found in the diamond-like structures of these elements, and the resulting materials are referred to as *covalent clathrates*.

According to the charge distribution between the host and guest species, covalent clathrates can be classified in three classes, namely, cationic, neutral, and anionic clathrates.

- In cationic clathrates, the host plays the role of a polycation and the guest is an anion, for example, the electronegative P atom in the $P_8Te_8Ge_{38}$ compound.
- In neutral clathrates, the host is formed by hydrogen-bonded molecules and the guests are neutral atomic or molecular species (e.g., noble gases, methane) as in the $(Cl_2)_8(H_2O)_{46}$ compound.
- In anionic clathrates, the host has a negative charge (e.g., group 14 elements such as Si, Ge, or Sn) with respect to the guest atoms such as alkaline metals (Na, K, Rb, Cs), alkaline-earth (Ba, Sr) or rare earth (Eu, Yb) atoms, as in the $Na_{24}Si_{136}$ compound. In anionic clathrates the framework atoms can be partially substituted by group 13 elements or transition metals, as it occurs in the $Ba_8Ga_{16}Sn_{30}$ or $Ba_8Cu_{16}P_{30}$ compounds.

In what follows we will focus on covalent anionic clathrates-based materials. The crystal structures of anionic clathrates belong to five basic types, which are denoted as the I, II, III, VIII, and IX clathrate types. The common structural feature of the frameworks I, II, III, and VIII is the presence of tetrahedrally bonded species only and the formation of large cages with 20, 24, 26, or 28 atomic vertices filled by the electropositive element. The most common forms of covalent anionic clathrates are types I and II, whose structural framework can be described in terms of two different kinds of face-sharing polyhedra (Fig. 3.18):

- The type I structure has a cubic cell (space group $Pm\bar{3}n$) and holds $n_U = 46$ atoms in the unit cell. There are two basic polyhedral building blocks in the structure, namely, pentagonal dodecahedra (containing 20 atoms) and tetrakaidecahedra (a polyhedron with two hexagonal faces and eight pentagonal faces containing 24 atoms). There are eight polyhedra in the unit cell, two dodecahedra and six tetrakaidecahedra, and the general

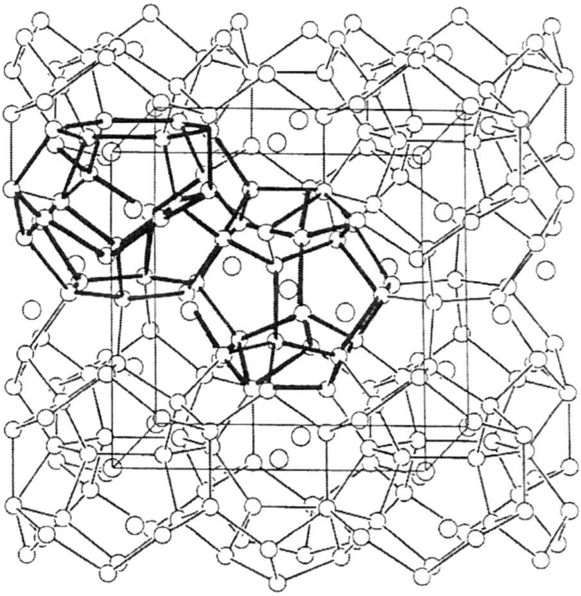

Figure 3.18 The type I covalent anionic clathrate ideal structure. Only the group 14 elements are shown. Outlined are the two different polyhedra that form the unit cell: tetrakaidecahedra (on the left) and pentagonal decahedra (on the right) [11]. Reprinted with permission from G. S. Nolas, T. J. R. Weakley, and J. L. Kohn, *Chem. Mater.* **11**, 2470, Copyright 1999, American Chemical Society.

formula is M_8E_{46}, where M represents an alkaline or alkaline-earth metal and E represents a group 14 element, although Zn, Cd, Al, Ga, In, As, Sb, or Bi atoms can also occupy the E positions.

- The type II structure has also a cubic cell holding n_U = 136 atoms in the cubic cell. These atoms are arranged in 16 pentagonal dodecahedra and 8 hexakaidecahedra (containing 28 atoms) building blocks. There are 24 cavities in total in the unit cell, and the general formula reads $M_{24}E_{136}$.

A hypothetical empty framework of group 14 elements would be valence balanced: each element uses its four valence electrons for the four bonds, leading to the fulfillment of the octet rule for each atom. The presence of guest cations gives rise to excess of

Table 3.16 Room temperature transport coefficients of type-I clathrate compounds of TE interest

Sample	σ (Ω^{-1}cm^{-1})	S (μVK^{-1})	P (μWcm^{-1}K^{-1})	κ (Wm^{-1}K^{-1})	ZT
$Ba_8Cu_{16}P_{30}$	1000	+50	2.5	3.7	0.020
$Sr_8Zn_{16}Ge_{30}$	3030	−24	1.7	1.9	0.028
$Ba_8Ga_{16}Si_{30}$	490	−66	2.1	1.4	0.047
$Ba_8Ga_{16}Ge_{30}$	1250	−50	3.1	1.6	0.059
$Sr_8Ga_{16}Ge_{30}$	78	−313	7.6	1.0	0.240
$Eu_8Ga_{16}Ge_{30}$	397	−152	9.2	0.9	0.310
$Sr_4Eu_4Ga_{16}Ge_{30}$	1000	−88	7.7	1.3	0.175
$Yb_{0.5}Ba_{7.5}Ga_{16}Ge_{30}$	590	−60	2.1	1.5	0.043

Source: Refs. [41, 65, 68, 108, 110]

electrons, leading to a metallic character. To restore semiconducting behavior it is necessary to partially substitute E atoms by an electron acceptor atom. For instance, the partial substitution of Ge by Ga creates an electron-deficient framework in comparison to the crystal structure of pure Ge. If M is divalent, electronic compensation can be realized by replacing 16 E atoms by group 13 atoms (e.g., Al, Ga, In). This is the case of the compound $Eu_8Ga_{16}Ge_{30}$, for instance, where neutron scattering data indicate that both Ga and Ge are randomly distributed through the framework [184]. The compound is then valence balanced, as each guest atom donates two electrons and every Ga utilizes one of these electrons for bonding with the framework. Deviation from this stoichiometry results in doped semiconductor behavior, with excess Ge giving n-type samples and excess Ga resulting in p-type ones.

Depending on the nature of both M and E atoms the lattice parameter of clathrates can vary from $a = 10.509$ Å for $Ba_8Ga_{16}Ge_{30}$ to $a = 12.113$ Å for Cs_8Sn_{46}. Thus, the lattice dimensions of clathrates are significantly larger than those of both skutterudites and half-Heusler compounds (Fig. 3.12). The melting points of ternary clathrates vary over a relatively broad range from $T \simeq 750$ K ($Ba_8Ga_{16}Sn_{30}$) to $T \simeq 1400$ K ($Ba_8Ga_{16}Si_{30}$) [183].

In Table 3.16 we list the transport coefficients, power factor, and FOM values for some representative clathrate compounds of TE interest. The most promising materials belong to the $X_8Ga_{16}Ge_{30}$ family, with X = Ba, Sr, Eu. At room temperature $Sr_8Ga_{16}Ge_{30}$

samples exhibit relatively high negative Seebeck coefficients (from $S = -125$ to $S = -320$ μVK^{-1}), electrical conductivities in the range $\sigma = 70$–$500\ \Omega^{-1}\text{cm}^{-1}$, and thermal conductivities of about $\kappa = 0.8$–1.3 Wm^{-1}K^{-1}, yielding $ZT \simeq 0.25$ at room temperature in the most favorable case. We note that most FOM reported values at room temperature are not specially large, ranging from $ZT = 0.020$ for $Ba_8Cu_{16}P_{30}$ to $ZT = 0.31$ for $Eu_8Ga_{16}Ge_{30}$. At higher temperatures, however, clathrates possess a TE efficiency comparable to state-of-the-art materials such as PbTe and SiGe. The highest reported $ZT = 1.35$ was obtained at 900 K for a single crystal $Ba_8Ga_xGe_{46-x}$ (grown using the Czochralski method) with a composition with x somewhere between 15 and 16 [185]. This feature is similar to that previously reported in previous sections for half-Heusler and skutterudite compounds, though when comparing materials with similar σ values clathrates generally exhibit significantly lower thermal conductivity values (compare Tables 3.12, 3.13, 3.15, and 3.16), which ultimately accounts for the potential of clathrate compounds for TE applications.

The physical origin of the low κ values in covalent clathrates has been the subject of much discussion. In the first place, we note that the valence stabilization requires the presence of two different types of atoms in the host framework, which naturally reduce the thermal conductivity due to alloying effects. In the second place, we have a contribution coming from guest atoms moving inside the cages in the clathrate open structure. The Einstein temperatures derived from atomic displacement data (e.g., $\Theta_{E1} = 74$ K and $\Theta_{E2} = 103$ K for $Sr_8Ga_{16}Ge_{30}$; $\Theta_{E1} = 53$ K and $\Theta_{E2} = 82$ K for $Eu_8Ga_{16}Ge_{30}$, respectively) are similar to those reported for skutterudites [184, 186]. Neutron scattering studies and molecular lattice dynamical studies have demonstrated that localized guest vibrations strongly interact with the host acoustic phonon branches in clathrates, so that the rattling motion is dominated by the weak electrostatic interaction between the host cage and guest atoms, which, in turn, is influenced by rattling. The highest frequency optic phonons in the clathrate structure have very low group velocities and contribute little to the total thermal conductivity, whereas the low-frequency acoustic phonons have a somewhat larger group velocity. Thus, the resonant scattering of these low-frequency phonons by vibrations of the guest ions in their polyhedral cages results in very low thermal conductivities. The guest atoms vibration frequencies increase as the size difference between guest atom radius and host cage size increases, a natural consequence of stronger restoring forces experienced by the guest atoms. As the space available to the guest atoms in the host cage becomes larger than their ionic radii, the restoration forces acting on them become increasingly weak. Their phonon energies decrease accordingly and they start vibrating in a more localized manner. The guest phonon spectrum becomes a low-lying, non-dispersive optical mode, which can disturb the flow of heat-carrying acoustic phonons propagating through the framework. Accordingly, one expects the magnitude of the low temperature thermal resistivity will correlate with the relative guest radius/host-cage mismatch (**Exercise 3.12**).

In clathrates the guest atoms are ionically bonded to the host framework. The electrostatic potential within the polyhedral cages are not the same everywhere and different points may be energetically preferred. Indeed, electrochemical interactions with the cage sides leads to a broadening deformation of the guest potential well until, at some point, the minimum energy potential is no longer at the cage center, being replaced by a fourfold split, off-center minima, which, in the most simple approach, can be modeled as $U = Ax^2 + Bx^4$, containing a harmonic plus an anharmonic (quartic) terms. This suggests that the guest atom can tunnel among these positions and its lattice site in the equilibrium configuration is no longer centered in a D_{2d} crystallographic point symmetry but is disordered in four equivalent off-center positions, still preserving, however, an average $Pm\bar{3}n$ crystal symmetry. Accordingly, the measured atomic displacement positions of rattling atoms will contain a static and a dynamic component, both mechanisms having an effect on the resulting lattice contribution to the thermal conductivity [184]. In addition, the largely deformed potential well produces strongly anharmonic vibrations and the Einstein model fails to properly account for the experimental data [187]. As we mentioned in the previous section, the early picture of isolated guest atoms in a host cage responsible for the scattering of heat carriers was revisited by inelastic neutron scattering studies in skutterudite crystals, were it was concluded that the guest atoms coherently coupled with the host-lattice dynamics and hybridized with low-energy optical phonon modes characterized by small group velocities. These low-frequency phonon branches open new paths for the occurrence of Umklapp scattering of heat-carrying acoustic phonons. At a microscopic level this implies the existence of a coupling between the nondispersive phonon modes of the isolated resonators and the acoustic phonon branches stemming from delocalized lattice phonons. Recent experiments have confirmed that this physical mechanism also operates in the case of clathrate compounds [188]. We note that this mechanism obeys momentum and energy conservation, hence it is in contrast with inelastic resonant scattering associated with energy dissipation at isolated atoms.

A widespread use of TE generators based on clathrate compounds would require both n- and p-type materials being of comparable TE performance and compatible in terms of thermal expansion to minimize stress effects and enhance their chemical compatibility within the module. However, even though several clathrates have been discovered to exhibit a p-type conduction (e.g., $Ba_8Ga_{16}Ge_{30}$, $Ba_8Ga_{16}Zn_xGe_{30-x}$, $Ba_8Ga_{16}Al_xGe_{30-x}$) most of the clathrates studied so far are characterized by n-type behavior, since the available p-type clathrates exhibit lower ZT values than the n-type counterparts. Indeed, in these materials it is not straightforward to tune the charge carrier concentration to the optimum value for both the n- and p-type legs. It is thus urgent to search for p-type clathrates for practical applications. This has spurred a search for TE-efficient clathrates containing transition metal elements. In these materials the charge carrier concentration can be systematically varied, for both n- and p-type conduction, by changing the transition metal content. Another issue restraining the actual commercial applications of these materials is the high cost of gallium and germanium. Consequently, efforts aimed at replacing these elements with cheaper silicon and tin are currently going on.

3.5.4 *Chevrel Phases*

Several compounds in the family $M_xMo_6Se_8$ (with M = Cu, Fe, Ti) have been reported to have good TE properties at high temperatures

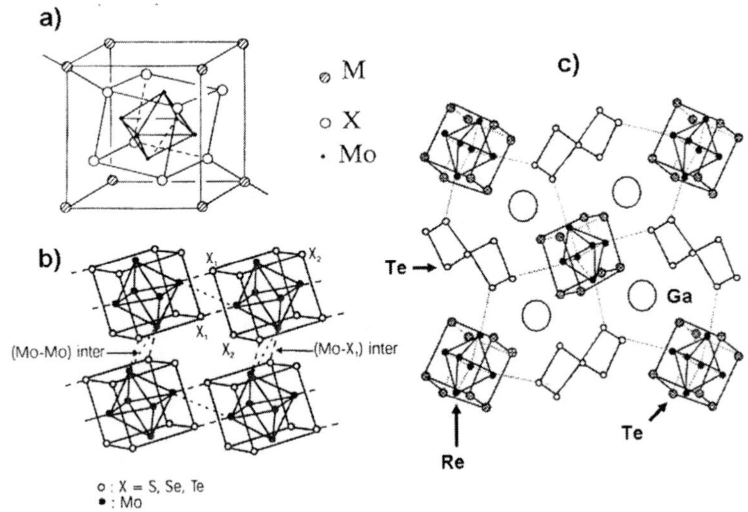

Figure 3.19 (a) Polyhedral nested structure of the Mo_6Se_8 basic building block. M stands for a metallic cation, and X is a chalcogenide element. (b) Spatial arrangement of the basic building blocks in the $M_xMo_6Se_8$ crystal. Mo–Mo and Mo–X intercluster bonds are represented by dashed lines. (c) Structure of the $Re_6Ga_xTe_{15}$ compound. Quasi-planar heptagonal Te clusters are linked to Re_6Te_8 cubic-like clusters. Large white spheres represent voids where Ga atoms are inserted into the structure. Reprinted from Ref. [50], by permission from Macmillan Publishers Ltd: *Nat. Mater.* (C. B. Vining, *Nature Mater.* **8**, 83), Copyright (2009), and from Ref. [189], with permission from D. S. Dalafave, 2010 *Mats. Chem. Phys.* **119**, 195, Copyright 2010, Elsevier.

(e.g., $ZT = 0.6$ for $M_{3.1}Mo_6Se_8$ at $T = 1150$ K). These compounds belong to the Chevrel phase class, which, in turn, can be regarded as another representative of the inclusion compounds broader class, since they are built from a three-dimensional network of pseudo-cubic clusters forming cavities or channels where various elements can be intercalated. The basic atomic arrangement is the same for all MMo_6Se_8 compounds. It consists of a stacking of Mo_6Se_8 building blocks, where each building block is a cube formed by eight Se atoms which contains an octahedron of tightly packed Mo atoms inside (see Fig. 3.19a). These Mo atoms, however, are slightly outside the middle of the faces of the Se atoms cube, so that the Mo_6Se_8 block has almost cubic symmetry. The arrangement of these building blocks leaves a certain number of cavities in the chalcogen atom network, which are empty in the binary Mo_6Se_8 compound but can be filled with cations in the ternary $M_xMo_6Se_8$ compound. In the $M_xMo_6Se_8$ crystal, the corners of each Mo_6Se_8 block lie directly opposite to the face centers of adjacent blocks. That means that there are close contacts between the Mo atoms of one block and the Se atoms of the other neighboring blocks. The $M_xMo_6Se_8$ resulting unit cell is then constructed as shown in Fig. 3.19b. The M cation is located in the origin (0,0,0) position. The other inserted cations are distributed on the corners of the rhombohedral unit cell which has an angle close to 90°. The size and shape of the cavities depend on the nature of the inserted cation and the limit concentration of these cations is determined by electronic factors. Indeed, as one adds M atoms the number of valence electrons increases due to the charge transfer from the M atoms to the Mo atoms. The Mo_6 then contracts and becomes more regular. The number of valence electrons per Mo atom is therefore, the parameter of interest. In this way, the resulting solid can be modified from metallic to semiconducting behavior for a valence electron concentration of 24 [191].

The Re_2Te_5 compound is a cluster type telluride with a structure closely related to that of the molybdenum chalcogenide Chevrel phases, where Re_6Te_{15} clusters play a role similar to that played by Mo_6Se_8 building blocks (see Fig. 3.19c). Thus, the resulting Re_6Te_{15} network structure includes voids with radius of 2.754 Å, large enough to host additional metal atoms (e.g., Ga, In, Ag) within. The crystal unit cell contains 84 atoms. The TE properties of selenium substituted $Re_6Ga_xSe_yTe_{15-y}$ compounds were studied in the temperature range 90–320 K. The measurements revealed p-type semiconductivity in all the considered samples, exhibiting high Seebeck coefficient values (500–600 μVK^{-1} at room temperature). Unfortunately, exceedingly large electrical resistivity values ($\sim$100 Ωcm at room temperature) render too small power factor values [189].

3.6 Exercises

(1) A comparative study of the thermoelectric properties of amorphous $Zn_{41}Sb_{59}$ and crystalline Zn_4Sb_3 compounds yielded the following room temperature transport coefficients values:

Table 3.17 Room temperature structural and transport coefficients data for two skutterudite samples (After Refs. [169, 175])

Sample	a (Å)	v_t (ms^{-1})	v_l (ms^{-1})	κ ($Wm^{-1}K^{-1}$)	σ ($\Omega^{-1}cm^{-1}$)
$CoSb_3$	9.037	2788	4623	8.7	40
$LaFe_3CoSb_{12}$	9.140	2661	4484	1.6	1000

$\rho = 0.7\ \Omega$ m, $S = +275\ \mu VK^{-1}$, $\kappa = 0.19\ Wm^{-1}K^{-1}$ for the amorphous sample and $\rho = 1.8\ m\Omega$ cm, $S = +108\ \mu VK^{-1}$, $\kappa = 0.9\ Wm^{-1}K^{-1}$ for the crystalline sample [190]. Compare their respective ZT values.

(2) Making use of the transport data listed in Table 3.1 determine the charge carrier contribution to the thermal conductivity of elemental lead, bismuth and tellurium. Comment on the obtained results.

(3) Making use of the transport data listed in Table 3.12 compare the lattice contribution to the thermal conductivity in the half-Heusler compounds $Zr_{0.5}Hf_{0.5}NiSn$ and $Zr_{0.5}Hf_{0.5}NiSn_{0.98}Bi_{0.02}$. Comment on the obtained results.

(4) In Table 3.17 we list the room temperature lattice parameter a, the transverse (v_t) and longitudinal (v_l) sound velocities, the thermal (κ) and the electrical (σ) conductivities for a filled and an unfilled skutterudite antimonide. Making use of these data determine their (a) lattice (κ_l) and (b) minimum (κ_m) thermal conductivities (assume the Lorenz number value $L = 2 \times 10^{-8}\ V^2K^{-2}$). Discuss the role played by La fillers within the framework of the electronic crystal-phonon glass approach.

(5) Obtain the room temperature FOM for the $CeFe_4Sb_3$ filled skutterudite from the measured values of the Seebeck coefficient ($+70\ \mu VK^{-1}$), electrical resistivity ($\rho = 1.3\ m\Omega$ cm), and thermal conductivity ($0.9\ Wm^{-1}K^{-1}$) [174].

(6) Show that the structural formula of an unfilled skutterudite can be expressed as $\square M_4Sb_{12}$.

Table 3.18 Room temperature lattice thermal conductivity values for a series of Tl-filled skutterudites

Sample	κ_l ($Wm^{-1}K^{-1}$)
$Tl_{0.1}Co_4Sb_{12}$	3.4
$Tl_{0.22}Co_4Sb_{12}$	2.9
$Tl_{0.5}Co_4Sb_{11.5}Sn_{0.5}$	2.3
$Tl_{0.75}Co_4Sb_{11.25}Sn_{0.75}$	2.1
$TlCo_4Sb_{11}Sn$	1.9

Source: Ref. [164]

(7) The distance between the Ce position (cage center) and the Sb position (cage border) in the $CeFe_4Sb_{12}$ skutterudite is 3.39 Å. The covalent radius of Sb is 1.46 Å, and the ionic radius of Ce^{+3} for a coordination number 12 is 1.34 Å. Determine the average available distance for the Ce atom thermal oscillations inside the cage.

(8) In the general case, assuming each atom contributes three normal modes, the mean atomic displacement is given by

$$\langle u_i^2 \rangle = \frac{3\hbar}{2M} \int \frac{Z_i(\omega')}{\omega'} \coth\left(\frac{\hbar\omega'}{2}\beta\right), \tag{3.2}$$

where M is the atom mass and $Z_i(\omega')$ (satisfying the normalization condition $\int Z_i(\omega')d\omega' = 1$) denotes the partial contribution to the vibrational DOS due to the i-th atom. Show that Eq. (3.2) reduces to Eq. (3.1) in the particular case $Z_i(\omega') = \delta(\omega' - \omega)$.

(9) Making use of the lattice thermal conductivity data listed in Table 3.18, (a) plot the thermal resistivity as a function of the Tl atoms filling fraction, (b) show that these data can be fitted to a power law curve of the form $\kappa_l^{-1} = a + bx^{1/3}$, and (c) give a physical picture of the obtained results.

(10) It was early suggested that the decrease in lattice thermal conductivity may be proportional to the ratio between the void and filling atom radii as well as to the mass of the guest atoms, that is, the smaller and the heavier the filling atom, the lower

Table 3.19 Mass (M), ionic radii (r) taken from ref. [194], coordination number (CN), and lattice thermal conductivity (after ref. [168] if not otherwise indicated) for a number of ion fillers in skutterudite compounds

Ion	M (uma)	r (Å)	CN	κ_l ($Wm^{-1}K^{-1}$)
Ca^{+2}	40.1	1.34	12	1.61
Sr^{+2}	87.6	1.44	12	2.07
In^{+3}	114.8	0.92	8	2.10 [162]
Sn^{+2}	118.7	1.10	8	1.97 [192]
Ba^{+2}	137.3	1.61	12	2.15
La^{+3}	138.9	1.36	12	1.63
Ce^{+3}	140.1	1.34	12	1.28
Pr^{+3}	140.9	1.34	9	1.09
Nd^{+3}	144.2	1.27	12	0.96
Eu^{+2}	152.0	1.35	10	1.79
Yb^{+2}	173.0	1.37	8	1.18
Tl^{+}	204.4	1.70	12	1.90 [164]

Table 3.20 Room temperature lattice thermal resistivity and carrier concentrations values for doped and undoped CoSb3 skutterudites

Sample	κ_l^{-1} ($W^{-1}mK$)	n (cm^{-3})
$CoSb_3$	0.109	1.2×10^{17}
$Co_{0.95}Fe_{0.05}Sb_3$	0.125	3.8×10^{19}
$Co_{0.95}Ni_{0.05}Sb_3$	0.175	5×10^{20}

Source: Ref. [167]

the lattice thermal conductivity [193]. Discuss the validity of this physical scenario making use of the experimental data listed in Table 3.19.

(11) In order to investigate the importance of electron–phonon scattering processes two-doped $CoSb_3$ skutterudites were synthesized and compared to the native one. Making use of the data listed in the Table 3.20 estimate the relative importance of this scattering mechanism for n-type and p-type doped samples.

Table 3.21 Low temperature ($T = 10$ K) lattice thermal conductivity for several clathrate compounds. (After Refs. [108, 184])

Sample	κ_l (Wm^{-1}K^{-1})
$Ba_8Ga_{16}Si_{30}$	1.20
$Sr_8Ga_{16}Ge_{30}$	1.10
$Eu_8Ga_{16}Ge_{30}$	0.22

(12) Making use of the ionic radii listed in Table 3.19, and transport data given in Table 3.21, show that the lattice thermal resistivity correlates with the guest/host-cage mismatch, $\Delta r \equiv r_C - r_I$, for the considered type I clathrates, where r_C measures the averaged cage radius. The cage radii for different structural polyhedra are estimated to be 2.05 and 2.23 Å, for Si_{20} and Si_{24} in Si-bearing clathrates [108], 2.197 and 2.377 Å for Ge_{20} and Ge_{24} in $Sr_8Ga_{16}Ge_{30}$ clathrate, and 2.183 and 2.365 Å in $Eu_8Ga_{16}Ge_{30}$ clathrate, respectively [183, 184].

3.7 Solutions

(1) In the first place, we note that the electrical conductivity of the amorphous sample, $\sigma = 1.4\ (\Omega\text{m})^{-1}$ is more than four orders of magnitude lower than that of the crystalline sample, $\sigma = 5.5 \times 10^4\ (\Omega\text{m})^{-1}$, as expected. On the other hand, the Seebeck coefficient of the amorphous sample doubles that of the crystalline one, and its thermal conductivity value is about five times lower, as well. The latter properties are both contributing to enhance the corresponding FOM value. Therefore, one can not decide which sample will ultimately render a better TE performance on merely qualitative basis. To this end, we will explicitly evaluate the FOM to obtain

$$ZT = 1.43\ \Omega\,\text{m}^{-1}\frac{275^2 \times 10^{-12}\ \text{V}^2\text{K}^{-2} \times 300\ \text{K}}{0.19\ \text{Wm}^{-1}\text{K}^{-1}} = 1.7\times 10^{-4},$$

for the amorphous $Zn_{41}Sb_{59}$ alloy, and

$$ZT = 5.5 \times 10^4\ \Omega\,\text{m}^{-1}\frac{108^2 \times 10^{-12}\ \text{V}^2\text{K}^{-2} \times 300\ \text{K}}{0.9\ \text{Wm}^{-1}\text{K}^{-1}} = 0.21,$$

for the crystalline Zn_4Sb_3 alloy. Thus, the TE performance of the crystalline phase is definitively better than that corresponding to the amorphous one for this compound.

(2) According to the Wiedemann–Franz law (see Section 1.2.3) we have

$$\begin{aligned}\kappa_e &= L_0\sigma T \\ &= 2.44 \times 10^{-8}\ \mathrm{V^2K^{-2}}\ 51700\ \Omega^{-1}\mathrm{cm}^{-1}10^2\ \mathrm{m}^{-1}\mathrm{cm}\ 273\ \mathrm{K} \\ &= 34.4\ \mathrm{Wm^{-1}K^{-1}}\end{aligned}$$

for lead

$$\begin{aligned}\kappa_e &= 2.44 \times 10^{-8}\ \mathrm{V^2K^{-2}}\ 7400\ \Omega^{-1}\mathrm{cm}^{-1}10^2\ \mathrm{m}^{-1}\mathrm{cm}\ 273\ \mathrm{K} \\ &= 4.9\ \mathrm{Wm^{-1}K^{-1}}\end{aligned}$$

for bismuth, and

$$\begin{aligned}\kappa_e &= L_0\sigma T = 2.44 \times 10^{-8}\ \mathrm{V^2K^{-2}}\ 2\ \Omega^{-1}\mathrm{cm}^{-1}10^2\ \mathrm{m}^{-1}\mathrm{cm}\ 273\ \mathrm{K} \\ &= 0.001\ \mathrm{Wm^{-1}K^{-1}}\end{aligned}$$

for tellurium. Therefore, the thermal conductivity of lead is mainly determined by its electronic contribution, as expected for a metallic material. The electronic contribution amounts to about ~70% in the case of semimetallic bismuth, whereas this contribution is almost negligible in the case of semiconducting tellurium.

(3) Making use of the Wiedemann–Franz law to express the electronic contribution to the thermal conductivity we have $\kappa_l = \kappa - L_0\sigma T$, so that

$$\begin{aligned}\kappa_l = 4.4\ \mathrm{Wm^{-1}K^{-1}} - 2.44 \times 10^{-8}\mathrm{V^2K^{-2}}\ 121 \\ \times 10^2\ \Omega^{-1}\mathrm{m}^{-1}\ 300\ \mathrm{K} = 4.31\ \mathrm{Wm^{-1}K^{-1}}\end{aligned}$$

for $Zr_{0.5}Hf_{0.5}NiSn$, and

$$\begin{aligned}\kappa_l = 7.4\ \mathrm{Wm^{-1}K^{-1}} - 2.44 \times 10^{-8}\mathrm{V^2K^{-2}}\ 633 \\ \times 10^2\ \Omega^{-1}\mathrm{m}^{-1}\ 300\ \mathrm{K} = 6.94\ \mathrm{Wm^{-1}K^{-1}}\end{aligned}$$

for $Zr_{0.5}Hf_{0.5}NiSn_{0.98}Bi_{0.02}$. Therefore, the charge carrier contribution to the thermal conductivity only amounts to about 2% (6.2%) of the total thermal conductivity value in the $Zr_{0.5}Hf_{0.5}NiSn$ ($Zr_{0.5}Hf_{0.5}NiSn_{0.98}Bi_{0.02}$) compounds, respectively. Whereas, this three-fold increment in κ_e is expected from doping, we realize that the thermal conductivity is yet governed by the lattice dynamics. We also see that the introduction of bismuth atoms has an important effect on the dynamical properties of the Sn sublattice, leading to a substantial (unfavorable) *increase* of the resulting κ_l conductivity.

(4) The lattice and the minimum thermal conductivities are given by $\kappa_l = k - L\sigma T$, and $\kappa_m = 0.4k_B\rho_a^{2/3}(v_l + 2v_t)$, respectively. For $CoSb_3$,

$$\kappa_l = 8.7\ \mathrm{Wm^{-1}K^{-1}} - 2 \times 10^{-8}\ \mathrm{V^2K^{-2}}\ 40 \times 10^2\ \Omega^{-1}\mathrm{m^{-1}}\ 300\ \mathrm{K} = 8.676\ \mathrm{Wm^{-1}K^{-1}},$$

$$k_m = 0.4 \times 1.38 \times 10^{-23}\ \mathrm{JK^{-1}}(4623 + 2 \times 2788)\ \mathrm{ms^{-1}} \left(\frac{16}{(9.037)^3}\right)^{2/3} 10^{20}\ \mathrm{m^{-2}} = 0.438\ \mathrm{Wm^{-1}K^{-1}}$$

For $LaFe_3CoSb_{12}$,

$$\kappa_l = 1.6\ \mathrm{Wm^{-1}K^{-1}} - 2 \times 10^{-8}\ \mathrm{V^2K^{-2}}\ 1000 \times 10^2\ \Omega^{-1}\mathrm{m^{-1}}\ 300\ \mathrm{K} = 1.0\ \mathrm{Wm^{-1}K^{-1}},$$

$$k_m = 0.4 \times 1.38 \times 10^{-23}\ \mathrm{JK^{-1}}(4484 + 2 \times 2661)\ \mathrm{ms^{-1}} \left(\frac{17}{(9.14)^3}\right)^{2/3} 10^{20}\ \mathrm{m^{-2}} = 0.428\ \mathrm{Wm^{-1}K^{-1}}.$$

The values of the minimum thermal conductivities are pretty similar for both compounds, but only the filled skutterudite exhibits a lattice thermal conductivity relatively close to κ_m, hence indicating the important role played by the guest atoms in order to significantly reduce κ_l. In addition, the substitution of Fe on the Co sites gives rise to a 25-fold increase in the electrical conductivity of the filled skutterudite as compared to the unfilled one. In this way, the electronic crystal–phonon glass scenario is reasonably agreed [175]. Note that the Wiedemann–Franz contribution is negligible for the highly resistive $CoSb_3$ sample.

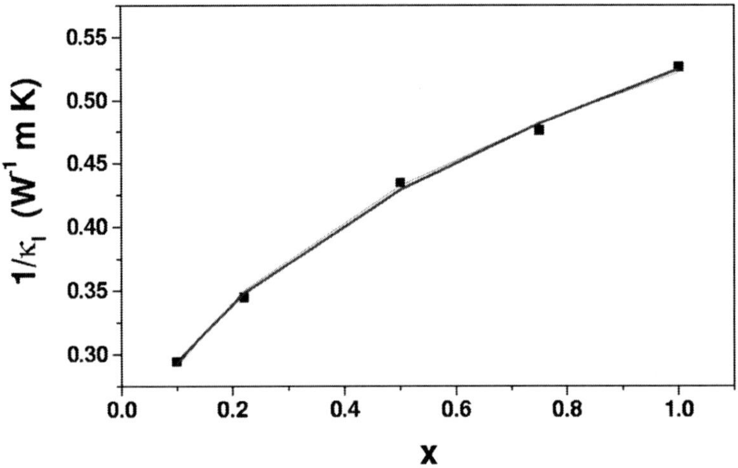

Figure 3.20 Thermal resistivity as a function of the Tl atoms filling fraction plot obtained from the data listed in Table 3.18.

(5) Making use of Eq. (1.71) we have

$$ZT = \frac{\sigma S^2 T}{\kappa} = \frac{10^3 \quad \Omega^{-1}\text{cm}^{-1}}{1.3\ 10^{-2}\ \text{m cm}^{-1}}$$

$$= \frac{70^2 \times 10^{-12}\ \text{V}^2\text{K}^{-2} \times 300\ \text{K}}{0.9\ \text{Wm}^{-1}\text{K}^{-1}} = 0.126$$

(6) By inspecting the skutterudite unit cell shown in Fig. 3.14a we count one metal (M) atom in the center, six M atoms sharing the faces, twelve M atoms sharing the edges, and eight M atoms sharing the corners, which amounts $1 + 6(1/2) + 12(1/4) + 8(1/8) = 8$ M atoms in the unit cell. In addition, we have six rings with four Sb atoms each, plus two voids per unit cell. Therefore, the structural formula reads $\square_2 M_8 Sb_{24} = \square M_4 Sb_{12}$.

(7) 3.39 Å−(1.46 + 1.34) Å= 0.59 Å.

(8) Making use of Eq. (2.52) the expression given by Eq. (3.2) can be readily integrated to get Eq. (3.1).

(9) (a) Making use of the data listed in Table 3.18 we obtain the plot shown in Fig. 3.20. Thus, within experimental error, there is no clear maximum in the thermal resistance data as

a function of the Tl concentration, and the maximum thermal resistance occurs near complete filling.

(b) From a fitting analysis of the plot shown in Fig. 3.20 one gets $\kappa_l^{-1} = 0.14 + 0.043x^{1/3}$.

(c) If the Tl atoms are treated as localized Einstein oscillators, then the phonon mean free path d should be a function of the distance between the Tl atoms in the crystal. The simplest estimate of d is therefore, the average distance between the Tl atoms. This simple argument suggests that, if the role of other scattering mechanisms can be neglected in Eq. (1.39) as a first approximation, κ_l^{-1} should vary as $x^{1/3}$, since the average spacing between Tl varies as $x^{-1/3}$ [164].

(10) By inspecting the data shown in Fig. 3.21 we realize that no clear trend relating the atomic mass and ionic radii with the thermal conductivity can be appreciated. Furthermore, we see that whereas the thermal conductivities of Ce^{+3} and Pr^{+3} containing skutterudites differ by about 20% these cations have almost the same masses and ionic radii. In addition, the samples exhibiting the lower thermal conductivity values are those filled with Nd^{+3}, Pr^{+3}, and Yb^{+2} cations, which significantly differ in valence, mass, and radii.

(11) By inspecting Table 3.20 one sees that the carrier density increases by more than two (Fe p-doping) and three (Ni n-doping) orders of magnitude. Since the amount of doping element is just of 5% one can reasonably assume that the lattice contribution to the thermal conductivity would be similar for the three compounds, whereas the electron–phonon scattering should be significantly stronger in the two-doped samples. The intensity of this electron–phonon interaction can then be roughly estimated from the ratios

$$\frac{0.125 - 0.109}{0.125} \rightarrow 12.8\%$$

for p-type $Co_{0.95}Fe_{0.05}Sb_3$, and

$$\frac{0.175 - 0.109}{0.175} \rightarrow 37.7\%$$

for the n-type $Co_{0.95}Ni_{0.05}Sb_3$ skutterudites.

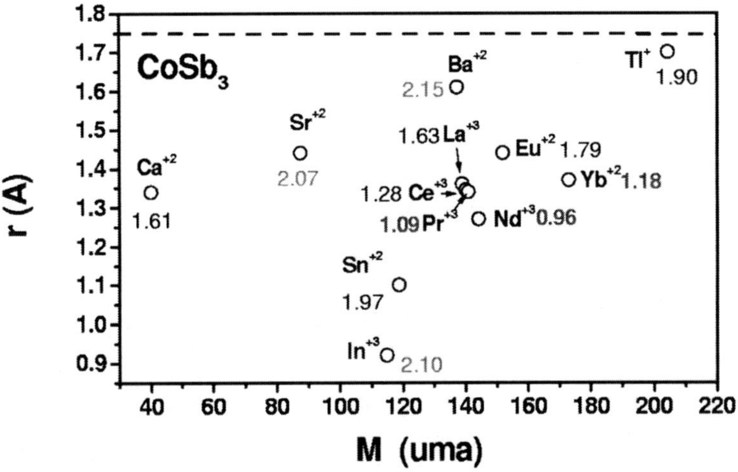

Figure 3.21 Ionic radius versus atomic mass chart derived from the data listed in Table 3.19. The numbers indicate the thermal conductivity value for the corresponding filled skutterudite. The dashed horizontal line marks the void radius in the $CoSb_3$ unfilled skuterudite.

(12) The average cage radii for Si- and Ge-based clathrates are

$$r_C = \frac{2 \times 2.05 + 6 \times 2.23}{8} = 2.19\ \text{Å},$$

for $Ba_8Ga_{16}Si_{30}$,

$$r_C = \frac{2 \times 2.197 + 6 \times 2.377}{8} = 2.33\ \text{Å}$$

for $Sr_8Ga_{16}Ge_{30}$, and

$$r_C = \frac{2 \times 2.183 + 6 \times 2.365}{8} = 2.32\text{Å}$$

Table 3.22 Thermal resistivity, average polyhedral radii, guest ionic radii, and guest/polyhedral size mismatch for the listed clathrate compounds

Sample	κ_l^{-1} ($W^{-1}mK$)	r_C (Å)	r_I (Å)	Δr (Å)
$Ba_8Ga_{16}Si_{30}$	0.83	2.19	1.61	0.58
$Sr_8Ga_{16}Ge_{30}$	0.91	2.33	1.44	0.89
$Eu_8Ga_{16}Ge_{30}$	4.55	2.32	1.35	0.97

for $Eu_8Ga_{16}Ge_{30}$, respectively. Note that we obtain essentially the same values for both Ge-based clathrates, indicating that the cage size mainly depends on the nature of the framework atoms and not on the filling atoms in these materials. Making use of the obtained r_C values, along with the data listed in Tables 3.19 and 3.21, we derive the data shown in Table 3.22. The obtained results indicate a remarkable non-linear growth of the lattice thermal resistivity with the guest-polyhedral size mismatch.

Chapter 4

The Electronic Structure Role

4.1 General Remarks

In the previous chapter we have focused on structural aspects related to the arrangement of atoms through the space in a given solid, and we saw that such an arrangement plays an essential role in controlling the lattice contribution to its thermal conductivity. One must note, however, that the arrangement the atoms adopt in a given material is mainly determined by its predominant bonding style (i.e., ionic, covalent, metallic, or van der Waals), stemming from the interaction among the solid constituents.

Indeed, the resulting crystal structure represents the balance of all the forces acting within both the electrons system and the atoms system, as well as between the two of them. Quite often this balance is very delicate and can be upset simply by changing the sample's temperature or pressure, so that the atoms reorganize themselves to form a new structure. In certain materials an important contribution to this delicate balance comes from their electronic band structure, which can even play a crucial role in some instances. Noteworthy, this is the case of several TEMs of current interest belonging to the family of the so-called *electronically stabilized* structures.[a]

[a]An important and representative class is provided by alloys obeying the so-called Hume-Rothery electron concentration rule (see Section 5.3.2).

Thermoelectric Materials: Advances and Applications
Enrique Maciá-Barber

ISBN 978-981-4463-52-2 (Hardcover), 978-981-4463-53-9 (eBook)
www.panstanford.com

Nevertheless, the electronic structure is important not only regarding cohesion energy issues, but as far as most transport coefficients of interest in TE research (namely σ, S, and κ_e) are mainly determined by the topology of the electronic energy bands close to the Fermi level. For instance, according to Eq. (2.51) the Seebeck coefficient can be regarded as a measure of the variation of the spectral conductivity function above and below the Fermi level, quantitatively expressed through the logarithmic derivative of $\sigma(E)$ with E. In this way, the Seebeck coefficient value depends upon the asymmetry of both the density of states (DOS) and the carriers diffusivity near the Fermi level (see Sections 2.4.2 and 2.4.3). In fact, one of the least understood problems in current TE research is how to increase the Seebeck coefficient of a material without depressing its electrical conductivity and to precisely predict which materials will have a high enough power factor. Of course, the electronic structure finer details are intimately tied to these goals achievement. Accordingly, a plausible approach in obtaining novel TEMs should be to engineer appropriate electronic structures by design, a goal which is, in turn, closely related to the synthesis of complex enough lattices discussed in the previous chapter.

Before proceeding with a detailed description of the main characteristics of the electronic structures of several materials of TE interest, some words are in order to recall a number of basic notions regarding the concept of energy bands in solid state physics, as well as other closely related topics.

The formation of energy bands in a solid can be understood in terms of the diffraction of the electrons by the regular planes of atoms in the lattice. In fact, being quantum particles the electrons have wavelike properties, and the wavelength of itinerant electrons with a kinetic energy relevant for transport processes turns out to be typically of a few angstroms. This is just the necessary order of magnitude to be diffracted by the atomic planes in common lattices (**Exercise 4.1**). As a consequence, if an electron is imagined to be accelerated along a given direction along the lattice, its kinetic energy will progressively increase (and its wavelength λ progressively shorten) until the Bragg reflection condition $2d \sin\theta = n\lambda$, $n \in \mathbb{N}$, is satisfied. This condition depends upon the spacing of the lattice planes, d, and the angle θ at which

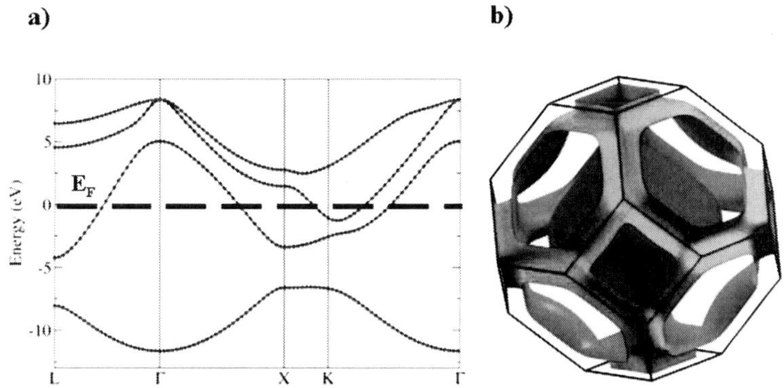

Figure 4.1 (a) Band structure of elemental lead. The lettering in the abscissas refers to the main symmetry directions in the Brillouin zone of an fcc lattice (see Fig. 4.4d), (b) constant energy surfaces (shadowed) on the fcc Brillouin zone (thick lines).

the electron strikes them, so that it generally depends upon the particular direction of the electron's motion through the lattice. Since Bragg reflections occur for certain values of the electron's wave vector ($k = 2\pi/\lambda$) it is convenient to describe this process in the momentum k space (also referred to as reciprocal space) by constructing all the planes upon which the wave vector has the appropriate value. These planes enclose a polyhedron in reciprocal space known as the *Brillouin zones*.[a]

On the other hand, one can plot the relationship between the electron energy E and its wave vector, the so-called dispersion relation $E(\mathbf{k})$, to obtain the energy band diagram. A typical example, corresponding to the elemental lead, is shown in Fig. 4.1a. The slope of the dispersion relation curves measures the group velocity of the charge carriers,

$$v_j = \frac{1}{\hbar}\frac{\partial E}{\partial k_j}, \tag{4.1}$$

[a]In what follows we will only consider the so-called first Brillouin zone, which can be defined as the set of points in reciprocal space than can be reached from the origin Γ without crossing any Bragg planes. The $(n+1)$-th Brillouin zone is the set of reciprocal space points not belonging to the $(n-1)$-th zone that can be reached from the nth zone by crossing only one Bragg plane [9].

whereas its curvature provides a measure of the so-called *band effective mass*, defined by the expression

$$\frac{1}{m_{ij}^b} = \frac{1}{\hbar^2}\frac{\partial^2 E}{\partial k_i \partial k_j}, \tag{4.2}$$

which, in general, defines a tensor magnitude. Thus, flat, narrow bands are related to heavy, slow carriers, and vice versa. Therefore, the charge carrier mobility, determined by both the group velocity and the effective mass, significantly depends on the dispersion relation topology (see Section 4.4). In band diagrams the Fermi energy level E_F (which is usually set to zero) indicates the energy of the electrons located at the topmost occupied band in the fundamental state of the system. The charge carriers located at the energy levels closer to E_F play the more significant role in the transport properties of the solid. In metallic systems, like lead, the Fermi level crosses through the bands, so that electrons can readily increase their energy in response to outside influences, such as an electric field or temperature gradient. It is this availability to electrons of easily accessible vacant, excited levels in an unfilled band that is at the heart of metallic behavior. On the contrary, insulators and semiconducting materials are characterized by the presence of an appreciable *energy gap* separating the top occupied band (referred to as the valence band) from the lowest unoccupied band (referred to as the conduction band, see Fig. 4.6). The main difference between both kinds of materials is related to the magnitude of the energy band gap, E_g, which ranges from a few tenths of eV for narrow-gap semiconductors, to about 1–2 eV for broadband semiconductors, up to 5–10 eV for typical insulating materials.

If the conduction band minimum occurs at the same point in reciprocal space as the valence band maximum then it is said that we have a *direct* energy gap, because in this case the energy gap value can be directly determined from optical absorption processes involving photons with an energy $\hbar\omega \geq E_g$. On the contrary, as is often the case, if the minimum and maximum occur at different points in k-space, then for crystal momentum $\mathbf{p} = \hbar\mathbf{k}$ to be conserved a phonon must also participate in the process, which is then known as an indirect transition. Accordingly, in that case it is said that we have an *indirect* gap (see Fig. 4.6).

With increasing temperature, electrons in the valence band begin to be thermally excited. Once the electrons are excited into the conduction band an equal amount of holes is left behind in the valence band. The material of this type is called an *intrinsic* semiconductor. The Fermi level is located in the middle of the forbidden gap, as long as the mass of the electrons in the conduction band is equal to that of the holes in valence band. The temperature dependence of the electrical conductivity in these materials is almost linear on a $\ln \sigma$ versus $1/T$ plot over a wide range of temperatures, and the slope provides a measure of the gap energy value. The band gap size plays a very important role in determining the TE performance of a semiconductor. This is because for a given band gap energy value there is a temperature at which thermally activated cross-gap carrier excitations occur to generate carriers of opposite sign, which decrease the Seebeck coefficient value according to the expression

$$S = \frac{\sigma_e S_e + \sigma_h S_h}{\sigma_e + \sigma_h}, \tag{4.3}$$

where $\sigma_{e,h}$ $(S_{e,h})$ respectively denote the electrical conductivity (Seebeck coefficient) of electrons and holes. In fact, since $S_e < 0$, whereas $S_h > 0$, Eq. (4.3) prescribes that the contribution to the Seebeck coefficient arising from each kind of charge carrier tends to cancel out each other. Thus for cooling applications narrow–band gap materials are preferable, whereas for high-temperature power generation larger gaps are necessary.

Semiconducting properties are known to be very sensitive to the addition of very small, controlled amounts of intentionally added "impurities" or alloying elements. Such alloys are called extrinsic semiconductors and the added atoms are called dopants. When the impurity atoms' energy levels are formed just below the bottom of the conduction band an electron in the impurity level can be easily excited into the conduction band even near room temperature. In that case the impurity level is called a donor level, since it donates electrons to the conduction band. Therefore, the semiconductors doped in this way are called n-type semiconductors. On the contrary, when the impurity levels are formed just above the valence band these levels can accept electrons coming from the valence band, thereby leaving holes in the valence band. In that case the impurity level is called an acceptor level. Holes in the valence band now serve as charge carriers with a positive charge and the resulting materials are called p-type semiconductors.

Due to symmetry reasons several bands may have the same energy value in the energy spectrum. It is then said that these bands are degenerate. In the case of semiconducting materials the degeneracy of band extrema gives the number of valleys in the conduction band (the relevant feature to n-type materials) or peaks in the valence band (the relevant feature to p-type materials). In that case, each band extremum, if occupied by carriers, contributes to a certain Seebeck coefficient and electrical conductivity. Accordingly, the presence of a large number of degenerate features in the band structure could lead to increased figure of merit (FOM) because the total power factor for a given material depends on the summation of contributions from all extrema, provided that *only one* type of carrier is present in the system (otherwise the algebraic summation could lead to a vanishing power factor). Now, as we will see through this chapter, a natural requirement for obtaining a high band degeneracy is that the material possess a high crystal symmetry, a property which is not generally compatible with the requirement of a complex enough lattice structure necessary to achieve the required low lattice contribution to the thermal conductivity.

In summary, we conclude that most relevant parameters to be considered when selecting or designing good TEMs are (1) the shape and width of the bands near the Fermi level, (2) the charge carriers' effective masses and mobilities, and (3) the band gap size.

4.2 Electronic Structure of Elemental Solids

In this section we shall review the electronic structure of those elemental solids whose transport properties were presented in Section 3.2. In Fig. 4.2 we indicate the electronic configuration of their constituent atoms, as well as their transport behavior, namely, metallic, semimetallic, or semiconducting. In the latter case, the band gap energy value and the direct (d) or indirect (i) character of the gap is also given. Attending to their electronic structure these elements belong to three different classes: (1) the tetravalent

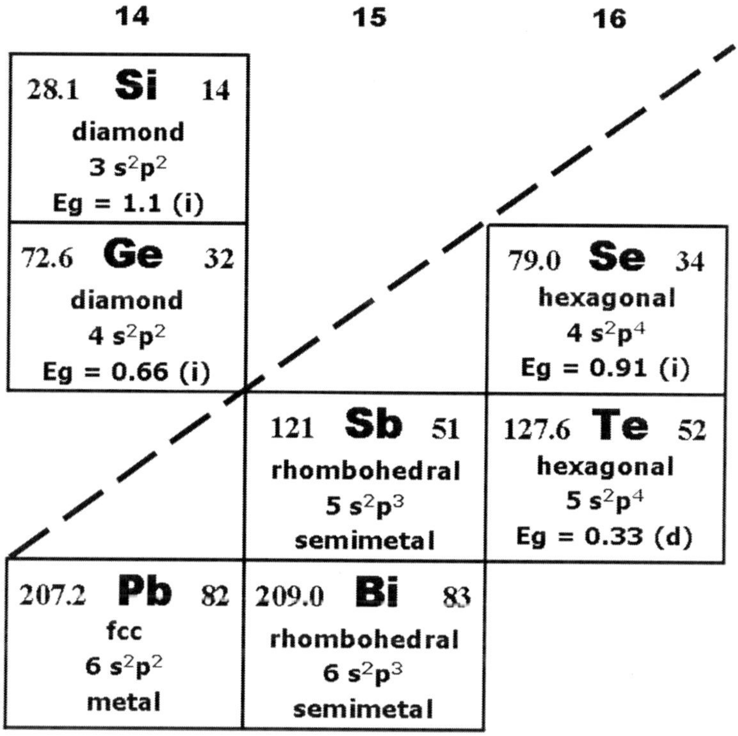

Figure 4.2 Chart showing the atomic number (Z, right upper corner), atomic mass M_a (in uma, left upper corner), crystalline structure, electron configuration in the fundamental state, and band gap energy (in eV) of direct (d) and indirect (i) semiconducting materials.

metal lead, (2) the trivalent antimony and bismuth, which behave as semimetals, and (3) the divalent selenium and tellurium, along with the tetravalent silicon and germanium, which are semiconductors.

Semimetallic materials (Fig. 4.3) are characterized by the fact that for some direction/s in reciprocal space their valence and conduction bands slightly overlap (showing metallic-like behavior) while in other directions the two kinds of bands do not overlap (showing insulating-like behavior). The resulting charge carrier density may be depleted by several orders of magnitude, and thus becomes substantially smaller than those of common metals. Another consequence of the semimetallic character is the presence

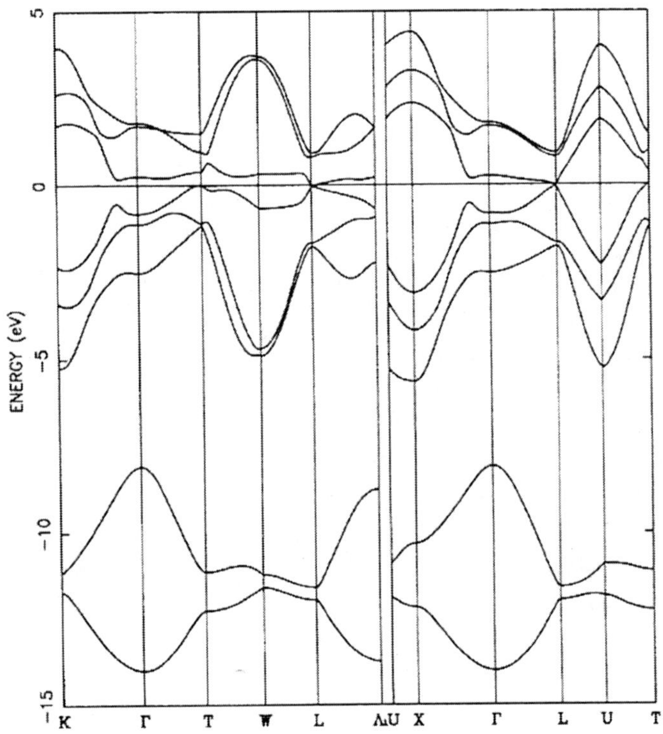

Figure 4.3 Semimetallic bismuth band structure. Note that the valence and conduction bands overlap close to the Fermi level (set as zero energy) at the L and T symmetry points (the lettering in the abscissas refers to the different symmetry lines in the Brillouin zone, see Fig. 4.4d) [195]. Reprinted with permission from A. B. Shick, J. B. Ketterson, D. L. Novikov, and A. J. Freeman, *Phys. Rev. B* **60**, 15484 (1999), Copyright 1999, American Physical Society.

of both electrons and holes in the electronic structure close to E_F. This leads to a significant reduction of the resulting Seebeck coefficient, as prescribed by the Eq. (4.3).[a] Accordingly, albeit antimony and bismuth are the best TEMs among elemental solids

[a]Equation (4.3) can also account for the difference in the Seebeck coefficient sign between bismuth and antimony (see Table 3.1). For the case of bismuth the mobility of electrons (3.8×10^5 cm^{2}V^{-1}s^{-1}, measured at 77 K along the trigonal direction), is much higher than that of the holes (2.1×10^4 cm^{2}V^{-1}s^{-1}), so that $\sigma_e \gg \sigma_h$ and $S < 0$.

(see Table 3.1), semimetallic behavior is not very convenient for good TE performance in general, and most promising TEMs are to be found among narrow-gap semiconducting materials instead. Indeed, as we will see through this chapter, most binary and ternary compounds currently used in TE industry belong to this class.

At this point, it is also worth mentioning that in semiconductors and semimetals the simultaneous presence of electrons and holes gives rise to the presence of the so-called *ambipolar* electronic thermal conductivity. In the particular case of semimetals a substantial amount of heat can be carried by this term. For example, up to 60% of the electronic thermal conductivity in bismuth is of this origin. The ambipolar term is given by

$$\kappa_b(T) = \frac{\sigma_e \sigma_h}{\sigma_e + \sigma_h}(S_h - S_e)^2 T, \tag{4.4}$$

so that it identically vanishes in compensated semimetals having $S_e(T) = S_h(T)$.

Most of the elemental solids of TE interest shown in Fig. 4.2 contain heavy atoms. In the previous chapter we saw that this is convenient in order to reduce their thermal conductivity. At this point, we realize that this feature implies they have a large atomic number Z as well. A large Z number gives rise to a strong electrostatic potential value, thereby enhancing the importance of spin–orbit splitting effects. As a result, relativistic effects must be explicitly taken into account in order to get accurate enough band structure calculations. For instance, a number of experiments showed some time ago that both the lowest conduction band and the highest valence band of the well-known TEM Bi_2Te_3 have sixfold degenerate band edges, but only recently has this important feature been reproduced by band structure calculations, and has the energy gap value also been correctly determined (see Section 4.3.2) [196, 197].

4.2.1 *Bismuth and Antimony*

Bismuth was one of the first elements whose electronic structure was explored both theoretically and experimentally as early as the 1930s. In spite of its electronic configuration ($6s^2p^3$) renders an odd number of electrons per atom (a feature that yields good electrical transport properties in common metals), its electrical conductivity is quite low (see Table 3.1). Indeed, the electrical conductivity of antimony and bismuth is usually ten to one hundred times smaller than that of typical metals, and it has a different temperature dependence. This low σ value is due to the low density of charge carriers arising from their typically semimetallic electronic structure, namely, $n_e = n_h \simeq 5 \times 10^{19}$ cm^{-3} and 2.5×10^{17} cm^{-3} for antimony and bismuth at room temperature, respectively. On the other hand, Eq. (4.3) indicates that both antimony and bismuth also have reduced Seebeck coefficients due to their semimetallic nature. Quite interestingly, these peculiar transport properties can be traced back to the crystal structure of these elements, which ultimately determines their resulting electronic structure.

Elemental crystals of bismuth, antimony, and arsenic crystallize into the so-called α-Arsenic structure (A7) with rhombohedral space group $R\bar{3}m$. This structure can be described as a distorted simple cubic lattice, resulting from the shift of atoms along one of the cube diagonals, accompanied by a relative shear (see Figs. 3.3 and 4.4a). Such a deformation is not a mere geometrical construction but it actually occurs in order to stabilize the structure on the basis of which is known as the *Jones–Peierls distortion*.[a] In the case of bismuth this lattice distortion from cubic to rhombohedral geometry can be described in two steps as follows. One first shifts each second (111) plane of the reference cubic structure along the [111] direction, resulting in an alternating displacement of atomic planes located at distances $d_{\pm} = a \pm \gamma$, successively (see Fig. 4.4b). Then one introduces a relative slip of these planes, ultimately leading to the shearing deformation. The corresponding bismuth first Brillouin zone (Fig. 4.4c) is then formed by choosing the rhombohedral unit cell (containing two atoms) shown in Fig. 4.4a, and it is bonded by six $\{1\bar{1}0\}$ side planes, three $\{221\}$ top planes, and three $\{\bar{2}\bar{1}\bar{1}\}$ bottom planes. Now, within this Brillouin zone[b] one can accommodate the

[a]Peierls theorem asserts that an infinite one-dimensional system with a half-filled band will undergo a structural distortion in such a way as to open up a band gap at the Fermi level. Later on it was shown that the concept can be used in a similar fashion for three-dimensional structures with a half-filled valence band. It is possible to go one step further and transfer the idea to three-dimensional systems with a partially filled valence band occupancy [199].

[b]Also known as the Jones zone.

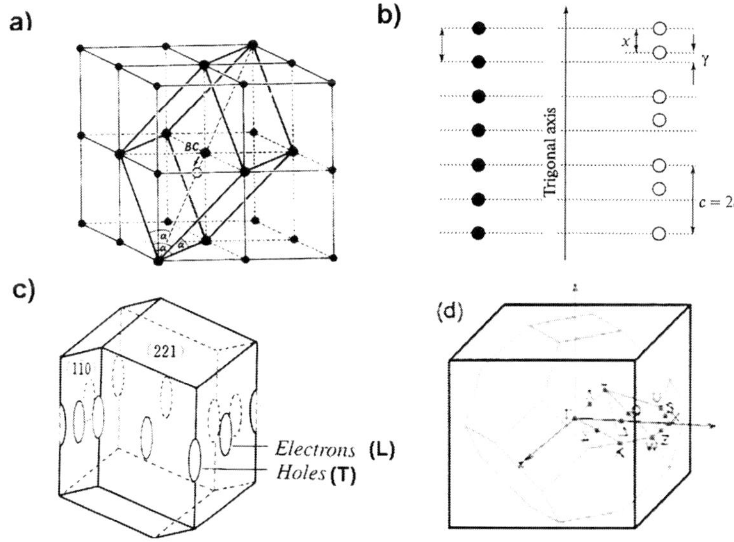

Figure 4.4 (a) Rhombohedral unit cell of bismuth, (b) the Peierls transition of the (111) planes of the cubic structure along the trigonal axis is shown schematically, (c) the bismuth Brillouin zone showing the electron (L points) and hole (T points) ellipsoidal pockets, (d) comparison between the Brillouin zones of the simple cubic and fcc cubic cells. Adapted from Ref. [10], U. Mizutani, *Introduction to the Electron Theory of Metals*, Copyright 2001, by permission of Cambridge University Press, and reprinted from Ref. [198], with permission from O. V. Misochko, *Bull. Rus. Acad. Sci.* **74**, 589 (2010), Copyright 2010, Russian Academy of Sciences.

required ten electrons per unit cell,[a] so that having an even number of conduction electrons per unit rhombohedral cell, one would expect bismuth would behave as an insulator, which is not the case. In order to account for its nonvanishing electrical conductivity, Jones noted in 1936 that the structural distortion from the simple cubic lattice in A7 compounds decreases the volume of the original simple cubic Brillouin zone by a factor of two, with the result that the free electron sphere intersects the Jones zone. As a consequence, a small number of electrons overlap outside this zone across the center of the $\{1\bar{1}0\}$ side planes (corresponding to the L points, see Fig. 4.4c),

[a]We recall bismuth contains two atoms per unit cell.

leaving a compensating number of holes at some corners of the zone (corresponding to the T points).

These electron and hole ellipsoids of revolution (usually referred to as electron and hole pockets, respectively) can also be recognized as stemming from band overlapping occurring at the Fermi level located at the L and T points in the band structure diagram shown in Fig. 4.3. Indeed, various band calculations agree that the dominant effect of the rhombohedral shear (which makes the L and T symmetry points unequal) is to lower the bands at the L point and to rise them at T. Hence, the formation of electron (hole) pockets around the $L(T)$ points naturally results. Therefore, the Peierls mechanism opens a gap over most of the Brillouin zone, leaving the electron and hole pockets only. This introduces a small band gap in the electronic structure over extended regions of the Brillouin zone which stabilize the structure, so that the elastic energy cost for the deformation is compensated by the electronic energy gain due to the opening of the gap.

Detailed first-principle studies of the total energy and phase stability of A7 structures disclosed that each one of the distortion processes involved can be respectively related to different energy gain contributions. Thus, the internal displacement distortion along the [111] direction leads to a metal–semiconductor transition accounting for a lowering of the total energy of about -66.4 meV per atom. This structure stabilization effect is supplemented by a further lowering, of about -0.53 meV per atom, due to the trigonal shear distortion, ultimately driving a semiconductor–semimetal transition [195]. Therefore, it is the presence of this small gap along with the conduction band overlapping upon the valence band by 18 meV, which is responsible for the semimetallic behavior of the A7 elemental compounds. This provides a striking illustration of the crucial importance of crystal structure in determining the electronic structure and related transport properties. Indeed, since bismuth has an odd number of electrons per atom, it would exhibit a typical metallic behavior instead of a semimetallic one, were it crystallized into a simple cubic lattice.

In this way, we realize that it is the band gaps arising from a very little deviation from simple cubic structure which ultimately deplete the effective number of carriers by a factor as large as 10^5 [9].

This fact is illustrated by the correlation between the electrical conductivity value and the rhombohedral shear distortion degree in bismuth and antimony. Indeed, on the basis of their structural parameters, the departure from cubic symmetry ($\alpha = 60^{\circ}$) is more pronounced in antimony ($\alpha = 57^{\circ}\ 14$) than in bismuth ($\alpha = 57^{\circ}\ 19$). Accordingly antimony has a larger electrical conductivity ($\sigma = 41000\ \Omega^{-1}\text{cm}^{-1}$) than bismuth ($\sigma = 7400\ \Omega^{-1}\text{cm}^{-1}$) at room temperature.

4.2.2 *Selenium and Tellurium*

In the previous section we learned that the A7 rhombohedral structure of the group 15 elements As, Sb, and Bi can be obtained from a simple cubic lattice under two separate distortions: (i) an internal displacement along a [111] diagonal of the two interpenetrating fcc lattices into which a simple cubic structure can always be resolved and (ii) a trigonal shear, associated with an extension along the same diagonal. As we will see below, an analogous Peierls distortion mechanism can also be applied in order to explain the hexagonal crystal structure of the group 16 elements selenium and tellurium.

From an electronic point of view, the Peierls mechanism can operate in this case because in selenium and tellurium atoms the s orbitals are completely filled and are therefore nonbonding, while the p bonding orbitals directed along the three orthogonal axes, would naturally lead to a simple cubic structure. Indeed, p-bonded systems would favor the simple cubic structure in order to maximize the directional p_{σ} bond formation. However, three mutually orthogonal linear atomic chains in a cubic pattern can be easily distorted by the Peierls mechanism when the corresponding energy band is partially filled. In the case of selenium and tellurium, having the atomic electron structure of the form s^2p^4, the p band is two-thirds filled, so that the lattice is unstable under a Peierls distortion dividing the six-nearest-neighbor p_{σ} bonds into two short and four long ones. This results in an energy gain achieved by the trimerization of atoms forming helicoidal chains along the c direction (see Fig. 3.4). The ratio of the intrachain (d_1) to the interchain (d_2) bond lengths provides a direct measure for the

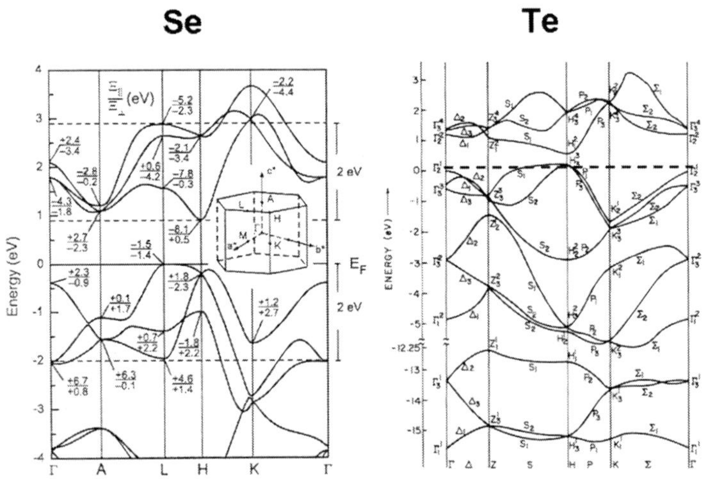

Figure 4.5 Band structure of the trigonal selenium (on the left) and tellurium (on the right) along some high-symmetry lines in the hexagonal Brillouin zone, which is shown in the inset. Adapted from Ref. [200], with permission from IOP Publishing Ltd., and reprinted from Ref. [201], with permission from A. Coke, T. Lee, and T. P. Das, 1980 *Phys. Rev. B* **22** 2968, Copyright 1980, American Physical Society.

strength of the Peierls distortion, which decreases with increasing atomic number ($d_2/d_1 = 1.45$ for Se and $d_2/d_1 = 1.23$ for Te) [202].

The band structures of selenium and tellurium are shown in Fig. 4.5. The first nine bands in tellurium spectrum constitute the valence bands that accommodate the 18 valence electrons arising from six valence electrons contributed by the three atoms within the unit cell. The next three bands are the conduction ones. A similar band arrangement is obtained for selenium (note the different energy scales in both diagrams). Both elements then behave as semiconducting materials: selenium has a relatively wide ($E_g = 0.91$ eV) indirect gap (the top of the valence band is located at the L point and the bottom of the conduction band is located at the H point), whereas tellurium has a narrower ($E_g = 0.33$ eV) direct gap at the H point.

Detailed band structure calculations have demonstrated that the stable crystal structure in 16 group elements is mainly determined by the competition between the relativistic spin–orbit splitting [which is more important in Te (with $Z = 52$) than in Se (with $Z = 34$)] and the crystal field splitting due to the structure's symmetry reduction induced by the Peierls transition. In the case of selenium the overall energy splitting between the bonding and antibonding states is so large that the system becomes more semiconductor-like, while tellurium becomes more semimetallic-like. In fact, from Table 3.1 we see that the electrical conductivity of tellurium is six orders of magnitude larger than that of selenium (**Exercise 4.2**). In this way, we one more time realize that the remarkable differences in the transport properties of two elemental solids made of close neighboring atoms in the same group of the periodic table are mainly determined by electronic structure features stemming from the Peierls mechanism, which stabilizes their crystal structure.

4.2.3 *Silicon and Germanium*

Silicon and germanium crystallize into the cubic diamond structure. This structure can be described in terms of two interpenetrating fcc lattices, so that it can be regarded as an fcc lattice but with two atoms in each unit cell. Thus, the first Brillouin zone is the same as that for the ordinary fcc structure (i.e., a truncated octahedron, Fig. 4.6a).

In silicon and germanium two 3s and two 3p electrons per atom are hybridized to form sp^3 directionally bonded states and the Brillouin zone is completely filled with eight electrons per unit cell (Fig. 4.6b). Therefore, the element becomes a semiconductor. The conduction band of silicon is characterized by six minima at equivalent Δ points located at $\frac{2\pi}{a}(0.85, 0, 0)$ along the (100) direction of the Brillouin zone, while band structure calculations indicate that silicon valence band maximum occurs at the Γ point, where two degenerate bands with different curvatures meet, giving rise to *light* holes and *heavy* holes (Fig. 4.6c). In addition, there is also a third band, only 0.044 eV below the valence maximum (which cannot be seen in the band diagram). This band is separated from the other two due to spin–orbit coupling effects. Conduction and valence bands are separated by an energy gap $E_g = 1.1$ eV.[a] As can

[a]The calculated energy gap of $E_g = 0.48$ is underestimated. This is a general shortcoming of most first-principle numerical calculations.

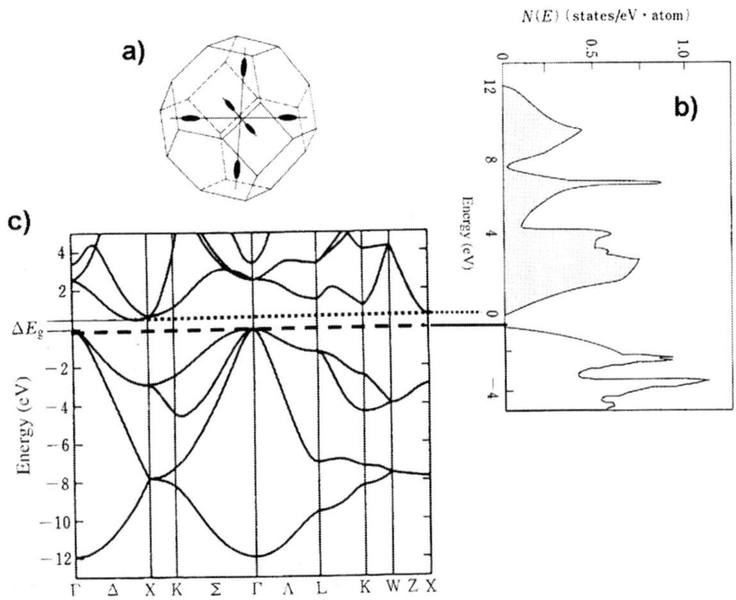

Figure 4.6 (a) Brillouin zone with the constant energy surfaces near the conduction band minima in silicon. There are six ellipsoidal pockets directed along the [100] directions (X symmetry point), (b) DOS, and (c) band structure of elemental silicon. Dashed area in the DOS indicates unoccupied conduction band states.

be seen, the top of the valence band is found at point Γ, but the bottom of the conduction band lies at $0.82X$ along the ΓX direction (indirect gap semiconductor). The conduction band minima along the directions [100] give rise to six pocket ellipsoids (Fig. 4.6a). These bands, which are formed from antibonding combinations of the orbital wavefunctions, are unoccupied.

The crystal structure and Brillouin zone of elemental germanium are as in silicon. There are again two degenerate valence bands, both with maxima at the Γ point, and four electron carrier pockets at the four equivalent L points $\frac{\pi}{a}(1, 1, 1)$ in the Brillouin zone. However, the conduction band minima now occur at the zone boundaries in the [111] directions (X point). The ellipsoidal constant energy surfaces are ellipsoids of revolution elongated along the [111] directions. The holes for both silicon and germanium are in a single, approximately spherical, carrier pocket at the Γ point of the Brillouin zone.

4.3 Electronic Structure of Binary Compounds

4.3.1 *BiSb Alloys*

In Section 4.2.1 we concluded that the semimetallic nature of bismuth, characterized by a conduction band overlapping upon the valence band along certain crystallographic directions, leads to a relatively low Seebeck coefficient because of the cancellation between the contributions to S from the electron and hole carriers. Therefore, one realizes that bismuth (being an n-type semiconductor, see Table 3.1) could be a better TEM if the hole carriers could be removed. A possible way to do this is through alloying with an isoelectronic element.[a] On this basis, the possible transition from the semimetallic to a semiconducting regime upon alloying of bismuth with the isoelectronic element antimony was proposed some time ago.

In the top panels of Fig. 4.7 we compare the band structures of bismuth and antimony semimetallic elemental solids. As we see the overall electronic structure is pretty similar in both solids, and it is characterized by the presence of electron (hole) pockets at the L (T) points of the Brillouin zone in both cases. A closer inspection of the band diagram around the Fermi level, however, reveals some important differences in the topology of bands close to the L and T symmetry points. Thus, the overlapping between the conduction and valence bands is significantly reduced at these points for antimony, which accordingly exhibits a more metallic-like behavior (namely, higher σ and κ values along with a lower Seebeck coefficient value, see Table 3.1), than the bismuth elemental solid. Accordingly, the substitution of Bi atoms by Sb atoms in the Bi lattice has a major effect on the electronic structure of bismuth, displacing

[a]Alloying does not necessarily work to this end when nonisoelectronic elements are considered. For instance, in Sb_2Te_3 compounds excess Sb atoms can occupy Te sites (the so-called antisite defects) creating acceptor states, which results in high hole concentrations.

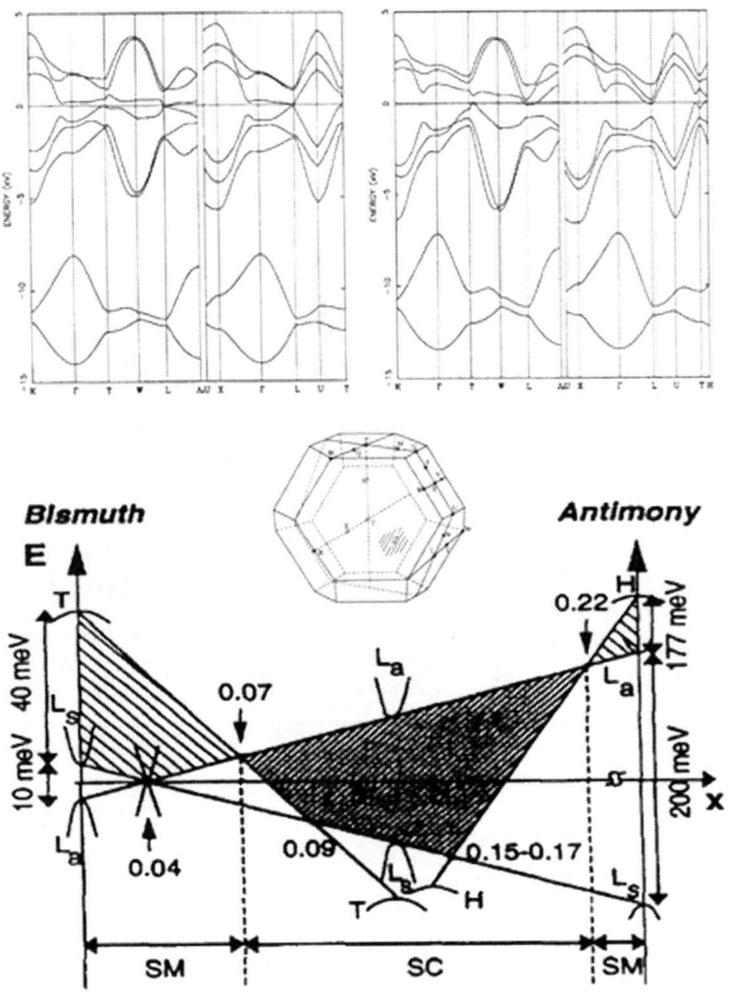

Figure 4.7 Band energy structure for bismuth (upper left panel) and antimony (upper right panel). The symmetry directions are labeled according to the Brillouin zone scheme shown in the middle frame. On the bottom frame a schematic diagram of the low temperature energy bands arrangement near the Fermi level for $Bi_{1-x}Sb_x$ alloys is shown as a function of the Sb composition, x. Depending on the Sb concentration, $Bi_{1-x}Sb_x$ alloys can be semimetals (SM) or narrow-band semiconductors (SC) [116]. Reprinted from B. Lenoir, A. Dauscher, M. Cassart, Yu. I. Ravich and H. Scherrer, Effect of antimony content on the thermoelectric figure of merit of $Bi_{1-x}Sb_x$ alloys, *J. Phys. Chem. Solids*, **59**(1), 129–134, Copyright (1998), with permission from Elsevier.

the heavy mass valence bands downward with respect to the light mass conduction band (see Fig. 4.7 bottom panel).

In fact, the progressive addition of antimony in the BiSb alloy results in a decrease in the overlapping between the conduction and valence bands at L and T points, which promotes a semimetallic–semiconductor transition. This process occurs when the overlapping between the L_s and T bands disappear at $x = 0.07$ Sb composition (see Fig. 4.7). Upon further addition of antimony $Bi_{1-x}Sb_x$ alloys become narrow-band semiconductors within the composition range $0.07 \leq x \leq 0.22$. The gap width progressively increases with antimony concentration, with the largest gap occurring around 12% Sb concentration. Indeed, within this composition, the interval range between $0.08 < x < 0.12$ is of special interest since the L_S band is located above the T and H bands, giving rise to the possibility of p-type carriers that are mirror images of the L-point electron carriers, rendering excellent TE properties. Therefore, the transport properties of $Bi_{1-x}Sb_x$ alloys show a strong dependence on composition (determining the E_g value) and temperature (controlling the charge carriers activation across the gap).

4.3.2 *Bismuth Chalcogenides*

The electronic band structures for the compounds Bi_2Te_3 and Bi_2Se_3 are shown in Fig. 4.8 The calculated electronic structure for Bi_2Te_3 indicates that it is a semiconductor with an indirect gap. The measured band gap is $E_g = 0.15$ eV at room temperature. The highest valence bands are located in the mirror planes of the Brillouin zone containing the trigonal and the bisectrix axes. They, therefore, contain six valleys ($g_v = 6$). Experiments also indicate the presence of six valleys in the conduction band minimum ($g_c = 6$), located on a mirror plane of the Brillouin zone. From symmetry considerations the energy near the valence band maximum can be written as

$$E = E_0 - \frac{\hbar}{2m_0}\left(\alpha_{11}k_x^2 + \alpha_{22}k_y^2 + \alpha_{33}k_z^2 + 2\alpha_{23}k_yk_z\right),$$

where $\mathbf{k} = (k_x, k_y, k_z)$ is the Bloch vector, m_0 is the free-electron mass, and the mass parameters α_{ij} are related to the effective-mass tensor referred to axes in the binary, bisectrix, and trigonal

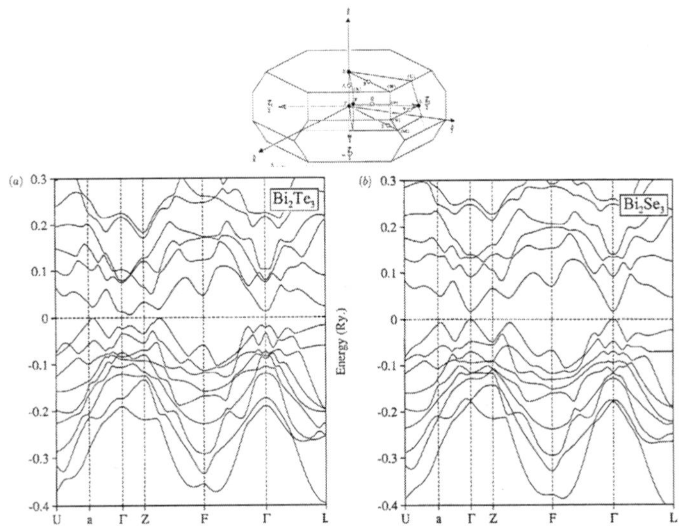

Figure 4.8 Electronic band structures for Bi_2Te_3 and Bi_2Se_3 obtained from linear muffin-tin orbital-local density approximation (LMTO-LDA) calculations with spin–orbit coupling included. Reprinted from Ref. [119], with permission from IOP Publishing Ltd. The band structures are referred to the first Brillouin zone illustrated in the top frame. The binary axis (with twofold symmetry) is along x, the bisectrix axis (occurring at the mirror planes) is along y, and the trigonal direction (with threefold rotation symmetry) is along z.

directions. The effective masses are highly anisotropic, as can be appreciated from the experimentally derived values $\alpha_{11} = 109.3, \alpha_{22} = 5.2$, $\alpha_{33} = 6.2$, and $\alpha_{23} = 3.1$ [119], which lead to the principal axes of the energy ellipsoid making an angle of about 35° with the crystalline axes. The similar values of α_{22} and α_{23} also show that the ellipsoids are nearly spheroidal and have a small effective mass normal to the reflection planes.

On the other hand, the calculated electronic structure for Bi_2Se_3 shows a direct band semiconductor with one-valley conduction band minimum and a one-valley valence band maximum occurring at the center of the Brillouin zone (Γ point), both valleys having an isotropic effective mass. These numerical results are in agreement with experimental data, and consistent with the measured band gap value $E_g = 0.2$–0.3 eV at room temperature.

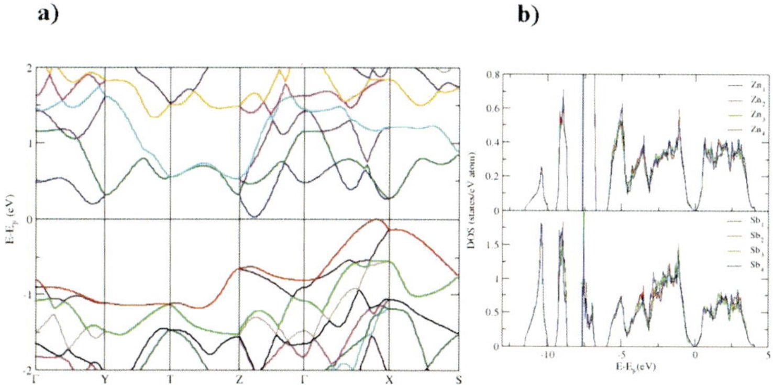

Figure 4.9 (a) Electronic band structure of ZnSb and (b) DOS of the four inequivalent Zn and Sb atoms in the unit cell [122]. Reprinted with permission from P. Jund, R. Viennois, X. Tao, K. Niedziolka, and J. C. Tédenac, *Phys. Rev. B* **85**, 224105 (2012), Copyright 2012, American Physical Society.

In both compounds the valence and conduction bands arise from the p atomic orbital, while the bands arising from the s orbitals occur about 0.8 Ry below the valence-band maximum. We note that, while elemental selenium (tellurium) has an indirect (direct) gap, the alloys Bi_2Se_3 (Bi_2Te_3), respectively have a direct (indirect) gap instead.

4.3.3 *Antimonides*

The calculated band structure of ZnSb shows a very narrow indirect band gap of about 0.05 eV (Fig. 4.9a), which is notably smaller than the experimental gap $E_g = 0.53$ eV. In Fig. 4.9b we show the total DOS of the eight Zn atoms and eight Sb atoms constituting the four Zn_2Sb_2 rings present in the orthorhombic unit cell (see Section 3.3.3). Although one may expect similar contributions coming from atoms occupying the same Wyckoff positions, one finds four different Zn atoms and four different Sb atoms, corresponding to the four different rings embedded in the unit cell (see Fig. 4.9). The role of these rings in the electronic band structure has been recently analyzed concluding that, in each ring, the contribution of the two Zn and Sb atoms to the DOS is similar, but they are slightly different from one ring to another. Therefore, from the electronic

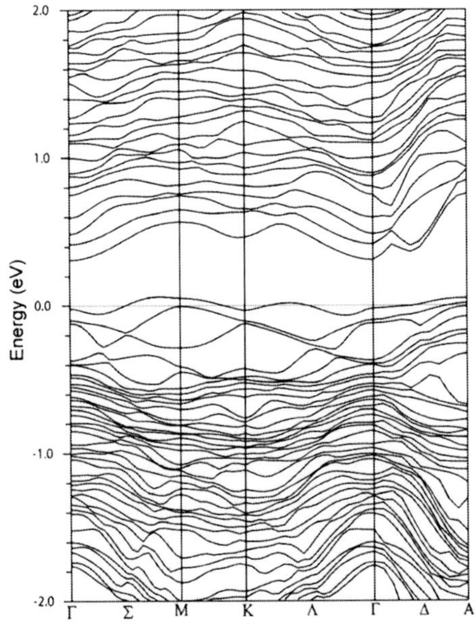

Figure 4.10 Electronic band structure of rhombohedral β-Zn_4Sb_3 [203]. Reprinted with permission from A. N. Quiu, L. T. Zhang, and J. S. Wu, *Phys. Rev. B* **81**, 035203 (2010), Copyright 2010, American Physical Society.

point of view, the four rings to which the 16 atoms in the unit cell belong are not strictly equivalent (see Section 3.3.3). This result nicely illustrates that, though closely related to each other, atomic and electronic structures have their own specific features, both of them contributing to the resulting complexity of the solid as a whole.

The band structure of the β-Zn_4Sb_3 compound is shown in Fig. 4.10 where we see that it is an indirect semiconductor with E_g = 0.26 eV. The conduction band minimum is located at the Γ point, whereas the valence band maximum is located about 4/5 of the ΓM direction. The maximum valence band crosses the Fermi level, resulting in a lightly p-type degenerate semiconductor.

4.3.4 *Lead Chalcogenides*

PbTe is a well-known narrow–band gap semiconductor (E_g = 0.31 eV at room temperature) with a direct band gap at the L point

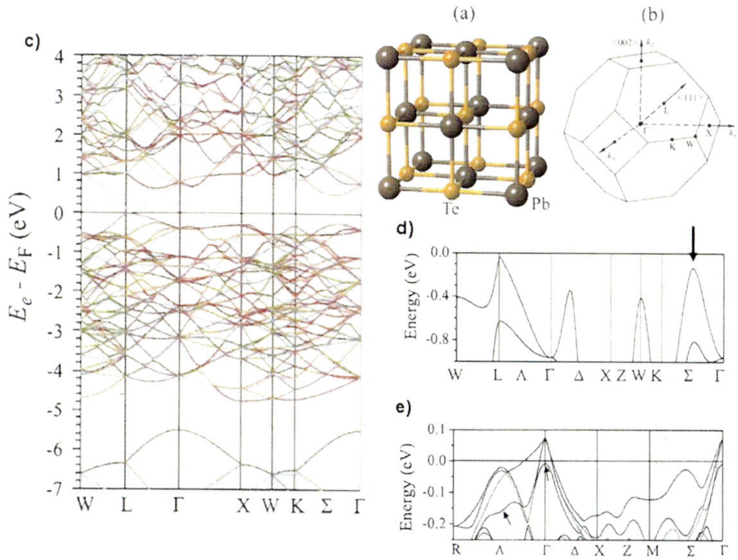

Figure 4.11 (a) Rock salt cubic structure of PbTe alloy, (b) the first Brillouin zone for PbTe and its high-symmetry k-points, (c) calculated electronic band structure of PbTe with an fcc unit cell using the optimized lattice constant $a = 6.557$ Å, (d) blow up of the band structure close to the Fermi level (top frame) compared with the corresponding band structure for a potassium doped PbTe alloy (bottom frame). Reprinted from Ref. [204], with permission from M. S. Lee and S. D. Mahanti, *Phys. Rev. B* **85**, 165149 (2012), Copyright 2012, American Physical Society, and Ref. [205], with permission from H. Kim and M. Kaviany, *Phys. Rev. B* **85**, 045213 (2012), Copyright 2012, American Physical Society.

(Fig. 4.11c). In contrast to most other semiconductors, due to lattice expansion the band gap of PbTe increases with temperature and decreases under pressure (due to lattice contraction). At the point of gap closure the band dispersion relations become linear (Dirac type) in the vicinity of the L point. As with all lead chalcogenides, the conduction band and the light valence band have been shown to be non-parabolic (see Fig. 4.11c), with transport properties that can be well characterized by the single-band Kane model (see Section 4.4). A number of experimental studies have shown the existence of a heavy second highest valence band maximum slightly below (0.2 eV) the principal (light) valence band, at a point close to Σ along the ΓK symmetry direction (arrow in Fig. 4.11d). This is important for charge transport since carriers around the Σ point can be introduced with rather small doping levels, as it is illustrated in Fig. 4.11e. Indeed, we see that after doping with potassium atoms, the valence band maximum shifts to the Γ point and second and third valence band maxima appear along the Σ (i.e., [110]) and Λ (i.e., [111]) directions. The DOS in the heavy lower valence band is much higher than that in the upper light valence band. This gives the lower valence band holes a higher Seebeck coefficient value than that of the upper valence band, and thus it is favorable for ZT to have the Fermi level in the lower valence band. Thus, the band structure is strongly affected by the presence of impurities, which in turn results in a substantial reduction of the Seebeck coefficient value (by about 20%) due to the removal of degeneracy of the valence bands maxima (see Section 4.4) [204].

PbSe and PbS (lower-cost analogs of PbTe) have band structures almost identical to PbTe, with a slightly larger direct gap at the L point ($E_g = 0.42$ eV for PbS at room temperature). The band extrema are located at the L point of the Brillouin zone and have four equivalent ellipsoid electron pockets that dominate the transport properties of n-type PbSe.

Originally it was thought that the TE performance of PbSe would be inferior to that of PbTe because of its higher thermal conductivity and smaller band gap size. However, this turns out not to be the case at high temperatures, where its ZT value is greater in fact. This is due to the gap width temperature dependence of PbSe, which increases rapidly with temperature, actually exceeding that of PbTe above 500 K. Thus, the FOM value $ZT = 1.1 \pm 0.1$ has been reported for bromine doped PbSe (n-type) samples at 850 K [57]. Even better FOM values have been reported for sodium doped PbTeS alloys (p-type) reaching $ZT = 1.55$ at 773 K for $x = 0.08$.[a]

[a]The highest band degeneracy ($g_v \geq 12$, see Section 4.4.1) at high temperatures as compared to that for PbTe ($g_v = 4$) also contributes to its good TE performance [206].

4.3.5 *SiGe Alloys*

Silicon and germanium form a complete range of solid solutions, which crystallize in the cubic diamond structure. The band structure of the SiGe alloys is characterized by the presence of two sets of valleys with their extrema at different levels. These sets are known as the X and L bands, respectively. Because of the complexities on the band structure, the power factor for p-type materials exhibits two maxima. One is associated with the X band and lies close to a carrier concentration of $n = 2 \times 10^{20}$ cm^{-3}, the other is related to the L band and lies near a concentration of $n = 2 \times 10^{21}$ cm^{-3}. At the high temperatures of interest, there is not much difference between the two maxima as far as the power factor is concerned, but one must keep in mind that the electronic contribution to the thermal conductivity also has an effect on the FOM value. Thus, it seems convenient to choose n-type materials with carrier concentration closer to the X band value.

4.3.6 *Pentatellurides*

The transition metal pentatellurides $HfTe_5$ and $ZrTe_5$ have been shown to possess interesting electrical transport properties in their bulk, single-crystal form. Their crystal structure is orthorhombic (space group $Cmcm$) with unit cell dimensions $a = 3.974$ Å ($a = 3.964$ Å), $b = 14.492$ Å ($b = 14.443$ Å), and $c = 13.730$ Å ($c = 13.684$ Å) for $ZrTe_5$ ($V_U = 790.7$ Å^3) and $HfTe_5$ ($V_U = 783.4$ Å^3), respectively, containing $n_U = 24$ atoms per unit cell. Their crystal structure is characterized by the presence of chains of MTe_6 (M = Zr or Hf) prisms ($n_E = 1$) running parallel to the a axis (Fig. 4.12a). These prismatic chains are linked along the c axis via zigzag chains of Te atoms (reminiscent of the tellurium elemental structure) to form two-dimensional sheets in the a-c planes (Fig. 4.12b). These sheets, in turn, stack to one another along the b axis through van der Waals weak bounds, in a way similar to that observed in Bi_2Te_3 or Se_2Te_3 compounds. The samples grow as ribbons with the preferred growth direction along the a axis and the thin part of the ribbon being the b axis.

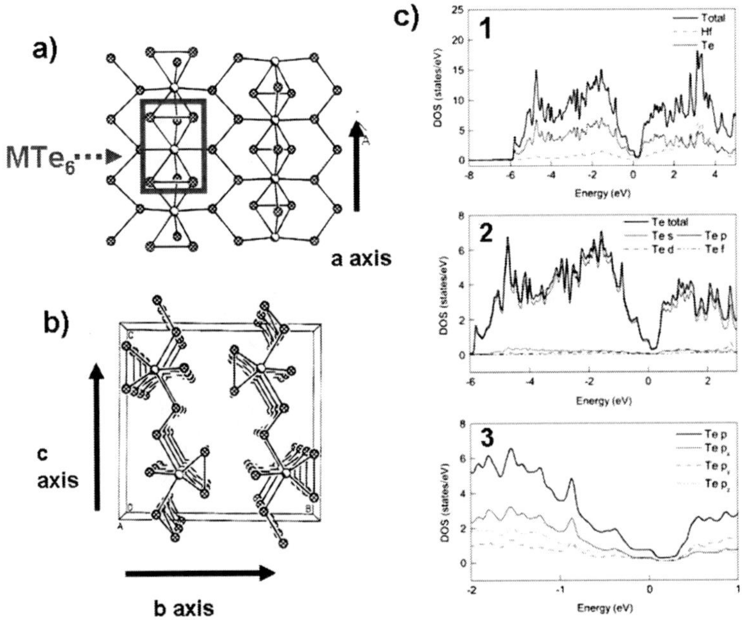

Figure 4.12 (a) Crystal structure of MTe_5 compounds (M = Zr or Hf) along the *a* axis. The open spheres are the metal (M) atoms and the cross-hatched spheres are the Te atoms. (b) A projection of the crystal structure as viewed down the *a* axis, (c) calculated partial density of states for $HfTe_5$ (including spin–orbit effects): (1) contributions to the total DOS from each atom, (2) contribution to the DOS from each orbital state of tellurium, (3) contributions to the total Te *p* orbitals from each *p* orbital state of Te. Reprinted from Ref. [47], with permission from R. V. Littleton IV, T. M. Tritt, J. W. Kolis, and D. R. Ketchum, *Phys. Rev. B* **60**, 13453 (1999), Copyright 1999, the American Physical Society, and Ref. [207], *Solid State Comm.* **146**, 454 (2008), M. W. Oh, B. S. Kim, S. D. Park, D. M. Wee, and H. W. Lee, Copyright 2008, with permission from Elsevier.

As a consequence of their anisotropic structure these materials do exhibit anisotropic transport properties as well, with the high conductivity axis being the preferential growth axis. As illustrated in Fig. 4.13a the electrical resistivity measured along the *a* axis exhibits an anomalous peak at around 75 K (145 K) in $HfTe_5$ ($ZrTe_5$). At room temperature the electrical resistivity is about $\rho = 0.6$ mΩ cm ($\sigma = 1666\ \Omega^{-1}\text{cm}^{-1}$) for $HfTe_5$ and $\rho = 0.7$ mΩ cm ($\sigma = 1429\ \Omega^{-1}\text{cm}^{-1}$) for $ZrTe_5$, and it decreases in both materials as the temperature

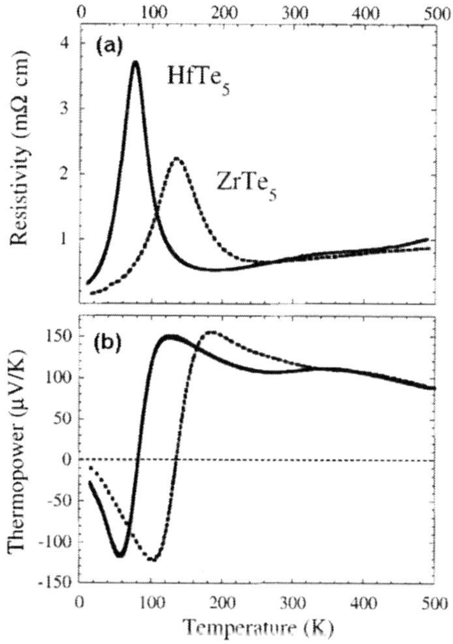

Figure 4.13 Temperature dependences of (a) the resisitivity, and (b) the Seebeck coefficient for $HfTe_5$ and $ZrTe_5$ compounds. Reprinted with permission from Ref. [47].

decreases before attaining the peak. From Fig. 4.13b we also see that the Seebeck coefficient is relatively large at room temperature ($S \simeq +125$ μVK^{-1}), and steadily increases as the temperature decreases until one reaches the critical temperature value at which the electrical resistivity attains its peak. When approaching this critical temperature the Seebeck coefficient rapidly drops to zero and changes its sign from large positive values above the peak to similarly large negative values below the peak.

It was originally thought that this resistivity peak may result from a charge density wave phase transition. However, typical evidences for structural phase transitions were not experimentally found. For example, the resistivity of ternary $Hf_{1-x}Zr_xTe_5$ alloys shows the resistivity peak to evolve smoothly as a function of Hf atomic content instead of becoming suppressed by the cation disorder, as it is observed in other compounds which are stabilized by charge

Table 4.1 Transport properties of pentatelluride compounds at $T = 150$–165 K arranged according to their FOM values. The values reported are measured along the a direction [47, 208, 210]

Compound	σ	S	P	κ	ZT
	$(\Omega^{-1}\text{cm}^{-1})$	(μVK^{-1})	$(\mu\text{Wcm}^{-1}\text{K}^{-2})$	$(\text{Wm}^{-1}\text{K}^{-1})$	
$ZrTe_5$	1539	+140	30.2	8.2	0.11
$HfTe_5$	1429	+120	20.6	4.5 − 6	0.10 − 0.14
Nd:$HfTe_5$	3333	+134	59.8	6.0	0.30
Sm:$HfTe_5$	3333	+121	48.8	6.0	0.24

density waves. Furthermore, detailed crystallographic studies failed to reveal any sort of extra diffraction peaks at low temperatures, and calorimetric studies showed no sign of any possible second order phase transition either. Accordingly, the reported transport anomalies suggest the possible presence of band structure effects.

Several experimental measurements indicated that the valence band structure of $ZrTe_5$ in the vicinity of the Fermi level consists of three bands: two electron bands and one hole band, in reasonable agreement with band structure calculations. More recent high-resolution angle-resolved photoelectron spectroscopy studies indicate that $ZrTe_5$ is semimetallic at low temperature ($T < 20$ K) with two strongly dispersing hole-like maximum valence bands close to the Fermi level. This semimetallic character arises from the van der Waals mediated interplane interactions. However, at $T > 20$ K these interactions become weak enough to allow for a semiconductor gap opening (with $E_g \simeq 50$ meV) [207]. In a similar way, the calculated DOS for $HfTe_5$ indicates a semimetallic behavior (Fig. 4.12c), where the conduction carriers mainly originate from tellurium atoms and have p orbital character [209].

In Table 4.1 we list the transport coefficient, power factor, and FOM values for several pentatellurides within the temperature interval 150–165 K. From these data we realize that, from about 120 to 170 K the effective power factor of $HfTe_5$ is higher than that of Bi_2Te_3. The selenium doped pentatellurides $ZrTe_{4.75}Se_{0.25}$ and $HfTe_{4.75}Se_{0.25}$ show some improvements over both Bi_2Te_3 and $HfTe_5$, with $ZrTe_{4.75}Se_{0.25}$ possessing an effective power factor of about $PT = 1.5\ \text{Wm}^{-1}\text{K}^{-1}$ at around 225 K. For $HfTe_5$:Sm a similar power factor was found, whereas for $HfTe_5$:Nd a substantial improvement, yielding $PT = 2\ Wm^{-1}K^{-1}$ at about 165 K, was reported [210]. This is more than a factor of two larger than the power factor of Bi_2Te_3 at low temperatures (**Exercise 4.3**).

4.3.7 *Rare-Earth Tellurides*

Lanthanum telluride $La_{3-x}Te_4$ was first considered for TEGs application during the late 1980s [30]. It is an n-type material with high TE conversion efficiency at temperatures above 1000 K ($ZT \simeq 1.2$ at 1275 K for $x \simeq 0.2$).[a] Nevertheless, synthesis issues related to the La-Te phase diagram and sensitivity to oxidation, as well as the absence of an efficient p-type material acting as a complementary leg in a TE couple, limited the advance of this material at that time. Nowadays, new interest has been spurred by the introduction of a low temperature mechanical alloying synthesis technique that provides reproducible control composition and optimization of TE response. Also, the discovery of $Yb_{14}MnSb_{11}$ Zintl phase as a suitable complementary p-type material has further contributed to the renewed interest in n-type rare-earth tellurides.

Rare-earth chalcogenides often adopt the cubic Th_4P_3 structure (space group $I\bar{4}3d$) with lattice parameters comparable to those of the skutterudites, namely, $a = 9.634$ Å ($a = 9.553$ Å) for La_3Te_4 (Ce_3Te_4), respectively, and they have $n_U = 28$ atoms in their unit cell. These compounds support a compositional range $R_{3-x}Te_4$ ($0 \leq x \leq 1/3$) associated to the existence of rare-earth vacancies. The interest in changing the rare-earth concentration of this compound is based on the consideration that rare-earth vacancies are expected to influence the charge carrier transport through both scattering effects and carrier concentration modification. The charge carrier concentration is directly proportional to the rare-earth content, with the number of free electrons per formula unit given by $n = 1 - 3x$ (**Exercise 4.4**). Accordingly, the transport properties vary from metallic to insulating as the vacancy count x increases from 0 ($n = 1$) to 1/3 ($n = 0$). Thus the inherently n-type TE performance of these

[a]This FOM value makes this compound one of the highest performing bulk n-type material above 1000 K.

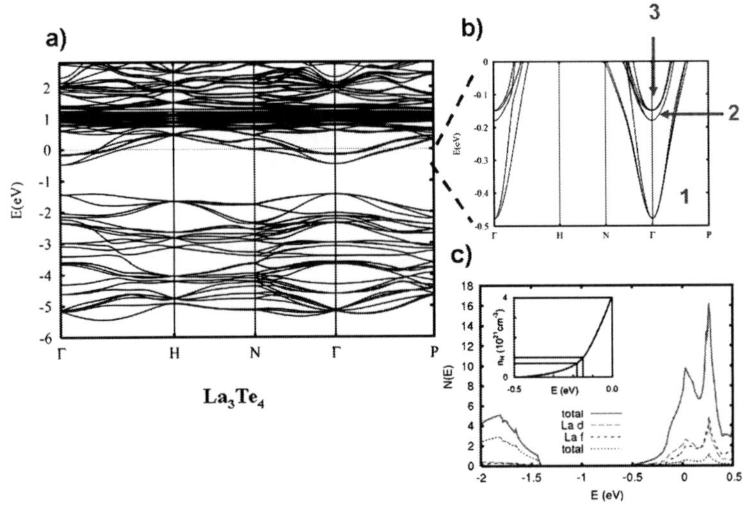

Figure 4.14 (a) Band structure of La_3Te_4 in the energy window from −6 to 3 eV. Note that the Fermi level (dashed horizontal line at 0 eV) is placed within the conduction band because La_3Te_4 has one free electron per formula unit, (b) close up of the conduction bands at a smaller energy scale, (c) the DOS showing a sharp increase near −0.16 eV due to the presence of heavy bands. This behavior is illustrated in the inset, which shows the Hall carrier concentration at 300 K. Solid lines indicate the location of the heavy band minima, which correspond to the optimum doping level [211]. Reprinted with permission from A. F. May, D. J. Singh, and G. J. Snyder, *Phys. Rev. B* **79**, 153101 (2009), Copyright 2009, American Physical Society.

binary compounds can be optimized via vacancy concentration without introducing additional alloying elements.

The band structure of $La_{3-x}Te_4$ has a direct gap ($E_g = 0.95$ eV) at the Γ point (Fig. 4.14a) and exhibits two light conduction bands close to the band edge along with three heavier bands roughly 0.3 eV higher in energy (Fig. 4.14b). The band-edge insulating composition corresponds to $x = 1/3$, and electrons progressively fill conduction states as La vacancies are filled. At full lanthanum occupancy there is one free electron per formula unit and the corresponding Fermi level is 0.48 eV above the conduction band edge. In Fig. 4.14b we show a close up of the parabolic conduction bands with band effective masses $m_3^b = 1.56m_e$, $m_2^b = 1.05m_e$, and $m_1^b = 0.34m_e$, from top to bottom. The corresponding degeneracy for each band is $g_3 = 2, g_2 = 1$, and $g_1 = 2$, respectively. Figure 4.14c shows that the conduction bands are primarily composed of La states, whereas the valence bands are tellurium derived.

For the sake of comparison, the band structure for the isostructural compound $Ce_{3-x}Te_4$ also has a direct gap of comparable size ($E_g = 1.06$ eV) at the Γ point, but the heavy bands are now located about 0.15 eV from the conduction band edge. Since cerium atoms [$4f\ 5d\ 6s^2$] have one $4f$ electron in addition to the $5d$ electrons that lanthanum atoms [$5d^1 6s^2$] have, one obtains a large peak in the DOS above the Fermi level in this case [212]. According to the Mahan–Sofo scenario discussed in Section 2.4.2, one may expect that the presence of this DOS peak could lead to a remarkably large TE efficiency for the $Ce_{3-x}Te_4$ compound. This is an interesting possibility, although recent experimental measurements indicate that electronic transport in $Ce_{3-x}Te_4$ compounds is not strongly influenced below room temperature by the presence of cerium f levels, which suggests that a TE efficiency similar to that in $La_{3-x}Te_4$ compounds may occur at high temperatures [213].

The impact of lanthanum (or cerium) vacancies on the thermal transport of these compounds is easily appreciated at high x values, where a low, nearly temperature independent lattice thermal conductivity is observed ($\kappa_l \simeq 0.6$ Wm^{-1}K^{-1} at 500 K). This is expected because randomly distributed vacancies are present in a very high concentration and these vacancies act as ideal point defects for phonon scattering because they provide the maximum mass contrast (see Section 1.2.2.3). However, we must note that a large number of randomly distributed cation vacancies can also promote Anderson localization of charge carriers, thereby reducing the electrical conductivity.

4.4 The Band Engineering Concept

In an ideal parabolic band model the energy dispersion relation is given by the expression,[a]

---

[a]For the sake of simplicity in the case of isotropic systems we will write $m_b \equiv m^b_{11} = m^b_{22} = m^b_{33}$, thereafter.

$$E(k) = \frac{\hbar^2 k^2}{2m_b}, \tag{4.5}$$

measured from the band edge. This relation provides an accurate enough prediction on transport properties for some TEMs. For instance, it is observed that the effective band mass of BaGaGe clathrates (see Section 3.5.3) is constant in the region of high ZT. According to Eq. (4.2) this is a clear indication that the corresponding band structure is parabolic. However, when the energy of the transport relevant charge carriers is high enough to be comparable with the energy gap value E_g, the dependence of energy on the wave vector is often non quadratic, and band nonparabolicity effects need to be taken into account. In that case, the energy dispersion relation given by the so-called Kane band model

$$E\left(1 + \frac{E}{E_g}\right) = \frac{\hbar^2 k^2}{2m_b}, \tag{4.6}$$

provides a more accurate description of the electrical transport properties. This is the case of skutterudites, lead chalcogenides, and other narrow–band gap TEMs. We note that when the charge carriers are mostly confined to the band edges (i.e., $E \ll E_g$) or the band gap becomes very large, Eq. (4.6) reduces to Eq. (4.5), as expected.

Band features such as energy band degeneracy, g_j, and band effective mass, m_{ij}^b, have an important impact on the S and σ coefficients, and large g_j and m_{ij}^b values are desirable for good TEMs. A closely related magnitude is the so-called *density-of-states effective mass*, which is defined as

$$m_{ij}^d \equiv g_j^{2/3} m_{ij}^b, \tag{4.7}$$

where g_j is the orbital degeneracy of the corresponding valence (holes) or conduction (electrons) bands.

For a single parabolic band n-type semiconductor the highest ZT is typically obtained at doping levels corresponding to a Fermi level positioned slightly above the minimum of the conduction band. The exact optimal location results from a precise balance between the Seebeck and the electrical conductivity coefficients values, which are given by the expressions

$$S = -\frac{2|e|L_0}{\pi^{4/3}\hbar^2}\left(\frac{g_v}{3n}\right)^{2/3} m_b T \quad (4.8)$$

$$\sigma = en\mu = e^2\frac{n\tau}{m_b} \quad (4.9)$$

where n is the charge carrier concentration, $\mu(E)$ is the mobility, and $\tau(E)$ is the relaxation time. Therefore, the electrical conductivity is typically small close to the band edge due to both low charge carrier concentration and low carriers mobility. This is because the mobility, $\mu = e\tau/m_b$, is inversely proportional to the effective mass, and because the relaxation time decreases with m_b when the carriers are predominantly scattered by phonons. Eqs. (4.8) and (4.9) clearly account for the well-known empirical fact that $\sigma(T)$ and $S(T)$ generally exhibit reciprocal behaviors. In fact, we see that the Seebeck coefficient increases (σ decreases) when m_b increases, whereas S decreases (σ increases) when n increases.

By comparing Eq. (4.9) with the spectral conductivity function given by Eq. (2.9), we see that the charge carrier concentration is proportional to the DOS (i.e., $n \sim N(E)$) and the diffusivity $D(E) \sim \tau(E)/m_b$. On the other hand, making use of Eq. (2.8), we can express the Mott's formula given by Eq. (2.49) in the more general form

$$\begin{aligned} S(T) &= -|e|L_0\left(\frac{d\ln N(E)}{dE} + \frac{d\ln\tau(E)}{dE} + 2\frac{d\ln v(E)}{dE}\right)_{E_F} T \\ &= -|e|L_0\left(\frac{d\ln n(E)}{dE} + \frac{d\ln\mu(E)}{dE}\right)_{E_F} T, \end{aligned} \quad (4.10)$$

where $N(E)$ is the DOS and $v(E)$ is the group velocity of the carriers. According to Eq. (4.10) it is possible to increase $|S|$ at a given charge carrier concentration value, by increasing the energy dependence of the DOS, for instance via the introduction of resonant states, as it has been reported for Tl doped PbTe compounds (see Section 3.3.4) or LAST materials (see Section 3.4.2).

The presence of a narrow peak close to the Fermi level should also lead to a high Seebeck coefficient value (see Section 2.4.1). As an illustrative example we shall consider the case of Ag_3AuTe_2 compound, whose electronic structure exhibits a very small (~4 meV) indirect gap between the valence band at the N point and an off-symmetry location near the Γ point in the conduction band.

The characteristics of the valence band centered at the N point are important for the electronic transport. In fact, this band is highly non-parabolic and quite flat, leading to a large carrier effective mass for holes and a high DOS value in the valence band near the Fermi level. Since the Fermi level is located on the right of this DOS peak, a large positive value of the Seebeck coefficient is expected for this sample, as it is experimentally confirmed (**Exercise 4.5**).

4.4.1 *The Thermoelectric Quality Factor*

Let us assume that a material can be doped in such a way that the dopant atoms do not appreciably change the electronic structure, the vibrational DOS, or the scattering mechanisms, but only affect the Fermi level location. In that case, the maximum ZT value for semiconductors with a single band can be estimated from the dimensionless expression [15]

$$B = \frac{k_B^2 h}{3\pi^2} \frac{g_v c_l}{m_I \, \Phi^2 \kappa_l} T, \tag{4.11}$$

where

$$c_l = \frac{1}{5}(3c_{11} + 2c_{12} + 4c_{44}), \tag{4.12}$$

is the average longitudinal elastic modulus, where c_{ij} are the matrix elements of the elastic tensor (in Nm^{-2} units), m_I is the inertial effective mass (which coincides with m_b for isotropic materials),[a] and Φ is the so-called *deformation potential coefficient*, which describes the change in energy of the electronic band due to the elastic deformation stemming from the coupling between carriers and phonons.

In fact, in a crystal, acoustic phonons would cause a fluctuation on the lattice deformation which, in turn, affects the energy of the conducting bands, thereby introducing a scattering source for the charge carriers. The key parameter determining the strength of

[a]Anisotropy that arises either from the crystal structure or the carrier pockets (valleys) affects the ratio m_I/m_b. For non cubic crystals the direction of lightest m_I is preferred for TE transport. In a cubic crystal, differing m_I/m_b relies on the different averaging methods for obtaining $m_I = 3(1/m_1^b + 1/m_2^b + 1/m_3^b)^{-1}$ and $m_b = (m_1^b m_2^b m_3^b)^{1/3}$ for anisotropic carrier pockets along three principal directions having effective masses m_1^b, m_2^b, and m_3^b, respectively.

carrier scattering by acoustic phonons is then given as the change of the band energy per relative volume change of the crystal, namely, $\Phi = \Delta E(\Delta V/V)^{-1}$. For semiconductors Φ is found in the range of 5 to 35 eV, without an obvious trend, hence suggesting that the deformation potential is independent of other parameters determining the TE quality factor [125].

Equation (4.11) is referred to as the *TE quality factor*, first introduced by Chasmar [215]. This quality factor can be split into two different contributions, namely, the band structure related contribution, $g_v m_I^{-1} \Phi^{-2}$, and the lattice-related contribution, c_l/κ_l. Whereas κ_l is generally independent of the electronic band structure (see Section 1.2.2.3), and ZT accordingly increases when κ_l decreases, it is not clear, however, whether increasing c_l in Eq. (4.11) is beneficial or not for TE performance, since the average longitudinal elastic modulus is directly related to the longitudinal speed of sound, v_l, which is also directly related to the lattice thermal conductivity, regardless of whether κ_l is dominated by Umklapp processes or has reached the minimum thermal conductivity value (see Section 1.2.2.3).

Anyway, the band structure component of the quality factor given by Eq. (4.11) provides a strategy to obtain high ZT materials by looking for compounds having (1) a large number of band valleys (pockets), (2) a low band effective mass, and (3) a low carrier-phonon scattering strength. Indeed, since the deformation potential coefficient Φ appears squared at the denominator of Eq. (4.11) it is an important factor determining the maximum performance of a TEM. For instance, silicon and germanium have remarkably low deformation potential values, which greatly improves their TE quality factor. Accordingly, large ZT values are expected for samples with small deformation potential coefficient values, due to high carrier mobilities resulting from weak electron–phonon scattering [57].

For instance, high ZT values are achieved in n-type PbSe with Cl or Br doping. This result can be traced back to the existence of a weaker carrier scattering in the conduction band as compared to that in the valence band. Experimental evidence for this transport asymmetry is provided by the observation that the carrier concentration dependence of Seebeck coefficient at room

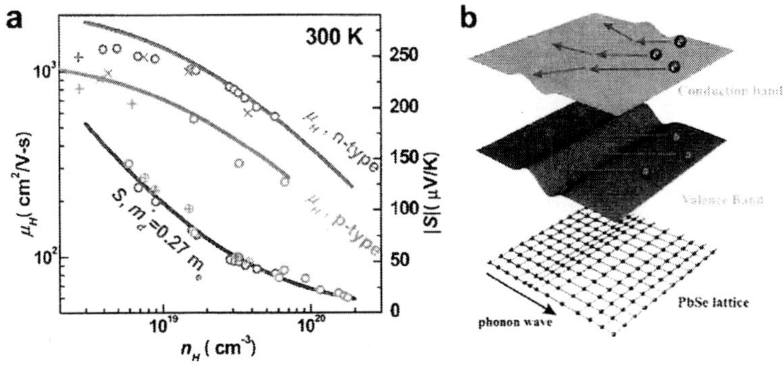

Figure 4.15 (a) Room temperature dependence of the Hall mobility and Seebeck coefficient for n-type and p-type PbSe samples, (b) sketch illustrating the different carrier–phonon coupling strength for electrons and holes in the conduction and valence bands, respectively [38]. Reprinted by permission from Y. Pei, H. Wang, and G. J. Snyder, *Adv. Mater.* **24**, 6125, Copyright 2012, John Wiley and Sons.

temperature is the same for both p-type and n-type PbSe alloys (**Exercise 4.6**), while n-type materials exhibit significantly higher mobility values than the p-type ones (Fig. 4.15).

4.4.2 *Band Convergence Effect*

Band degeneracy increases when multiple bands have the same (or comparable) energy within a few $k_B T$. This can occur when multiple bands have their band extrema with no (or insignificant) difference in energy or when multiple carrier pockets in the Brillouin zone are degenerate because they are symmetrically equivalent due to the crystal symmetry. The first case is referred to as *orbital degeneracy*, while the latter is dubbed *valley degeneracy*. High symmetry crystals can have very high carrier-pocket degeneracy when the band extrema are located at low symmetry points in the Brillouin zone. For instance, a number of well-known TEMs, such as SiGe, Bi_2Te_3 and skutterudites, are thought to have $1 < g_v < 6$. While it may not be possible to increase g_v by increasing the carrier pocket symmetry in classical periodic crystals,[a] one

[a]As we will see in the next chapter, this is not the case for quasicrystals having high order decagonal, dodecagonal, or icosahedral point group symmetries.

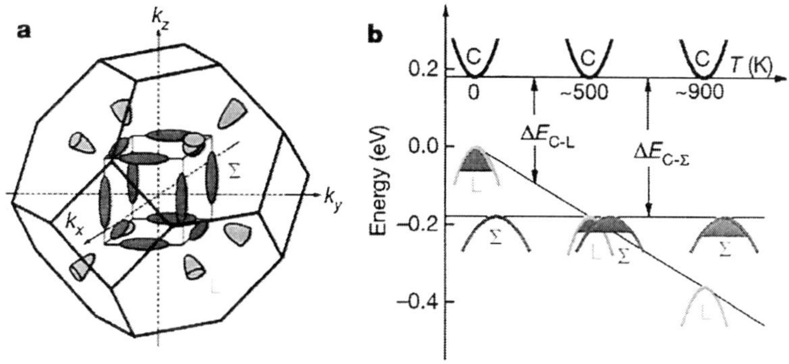

Figure 4.16 (a) Brillouin zone of PbTe showing the low degeneracy hole pockets (light gray) centered at the L point, and the high degeneracy hole pockets (dark gray). The figure shows eight half-pockets at the L point, so that the full number is $g_L = 4$, while $g_\Sigma = 12$. (b) Relative energy of the valence bands in $PbTe_{0.85}Se_{0.15}$. At about 500 K the two valence bands converge, resulting in transport contributions from both the L and Σ bands. C denotes the conduction band [76]. Reprinted by permission from Macmillan Publishers Ltd: *Nature* (Y. Pei, X. Shi, A. LaLonde, H. Wang, L. Chen, and G. Jeffrey Snyder, *Nature*, **473**, 66 (2011)), Copyright 2011.

effective way to increase the g_v value is to *converge* different bands (which are not originally degenerate) to have energies within a few k_BT of each other. This concept was at the basis of early carrier pocket engineering attempts to produce high band degeneracy, and therefore high ZT, by manipulating low-dimensional structures, though it becomes more useful in bulk materials, where the band degeneracy is already higher from the start due to their higher structural dimensionality[a].

For the sake of illustration, let us consider the case of PbTe alloys previously considered in Section 4.3.4. The valence band occurring at the L point (see Fig. 4.11) has a band degeneracy value $g_L = 4$, while the band along the Σ line has an exceptionally large $g_\Sigma = 12$. As the temperature increases the light L band lowers its energy, while the energy for the heavy Σ band remains roughly constant as measured from the conduction band edge (Fig. 4.16b).

[a]The effect of strain-induced band convergence in the TE properties of a single layer of black phosphorus (phosphorene) has been numerically studied recently reporting that a $ZT = 1.65$ value could be obtained at 300 K by applying a 5% strain along the zigzag direction [306].

This leads to an effective convergence of these two valence bands which increases the overall m_d value and, after Eq. (4.8), also the Seebeck coefficient at high enough temperatures, without any adverse effect on the carrier mobility [125]. Indeed, following this route the remarkable value $ZT = 1.8 \pm 0.1$ has been obtained for sodium doped $PbTe_{0.85}Se_{0.15}$ when optimal doping levels ($\sim$2%) are achieved [76]. In a similar way, the L and Σ bands converge in p-type PbSe to produce high effective g_v at 700 K, enabling a peak $ZT \simeq 1.2$ at 800 K.

4.4.3 *Band Gap Size Control*

In semiconductors the maximum ZT peak value is generally determined by the band gap size. Two main processes contribute to this important role of E_g. On the one hand, according to Eq. (4.3), minority carriers thermally excited across the band gap significantly decrease the overall Seebeck coefficient, causing it to level off with temperature. On the other hand, the onset of the bipolar transport also increases the electronic term contribution to the thermal conductivity κ_l (see Eq. (4.4)). Thus, with all other physical parameters remaining the same, the FOM at high temperatures would improve if the band gap could be increased in order to minimize the unfavorable effects stemming from bipolar transport. For instance, lead chalcogenides can be alloyed with a wider band gap compound, such as MgTe or CdTe to this end [125]. In the same vein, substitutions at the X site in half-Heusler compounds (see Section 3.5.1) also modify the size of the band gap.

As an illustrative example of band engineering, which includes both band convergence and gap size control let us consider the case of magnesium silicides. Some time ago, researches working at Ioffe Institute (see Section 1.3) showed that the compounds Mg_2X, with X = Si, Ge, or Sn have good properties to be considered as potential TEMs in order to replace PbTe and their derivatives in the mid-temperature (500–800 K) range because their constituting elements are abundant, inexpensive, environmentally harmless, and do not contain expensive tellurium or toxic lead. Mg_2Si ($T_m = 1375$ K) crystallizes into an fcc cubic lattice with a lattice constant $a = 6.338$ Å ($V_U = 254.6$ Å^3). It possesses the so-called antifluorite structure (space group $Fm\bar{3}m$) with Si^{-4} ions occupying the corners and face-centered positions of the unit cell and Mg^{+2} cations occupying eight tetrahedral sites in the cell. Mg_2Si and Mg_2Sn (T_m = 1051 K, a = 6.765 Å, V_U = 309.6 Å^3) are indirect band gap semiconductors (E_g = 0.77 and 0.35 eV, respectively) with good electronic mobility values (μ = 405 and 320 cm^{2}s^{-1}V^{-1}, respectively). However, these binary compounds have too high thermal conductivity values (κ_l = 7.9 and 5.9 Wm^{-1}K^{-1}, respectively) to be used as TEMs. Therefore, in order to increase the FOM one could alloy them, hence reducing their thermal conductivities. In fact, the lattice thermal conductivity is significantly reduced ($\kappa_l \simeq$ 2 Wm^{-1}K^{-1}) due to alloying effects in $Mg_2Si_{1-x}Sn_x$ solid solutions, although the electronic mobility is reduced as well ($\mu \simeq$ 50 cm^{2}s^{-1}V^{-1}).

The electronic structure of the resulting ternary compound is shown in Fig. 4.17a, and it has two low-lying conduction bands (referred to as the low mass, C_L, and high mass, C_H, conduction bands attending to their respective curvatures). As we see, by properly changing the Sn concentration the respective conduction band edges can be reversed (i.e., there is a band inversion) in these solid solutions. Indeed, around composition $x \simeq$ 0.65–0.70, $Mg_2Si_{1-x}Sn_x$ solid solutions display a convergence in energy of both heavy and light conduction bands in the energy spectrum and both bands give practically the same contribution to electrical conductivity [216]. This is a favorable situation for TE performance because in this case the DOS can increase without a related decrease of electron mobility. Transport measurements, carried out from 4 to 800 K in samples ranging the interval composition x = 0.2–0.8, reveal a progressively increasing Seebeck coefficient that peaks at x = 0.7. The peak FOM attains $ZT \simeq$ 1.3 near 750 K (850 K) for samples with x = 0.6 (x = 0.4), respectively with an average value of about $\langle ZT \rangle \simeq$ 0.8 in the temperature range 400–850 K (Fig. 4.17b) [217].

4.4.4 *Carrier Concentration Optimization*

From Eqs. (4.8) and (4.9) it is clear that the FOM depends on the carrier concentration, which can be optimized by controlling the doping level in order to properly locate the Fermi energy. For

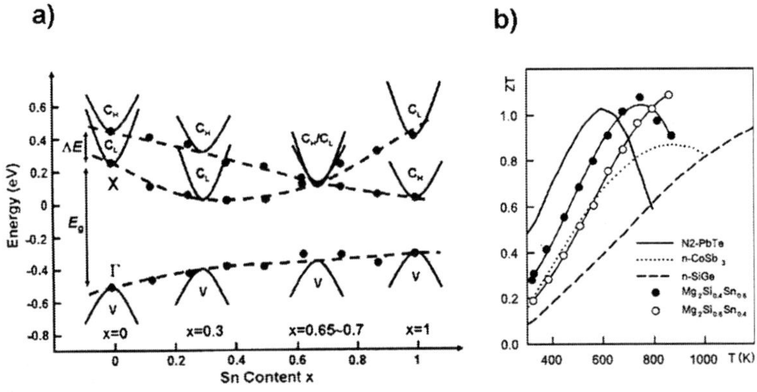

Figure 4.17 (a) Relative positions of the heavy (C_H) and light (C_L) conduction bands as well as the topmost valence band (C_V) as a function of the Sn content for $Mg_2Si_{1-x}Sn_x$ solid solutions. The dashed lines indicate the relative displacement of these bands as x is progressively increased as well as the resulting evolution of the bang gap size E_g and the band splitting between the two conduction bands ΔE. Solid dots indicate numerically calculated data for different Sn contents [216]. Reprinted with permission from W. Lu, X. Tn, K. Yin, H. Liu, X. Tang, J. Shi, Q. Zhang, and C. Uher, *Phys. Rev. Lett.* **108**, 166601 (2012), Copyright 2012, American Physical Society. (b) The FOM values of two $Mg_2Si_{1-x}Sn_x$ compounds are compared to those of other high temperature TEMs [217]. Reprinted with permission from V. K. Zaitsev, M. I. Federov, E. A. Gurieva, I. S. Eremin, P. P. Konstantinov, A. Y. Samubib, and M. V. Vedernikov, *Phys. Rev. B* **74**, 045207 (2006), Copyright 2006, American Physical Society.

instance, in the high temperature TEM SiGe, boron (with one less valence electron than Si or Ge) and phosphorus (with one more valence electron) substitutions are used to render the material p- and n-type, respectively. For most TEMs, the optimal doping level is obtained when the reduced Fermi level, βE_F, takes a value close to zero (where E_F is measured from the band edge). Using a classical statistics approximation, Ioffe estimated that the optimal carrier concentration for a single carrier type system is given by $n^* \sim (m_d T)^{3/2}$. Such a strong temperature dependence inevitably leads to a fairly narrow temperature range for attaining high FOM values. Thus, in order to maximize ZT in a wide enough temperature range, one would be able to achieve significantly different carrier concentrations at different temperatures. A straightforward strategy to achieve this goal is by functional grading of doping concentration, namely, using higher dopant concentration at the hot side of the TE element compared to the cold side.

This approach, however, presents challenges such as dopant solubility dependence with temperature and diffusion effects. Alternatively, one may try band engineering aimed to tune the DOS by exploiting the temperature dependence of the effective band degeneracy through band convergence. Indeed, making use of Eq. (4.7), Ioffe's optimal carrier concentration reads $n^* \sim g_v(T)(m_b T)^{3/2}$. Accordingly, one can stabilize the optimal carrier concentration, that is, to get an almost constant $g_v(T)(m_b T)^{3/2}$ value over a wide enough temperature interval by properly tuning the $g_v(T)$ function with temperature. For instance, alloying PbTe with MgTe or MnTe increases the energy separation between the L valence band and the conduction band faster than that of Σ valence band, leading to a higher effective g_v value (and therefore an increase on n^*) due to $L - \Sigma$ bands convergence (see Fig. 4.16) at low temperatures [125].

4.4.5 *Impurity-Induced DOS Peaks*

Another example of DOS engineering near the Fermi level was promoted by experimental studies showing that the Seebeck coefficient of TiO samples containing excess oxygen atoms exhibits a sharp increase by a factor of 80 compared to that of stoichiometric TiO, resulting in $ZT = 1.64$ at 1073 K [218]. This figure is among the largest FOM values reported for high temperature metal oxides TEMs (see Section 4.5). Similar improvements of TE properties upon the addition of certain doping elements has been also observed in more traditional TEMs. Thus, Tl doped PbTe reaches $ZT = 1.5$ at 773 K, a value which is twice as large as that of the best conventional p-type PbTe-based alloys. It has been confirmed that the reported improvement in ZT results from a Seebeck coefficient increase in this compound, caused by the presence of thallium-induced peaks in the DOS [127]. Indeed, dopant atoms introduce electronic states in the gap energy region, which can resonate with the states belonging to either the valence or conduction bands. In this way, the DOS can be effectively increased at certain narrow energy regions, opening a potential mechanism for electronic structure engineering based on the presence of convenient narrow features in the DOS close to the Fermi level (see Section 2.4). In just the same vein, the addition of silver and antimony atoms to the semiconductor PbTe parent alloy also introduces impurity-induced states near the band gap. Specifically, the addition of silver atom creates resonant states at the top of the valence band, which not only are resonant with the band, but also extend into the gap region, while the equivalent situation is found for antimony atoms and the PbTe conduction band.

Thus, a broad class of materials exists that is known to be capable of achieving wide control over the electronic structures and DOS. This control stems from the alloying of isovalent, but highly electronegative, mismatched elements at dilute concentrations, forming a system known as *highly mismatched alloys*. In these materials class, the hybridization between the extended states of the majority component and the localized states of the minority one results in a strong band restructuring, leading to the presence of new subbands near the original conduction or valence band edge. These narrow subbands have a heavy effective carrier mass and give rise to sharp DOS features. The presence of these spectral features gives rise to a significant improvement of ZT, in agreement with the Mahan–Sofo scenario (Section 2.4.2).

4.5 Oxide Semiconductors

The discovery of $NaCo_2O_4$ material in 1997 completely changed the traditional understanding of oxides in TE research, which have been almost entirely ignored for their potential use in TE field due to their extremely low electrical conductivity. However, $NaCo_2O_4$ oxide exhibits both high electrical conductivity ($\rho = 200$ μΩcm), a figure which is comparable to that of typical metallic alloys, and a moderate Seebeck coefficient ($S = +100$ μVK^{-1}) at room temperature in the in-plane direction, resulting in a remarkably high power factor of 50 $\mu Wcm^{-1}K^{-2}$ and $ZT = 1.2$ at 800 K [219]. It is worth noticing that this P value is even higher than that of state-of-the-art Bi_2Te_3 alloys at room temperature (see Section 3.3).

The layered $NaCo_2O_4$ material consists of ordered CoO_2 sheets and disordered sodium nanosheets. The CoO_2 layers form a strongly connected electron system that serves as an electron transport layer, leading to a high power factor, whereas the sodium ions layers in calcium cobalt oxide layers act as phonon scattering regions, providing low thermal conductivity. Consequently, p-type cobaltites have been the starting point of interest for TE oxides with Na_xCoO_2 as its most celebrated representative.

Compared with traditional TEMs, oxide-based TEMs have many advantages. They are nontoxic and nonpolluting, are made of inexpensive materials, can be simply prepared in air without vacuum protection, and can operate at high temperatures under an oxygen atmosphere for long periods. These merits make oxide TEMs promising candidates for high-temperature TE devices.

Motivated by these considerations other oxides families have also been explored for TE applications, including perovskites ($ACoO_3$ type), manganites ($AMnO_3$ type), ruthenates ($ARuO_3$ type), delafossites ($CuMO_2$), $YBa_2Cu_3O_{6+x}$ compounds, and $Zn_{1-x}Al_xO$. Among these compounds the best reported FOM values range from $ZT = 0.4$ (for the p-type representatives) and $ZT = 0.3$ (for the n type representatives) at 1273 K. Additional improvement with respect to the FOM values obtained for these conventional oxides has been reported for certain misfit $Ca_3Co_4O_{9-x}$ cobaltites (see Section 5.1.2).

In order to understand the relatively large Seebeck coefficient measured in cobaltites various mechanisms have been considered, including strong electron–electron correlation effects, charge valence fluctuations, orbital degeneracy, spin fluctuations due to the possibility for the Co ions to occur in different spin states, or effective mass enhancement in the proximity of a Mott transition. Of the listed mechanisms, electron correlation is considered to be one of the most probable explanations. In fact, in recent years the search for new TEMs, has focused on *strongly correlated* materials, going beyond traditional semiconductor materials.

Strongly correlated systems are characterized by the role of Coulomb interaction, forcing the electrons to locate as far apart as possible to each other, on the charge carriers dynamics. The main reason to expect a good TE performance in these systems is that electron correlation makes the relaxation time $\tau(E)$ strongly energy dependent. According to Eq. (4.10) this feature gives rise to a significant Seebeck coefficient enhancement. Kondo effect, describing the interaction of the electron's spin with the magnetic momenta of lattice ions, has been also invoked in order to explain the colossal Seebeck coefficient of about $S \simeq 45000$ μVK^{-1} measured in a $FeSb_2$ alloy (referred to as a *Kondo insulator*) at very low temperatures. Indeed, Kondo effect provides a mechanism for Seebeck coefficient enhancement via correlation effect-induced asymmetry in the spectral function $\sigma(E)$ close to the Fermi level, as it was discussed in Section 2.3. Nevertheless, one must keep in mind that correlated motion requires the presence of the so-called coherent regime, which can only develop at low enough temperatures. For instance, estimates for different materials give $T \simeq 580$ K for V_2O_3 oxide, $T \simeq 152$ K for FeSi silicide, and $T \simeq 100$ K for Na_xCoO_2. This restricts the use of these TEMs to refrigeration and cryogenic applications mainly.

4.6 Exercises

(1) The Fermi energy of the fcc elemental aluminum ($a = 4.04$ Å) is $E_F = 11.65$ eV. Assuming the validity of the free-electron model, determine the wavelength of the electrons with energy values close to E_F.

(2) Estimate the role of diffusivity and charge carrier concentration differences in the resulting electrical conductivity of selenium and tellurium, respectively.

(3) (a) Making use of the data given in the text determine the power factor of the neodymium doped $HfTe_5$ compound and compare it with that corresponding to Bi_2Te_3 at the same temperature. (b) Compare the obtained value with those listed in Tables 3.2 and 3.3 for other TEMs of interest.

(4) Taking into account the atomic electronic structures of lanthanum [$5d^1 6s^2$] and tellurium [$5s^2 p^4$] determine the number of electrons per formula unit in the $La_{3-x}Te_x$ compound.

(5) Determine the ZT value of the $Ag_{3.05}AuTe_2$ compound making use of the transport coefficient values $\rho = 0.01$ Ωcm, $S = +160$

μVK^{-1}, and $\kappa = 0.15$ Wcm^{-1}K^{-1} measured at $T = 500$ K [214]. Comment on the obtained result.

(6) Could you explain the physical basis for this result?

4.7 Solutions

(1) The kinetic energy of free electrons with momenta p and mass m is given by

$$E = \frac{p^2}{2m} = \frac{\hbar^2 k^2}{2m} = \frac{h^2}{2m\lambda^2}.$$

Equating this expression to the Fermi energy one gets

$$\lambda_F = \frac{h}{\sqrt{2mE_F}} = \frac{6.62 \times 10^{-34}\ \text{Js}}{\sqrt{2 \times 9.1 \times 10^{-31}\ \text{kg} \times 11.65\ \text{eV} \times 1.6 \times 10^{-19}\ \text{J(eV)}^{-1}}}$$
$$= 3.59 \times 10^{-10}\ \text{m}$$

Therefore, the electrons' wavelength $\lambda \simeq 3.6$ Å is comparable to the lattice parameter of the elemental aluminum crystal ($a = 4.04$ Å).

(2) Considering that the electrical conductivity of a semiconductor is given by the expression $\sigma(T) = \sigma_0 e^{-\beta E_g/2}$, where σ_0 depends on the charge carriers density and mobility, and $\beta \equiv (k_B T)^{-1}$, one obtains the room temperature electrical conductivity ratio

$$\frac{\sigma_{\text{Se}}}{\sigma_{\text{Te}}} = \alpha e^{(E_g^{\text{Te}} - E_g^{\text{Se}})\beta/2},$$

where $\alpha \equiv \sigma_0^{\text{Se}}/\sigma_0^{\text{Te}}$. Making use of the σ values listed in Table 3.1 and the band gap energies given in Fig. 4.2 we get

$$\alpha = \frac{2 \times 10^{-6}}{2} \exp\left(\frac{0.91 - 0.33\ \text{eV}}{2 \times 8.617 \times 10^{-5}\ \text{eVK}^{-1} \times 300\ \text{K}}\right) \simeq 0.07$$

at room temperature, which indicates that the charge carrier density and mobility of both materials differ by an order of magnitude at least, at this temperature.

(3) (a) Making use of the values indicated in the text we obtain

$$P = \frac{2\ \text{W}\ 10^{-2}\text{cm}^{-1}\text{K}^{-1}}{165\ \text{K}} = 121.2\ \mu\text{Wcm}^{-1}\text{K}^{-2}.$$

On the other hand, we have $PT = 0.88\ \mu\text{Wcm}^{-1}\text{K}^{-2}$ for Bi_2Te_3 at $T = 165$ K [210], so that

$$P = \frac{0.88\ \text{W}\ 10^{-2}\text{cm}^{-1}\text{K}^{-1}}{165\ \text{K}} = 53.3\ \mu\text{Wcm}^{-1}\text{K}^{-2}.$$

(b) By comparing the above figure with the data listed in Tables 3.3 and 3.2 we see that the HfTe:Nd doped pentatelluride exhibits a power factor about sixfold larger than that of $Bi_{0.88}Se_{0.22}$ alloys at 80 K ($P = 21\ \mu\text{Wcm}^{-1}\text{K}^{-2}$), and threefold larger than that reported for bismuth ($P = 74\ \mu\text{Wcm}^{-1}\text{K}^{-2}$) and antimony ($P = 66\ \mu\text{Wcm}^{-1}\text{K}^{-2}$) elemental solids at room temperature.

(4) In one formula unit $La_{3-x}Te_4$ lanthanum cations donate $3(3 - x)$ electrons of which 4×2 are required by tellurium anions in order to satisfy the valence octet rule. Thus, the free-electron concentration is given by $n = 3(3 - x) - 8 = 1 - 3x$ per formula unit.

(5) Making use of the experimentally measured transport data we have

$$\begin{aligned} ZT &= 100\ \Omega^{-1}\text{cm}^{-1}\ 10^2\ \text{m}^{-1}\text{cm}\ \frac{160^2 \times 10^{-12}\ \text{V}^2\text{K}^{-2} \times 500\ \text{K}}{1.5\ \text{Wm}^{-1}\text{K}^{-1}} \\ &= 0.085 \end{aligned}$$

which is an order of magnitude smaller than the values usually obtained for good TEMs. This exercise provides another instance (see Exercise 2.1) of a system exhibiting a relatively large Seebeck coefficient value, but a typically metallic ZT value due to its very low electrical conductivity.

(6) The independence of Seebeck coefficient with the carrier concentration can be explained as follows: according to the PbSe alloy band structure (see Fig. 4.11) close to the Fermi level only the symmetric bands at the L point in the Brillouin zone (conduction band for n-type and valence band for p-type,

respectively) are involved in the transport at room temperature. Since both bands have the same curvature and degeneracy ($g_L = g_c = 4$), according to Eqs. (4.2) and (4.7) the corresponding carriers (holes and electrons, respectively) will have the same DOS effective mass, which leads to similar $|S|$ values. On the other hand, having the same DOS effective masses, one would expect similar mobilities for both holes and electrons, which is not the case. This unexpected result has been explained as an indication of different scattering strengths for both electrons ($\Phi = 25$ eV, derived from $\mu(n_H)$ curve fit) and holes ($\Phi = 35$ eV) in this material, naturally leading to different deformation potential values, thereby different mobilities as well [57].

Chapter 5

Beyond Periodic Order

In Chapter 3 we learned that complex crystal structures, having low symmetry lattices, generally lead to low lattice thermal conductivities, which favors large figure of merit (FOM) values. Afterward, in Chapter 4 we saw that the presence of highly degenerate, relatively flat bands close to the Fermi level is a very convenient feature in order to obtain large Seebeck coefficient and electrical conductivity values, naturally leading to high power factors as well. Nevertheless, these degenerate bands usually occur in materials crystallizing into high symmetry lattice structures. Therefore, we face an apparent structural incompatibility, demanding a low enough crystal symmetry to guarantee the lattice complexity requirement along with a high enough lattice symmetry in order to satisfy optimal power factor conditions.

In this chapter we will see that this riddle can be properly addressed by relaxing a tacit assumption: that of limiting ourselves to the consideration of *periodic* structures only. In fact, well ordered arrangements of atoms are also possible in the so-called *aperiodic crystals*, which exhibit *complex* long-range spatial correlations, well beyond mere periodicity [32]. This property clearly distinguishes aperiodic structures from amorphous matter, the latter being able to display short-range correlations only. Therefore, aperiodic order

Thermoelectric Materials: Advances and Applications
Enrique Maciá-Barber

ISBN 978-981-4463-52-2 (Hardcover), 978-981-4463-53-9 (eBook)
www.panstanford.com

has nothing to do with disorder in any of its possible multiple forms. Etymologically aperiodic order means *order without periodicity*, and this sort of ordered matter shares very interesting properties with the more usual, periodically ordered one. For instance, arrays of atoms in aperiodic crystals are able to give rise to high-quality x-ray or electron diffraction patterns composed of a collection of discrete Bragg spots, just as periodic arrays of atoms do in classical crystals.

The three main classes of aperiodic crystals currently known will be described in Section 5.1. We shall start by considering the case of calaverite mineral, a remarkable gold–silver telluride, in Section 5.1.1, and then we will describe the so-called misfitted cobaltites and the manganese silicides in Section 5.1.2. The structural complexity present in the materials studied in these two sections is based on the existence of two basic structural sublattices in the solid, whose periodicities are incommensurate to each other. The third class of aperiodic structure to be considered exhibits a novel kind of long-range aperiodic order instead, which is based on the notion of a quasiperiodic arrangement of atoms through the space.[a] Accordingly, this phase of matter is referred to as *quasiperiodic* crystals (*quasicrystals*, for short) and they will be introduced in Section 5.1.3 and fully described in Sections 5.2 and 5.3. Indeed, these materials are of particular interest to our purposes, for quasicrystalline alloys possess crystal structures with very high lattice symmetries, characterized by the presence of fivefold, eightfold, tenfold, or twelvefold rotation axes, respectively related

[a] From a mathematical viewpoint quasiperiodic functions are a special case of *almost periodic functions*. The theory of these functions was developed by Harald Bohr (1887–1951, brother of the well-known physicist Niels Bohr) [5]. A function $f(\mathbf{x})$ is called almost periodic if for any arbitrary small number $\varepsilon > 0$, there are almost-periods $\mathbf{P}$ such that the shifted function differs less than ε from the unshifted one, namely, $|f(\mathbf{x}) - f(\mathbf{x}+\mathbf{P})| < \varepsilon$, for all $\mathbf{x} \in \mathbb{R}^n$. In general, the smaller the value of ε, the larger becomes the required translation $\mathbf{P}$, although these translations are relatively dense in $\mathbb{R}^n$. By this we mean that there are, for each ε, values R_1 and R_2 $(R_1 < R_2)$ such that every sphere of radius R_2 contains at least one $\mathbf{P}$ satisfying the above condition, and in every sphere of radius R_1 around any translation $\mathbf{P}$ satisfying the condition, there is no other translation but $\mathbf{P}$.

Almost periodic functions can be uniformly approximated by Fourier series containing a countable infinity of pairwise incommensurate frequencies. When the set of required frequencies can be generated from a finite-dimensional basis, the resulting function is referred to as a *quasiperiodic* one. Let us consider a quasiperiodic function given by its discrete Fourier decomposition to decagonal, octagonal, dodecagonal, or icosahedral point group symmetries. Thus, the diffraction patterns of quasicrystals are quite bizarre, unveiling the existence of unexpected symmetries which endow them with an impressive aesthetic appeal. Noteworthy, these high-order symmetries cannot be found in conventional periodic crystals, since they are forbidden by the restriction theorem of classical crystallography. In this way, the requirements of structural lattice complexity and high-order symmetry can be *simultaneously* met in a given piece of bulk matter. Afterward, in Section 5.1.4 we will present the properties of certain *complex metallic alloys* (some of them closely related to quasicrystalline phases), having huge unit cell containing up to thousand atoms.

5.1 Aperiodic Crystals

Aperiodic crystals include three classes of materials, namely, incommensurate composites, incommensurately modulated phases, and quasicrystalline alloys.[b] In all of these structures the atomic

---

$$f(\mathbf{x}) = \sum_{\mathbf{k}} \tilde{f}(\mathbf{k}) e^{i\mathbf{q}\cdot\mathbf{x}}, \tag{5.1}$$

where the reciprocal vectors are defined by

$$\mathbf{k} = \sum_{j=1}^{N} n_j \mathbf{b}_j. \tag{5.2}$$

If the minimal number of basis vectors $\mathbf{b}_j$ is larger than three (i.e., $N > 3$), then a higher dimensional description (see Section 5.2) is needed to describe the reciprocal lattice, and the related structure is an aperiodic crystal. Otherwise, we obtain a periodic crystal, which indicates that, in turn, periodic functions are just a particular case of quasiperiodic ones. The interested reader is referred to the excellent books [20, 23] for more details.

[b] The International Crystallographic Union approved, in April 1991, the establishment of a Commission on Aperiodic Crystals with the membership of J. M. Pérez-Mato (Chairman), G. Chapuis, M. Farkas-Jahnke, M. L. Senechal and W. Steurer. According to their terms of reference:

> In the following by "crystal" we mean any solid having an essentially discrete diffraction diagram, and by 'aperiodic crystal' we mean any crystal in which three-dimensional lattice periodicity can be considered to be absent [6].

In the new definition, the essential attribute of crystallinity is transferred from real space to reciprocal space. Consequently, within the crystalline family we can arrangements exhibit well-defined long-range order, which gives rise to high-quality discrete diffraction patterns, in spite of the absence of lattice periodicity. Attending to the x-ray, electron, or neutron diffraction patterns of different aperiodic crystals representatives, an important distinction can be made between incommensurate structures and quasicrystals. For the former there always exists a sub-pattern of high-intensity spots, the so-called main reflections, corresponding to an average periodic structure in physical space. In addition, there is a set of peaks, the so-called satellite spots, grouped around the high-intensity spots. The satellite reflections occur due to the presence of a displacive or other type of structural modulation characterized by a periodicity which is incommensurate with that of the average structure. Often the amplitude of this modulation is not too big and the overall structure can be adequately described in terms of a perturbed averaged periodic structure. Unlike modulated structures, there is no periodic reference structure in quasicrystals, so that their structures determination is more demanding in this case.[c]

In mathematical terms aperiodic crystals are characterized by the property that the Fourier transform of their atomic density consists of components belonging to a set of integral linear combinations of *more than three* reciprocal lattice vectors. This is at variance with the classical crystallography wisdom, which was used to employ just three reciprocal lattice vectors to fully describe the geometrical properties of crystal unit cells. At the same time, the necessity of using more than three reciprocal lattice vectors clearly highlights the higher structural complexity degree enclosed in aperiodic crystals.

In order to properly deal with these additional lattice vectors it is convenient to make use of the so-called superspace framework introduced by de Wolff in 1974 to describe certain incommensurate crystals [220]. The basic idea inspiring this approach is to embed

---

now distinguish between periodic crystals, which are periodic on the atomic scale, and aperiodic crystals, lacking lattice periodicity. This broader definition reflects our current understanding that microscopic periodicity is a sufficient but not necessary condition for crystallinity.

[c] The interested reader is referred to the excellent books [22] and [23] for a full description of this topic.

the complex aperiodic structure in a higher-dimensional space (the so-called superspace) in such a way that the periodic order in the atomic distribution is restored. Therefore, aperiodic crystals in our physical space become periodic crystals in the superspace. Thus, the positions of atoms in an aperiodic crystal can be regarded as a distorted pattern resulting from the projection of a periodic hyperlattice onto the 3D physical space.

In this way, incommensurate composites, incommensurately modulated phases, and quasicrystals can be described in terms of a unified formalism involving the generalization of the basic principles of classical 3D crystallography to higher-dimensional spaces. In fact, most incommensurate phases can be embedded in 4D superspaces, whereas quasicrystals require 5D (for the so-called decagonal phases) or 6D (for the so-called icosahedral phases) superspaces.

5.1.1 *The Calaverite Puzzle*

Calaverite ($Au_{1-x}Ag_xTe$, with $0 \leq x \leq 0.15$), together with krennerite ($Au_{1-x}Ag_xTe$, with $0.15 < x \leq 0.3$), sylvanite ($AuAgTe_4$), and muthmannite ($AuAgTe_2$) belongs to a class of related gold–silver telluride minerals. The first morphological descriptions of calaverite were published in the early 1900s, and the authors were puzzled by the rich variety of crystalline forms, most of which could not be satisfactorily indexed by three small integer numbers, as it was usual. Accordingly, they concluded that calaverite crystals apparently contradicted one of the basic rules of classical crystallography [221, 222].

The problem of the indexation of the calaverite crystal forms was again considered in the early 1930s by considering morphological data of 105 samples of the finest quality single crystals found in nature (Fig. 5.1a). After attempting to index most of the faces with four distinct and independent lattices (one monoclinic and three triclinic), the authors realized that the proposed solution was highly unsatisfactory and consequently declared that the law of rational indices probably was not generally valid [223]. However, the scientific community was apparently not ready to

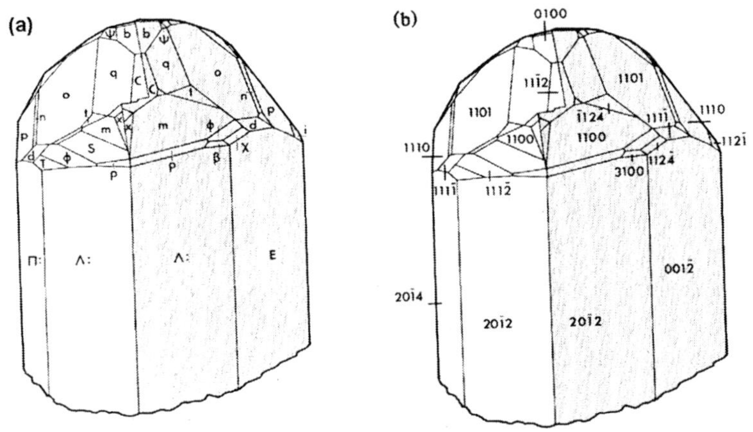

Figure 5.1 (a) Copy of a drawing of a calaverite twin crystal showing a rich variety of faces [223]. Reprinted with permission from V. Goldschmidt, Ch. Palache, and M. Peacock, *N. Jahrbuch f. Mineralogie, Beil. Abt. A* **63**, 1 (1931), Copyright 1931, E. Schweizerbart Science Publishers. (b) Superspace-based four-index scheme for the different faces [224]. Reprinted with permission from B. Dam, A. Janner, and J. D. H. Donnay, *Phys. Rev. Lett.* **55**, 2301 (1985), Copyright 1985, American Physical Society.

take this consideration too seriously, likely since it could deprive crystallography of one of its historical bases.[a]

Additional evidences of some kind of peculiar atomic arrangement in calaverite was provided by the study of its x-ray diffraction pattern in 1936, which showed the presence of some extra spots that could not be adscribed to the average structure lattice derived from the main diffraction spots, consistent with a monoclinic $C2/m$ space group, with lattice parameters $a = 7.1905$ Å, $b = 4.3975$ Å, and $c = 5.0613$ Å and $\alpha = 90.04°$ ($V_U \simeq 160.4$ Å^3). In this structure the vertices and centered positions of an almost orthorhombic unit cell are occupied by gold atoms, whereas the space between the gold planes is filled with two tilted zig-zag chains of tellurium atoms (Fig. 5.2a).

[a]During the XVIII and XIX centuries the determination of lattice parameters and crystal symmetry from the morphology of mineral crystals was a powerful research program. The subsequent development of X-ray crystallography, starting in 1913, made this branch of classical crystallography progressively go out of fashion.

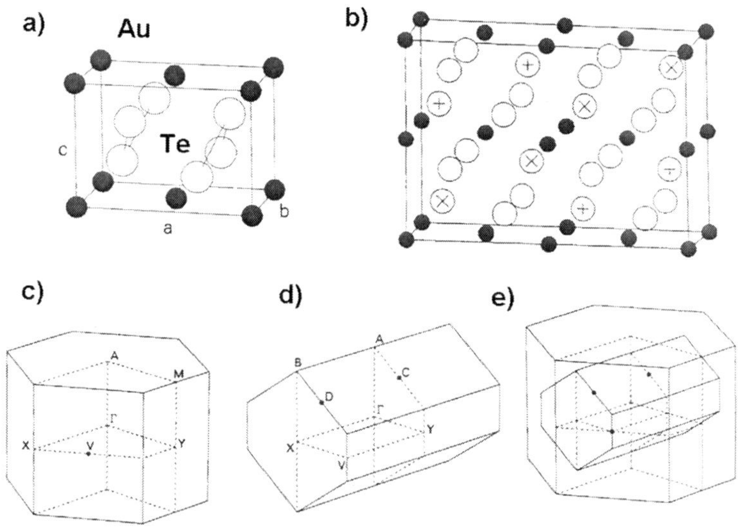

Figure 5.2 (a) The unit cell of the average monoclinic structure of calaverite containing six atoms ($n_U = 6$). Small full circles represent Au and large open circles Te atoms. The Te atoms form zig-zag chains ($n_E = 1$). The directions of the primitive lattice vectors **a**, **b**, and **c** are indicated. (b) Unit cell for the first-order approximant structure composed of four averaged structure unit cells. The phase of the modulation (if not zero) is shown inside the Te atoms by the symbols + (phase +1/4) and × (phase −1/4), respectively. The Au atoms in the centered position have a nearly linear coordination with the neighboring Te atoms (as in the average structure), whereas the Au atoms on the corners of the cell (and their equivalent atoms inside) have a nearly square coordination, because two Te atoms are shifted toward them, two away from them, and two others stay in place. The Brillouin zones corresponding to the average calaverite structure and its approximant structure are shown in (c) and (d), respectively. Their irreducible wedges are comprised by the broken lines. (e) The relative size and orientation of the first-order approximant structure Brillouin zone is compared to that of the average structure. Adapted from Ref. [225], with permission from IOP Publishing Ltd.

Further progress must await until 1979 when detailed electron microscope studies revealed a beautiful series of satellite spots in the electron diffraction patterns [226]. The presence of these satellite reflections around the main diffraction spots clearly signaled the modulated nature of calaverite. In particular, accurate measurements proved the incommensurate nature of the modulation describing the satellite reflections pattern and confirmed the presence of a displacive as well as an occupation modulation due to the presence of silver atoms in the (Au,Ag)Te_2 compound.[a] Both modulations can be expressed in terms of the same incommensurate wave vector $\mathbf{q} = \alpha\mathbf{a}^* + \gamma\mathbf{c}^*$, where $\mathbf{a}^*$ and $\mathbf{c}^*$ are reciprocal lattice vectors, and α and γ are real numbers, which take on the approximate values $\alpha \simeq -0.4076(16)$ and $\gamma \simeq 0.4479(6)$.[b] The introduction of this wave vector allows one to index the Bragg spots in the form $\mathbf{K} = h\mathbf{a}^* + k\mathbf{b}^* + l\mathbf{c}^* + m\mathbf{q}$, thereby extending the classical morphological laws in such a way that each crystal face can now be indexed in terms of *four* integers $\{hklm\}$, instead of the usual three Miller indices, as it is illustrated in Fig. 5.1b. Accordingly, it can be said that the law of rational indices still holds for aperiodic crystals provided that an appropriate number of integer indices is used.

The morphological importance of satellite-related faces (given by the $m \neq 0$ satellite spots in the diffraction pattern), as growth habit controlling factors, can be clearly appreciated in the enormous prismatic faces $\Lambda \rightarrow (20\bar{1}2)$, $\Pi \rightarrow (20\bar{1}4)$, and $E \rightarrow (001\bar{2})$ shown in Fig. 5.1b. This macroscopic feature highlights the sensitivity of crystal morphology to finer details in the long-range order present at the atomic level. Indeed, analysis of the calaverite structure refined in superspace indicates that as a result of a displacive modulation the coordination of the gold atoms change continuously from a linear coordination to a square–planar coordination with tellurium atoms, with a periodicity incommensurate to the other three fundamental periodicities of the lattice structure. This gold coordination modulation induces some strong Te displacements, of the order of 0.5 Å, which are able to break up Au–Te bonds and

---

[a]The occupation modulation leads to partial occupation of the gold positions, with a probability of finding Ag atoms on Au sites of at most 0.15. The displacive modulation is most important on Te sites, with an amplitude of approximately 0.76 au parallel to the twofold axis.

[b]The goniometric-based corresponding values, refined in a recent revision of the original measurements obtained by Goldshmidt, Palache, and Peacock in 1902, read $\alpha = -0.4095$ and $\gamma = 0.4492$, respectively. The remarkable agreement of these figures with the more accurate x-ray diffraction-based data confirms the high quality of their crystallographic work. The interested reader is referred to reference [23] (Chapter 7) for more details.

create isolated Te–Te pairs in the lattice. Therefore, the resulting modulation is presumably driven by Te–Te bonding interactions [227]. This chemical bonding arrangements naturally influence the resulting electronic structure of calaverite.

The calculations of the electronic structure of incommensurately modulated crystals is inhibited by the difficulty of handling their related infinitely large unit cell (as it is also the case with quasicrystals described in Sections 5.2 and 5.3). The usual way to deal with this problem is to study a specific series of periodic approximations to the crystal structure (Fig. 5.2b) and to extrapolate for the limiting case of the incommensurate structure from the trends in the approximants results. In Fig. 5.3a the electronic band structure corresponding to the average structure of calaverite, plotted along the symmetry directions of the irreducible wedge indicated in Fig. 5.2c, is shown. The overall picture of the band structure is determined by three contributions, stemming from Te $5s$, Au $5d$, and Te $5p$ states, respectively. The Te $5s$ bands extend from -13.4 to -10.0 eV (measured with respect to the Fermi level) and are well separated from the rest of the valence bands. The Au $5d$ bands extend from -6.5 to -4.0 eV, with a small extended tail up to $+1.5$ eV, which can be better appreciated in the DOS profile (Fig. 5.3c). The main contribution to the DOS at the Fermi energy then comes from the Te $5p$ states. In Fig. 5.3b we show the band structure corresponding to the first-order approximant of calaverite (Fig. 5.2b), plotted along the symmetry directions of the irreducible wedge indicated in Fig. 5.2d. As we can see, the main effect of increasing the lattice size is a significant reduction of the Brillouin zone size (Fig. 5.2e), which naturally leads to a remarkable splitting of the energy spectrum due to the band folding of the average structure-related bands (see Section 3.1). In addition, some signatures of orbital hybridization between $5p$ Te states and $5d$ Au states can be observed at the Γ, X, and A symmetry points. By comparing the DOS corresponding to the average structure and the approximant structure in Figs. 5.3c and d, respectively, we can appreciate a well-defined pseudogap located at about 2.5 eV on the right of the Fermi level, along with a smaller depletion of the local DOS at the Fermi level position. A close inspection of this feature in both DOS plots reveals that this depletion gets more pronounced in

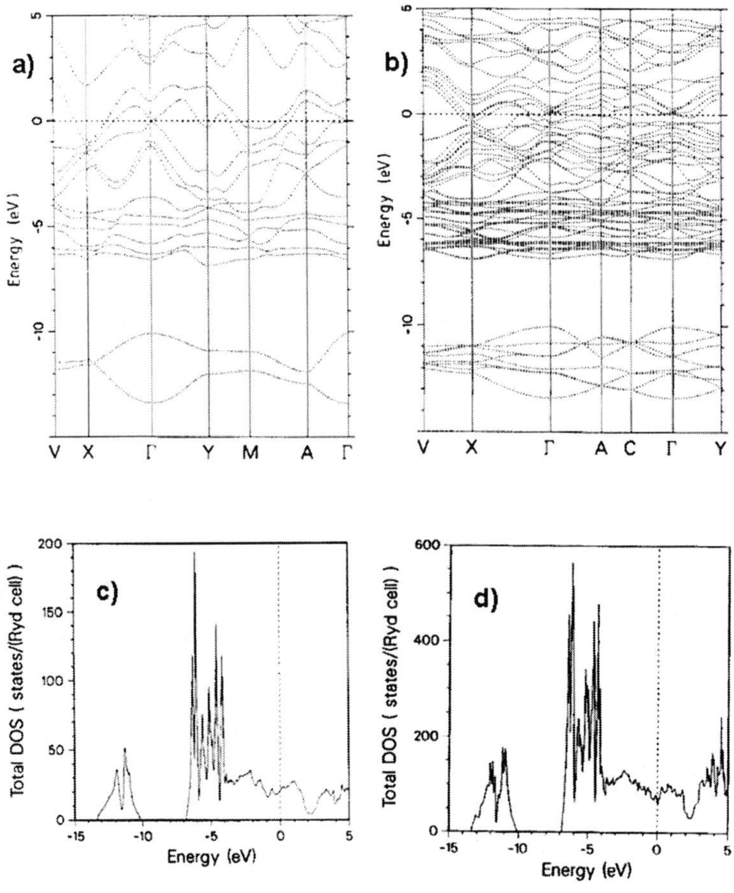

Figure 5.3 (a) Energy bands and (c) DOS of the average structure of calaverite along the symmetry directions indicated in the Brillouin zone shown in Fig. 5.2c. (b) Energy bands and (d) DOS of the approximant superstructure of calaverite along the symmetry directions indicated in the Brillouin zone shown in Fig. 5.2d. Adapted from Ref. [225], with permission from IOP Publishing Ltd.

the DOS corresponding to the approximant structure in agreement with recent specific heat measurements [270]. These features suggest that the structural modulation may play a significant role in the electronic stabilization of calaverite, which is confirmed by detailed calculations showing that the approximant structure is more stable than the average one by about 0.05 eV per atom, and the energy gain comes mostly from Te p - Au d states [225].

The transport properties of the $AuTe_2$ alloy ($T_m = 737$ K) indicate it is a poor metal with a room temperature resistivity value $\rho \simeq$ 300 μΩcm ($\sigma \simeq 3330\ \Omega^{-1}\text{cm}^{-1}$) [270], which moderately increases with temperature ($\rho \simeq 450$ μΩcm at 600 K). The room temperature Seebeck coefficient is moderately high ($S = -48\ \mu\text{VK}^{-1}$) and also slightly increases with temperature, attaining the maximum $S = -50\ \mu\text{VK}^{-1}$ at 450 K. The measured power factor exhibits its maximum value $P = 7\ \mu\text{Wcm}^{-1}\text{K}^{-2}$ at 320 K. The thermal conductivity is quite low ($\kappa = 2.6\ \text{Wm}^{-1}\text{K}^{-1}$) at room temperature and increases slowly with temperature ($\kappa = 3.6\ \text{Wm}^{-1}\text{K}^{-1}$ at 600 K), being dominated by the electronic contribution in the entire temperature range (**Exercise 5.1**). As we will see through the chapter, the lattice contribution to the thermal conductivity is very low in aperiodic crystals, due to their unusually complex atomic structure. Finally, the maximum value of the FOM is $ZT \simeq 0.09$ at around 450 K [228]. This low value is mainly due to the large room temperature carrier concentration of this material ($n \simeq 2.5 \times 10^{21}$ cm^{-3}), and one may reasonably expect a significant improvement of the TE performance by properly tuning the carrier concentration (**Exercise 5.2**). Indeed, Fig. 5.3d strongly suggests that shifting the Fermi level position toward the pseudogap located on the right position would be highly beneficial to this end.

5.1.2 *Incommensurate Structures*

In the recent years calcium cobaltite $Ca_3Co_4O_{9-\delta}$ compounds have been reported to show a moderately large Seebeck coefficient ($S = +140\ \mu\text{VK}^{-1}$) and considered to be promising TEM candidates, albeit their high electrical resistivity (typically $\rho \sim 0.01$ Ωcm), which is one order of magnitude higher than that of $(Bi,Sb)_2Te_3$ compounds (see Section 3.3), hence leading to power factor values of about 2 μWcm^{-1}K^{-2}.

The $Ca_3Co_4O_9$ compound is an oxide composite consisting of two basic building blocks: a CoO_2 layered subsystem (with a pseudo-hexagonal lattice) and a three-layered Ca_2CoO_2 subsystem (cubic NaCl structure), alternately stacking along the c axis. Detailed

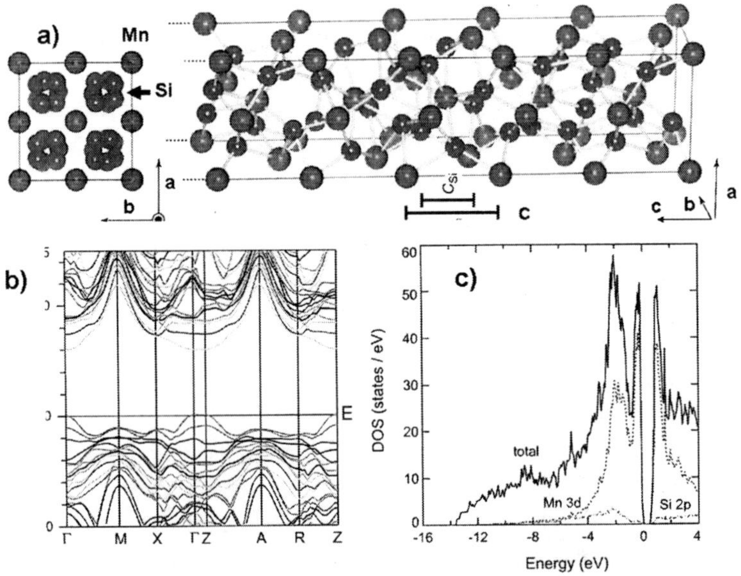

Figure 5.4 (a) The unit cell structure of the modulate composite structure $MnSi_\gamma$ at room temperature. The figure on the left is a c axis projection to illustrate the rotational arrangement of the Si atoms forming the chimney sublattice. The picture on the right depicts the atomic arrangement within a $5 \times c_{Mn}$ length in order to illustrate the Mn superstructure. The directions of the primitive lattice vectors **a**, **b**, and **c** are indicated. (b) The band structure and (c) the total DOS along with the contribution of Mn-3d and Si-2p bands of $MnSi_\gamma$ (with $\gamma = 1.7361$) periodic approximation ($a = 5.5271$ Å, $c = 4c_{Mn} = 17.4672$, $\alpha = \beta = \gamma = 90^\circ$) to the incommensurate [229]. Reprinted with permission from Y. Miyazaki, D. Igarashi, K. Hayashi, T. Kajitani, and K. Yubuta, *Phys. Rev. B* **78**, 214104 (2008), Copyright 2008, American Physical Society.

crystal structure studies, combining X-ray diffraction, electron diffraction, and high-resolution electron microscopy techniques, indicated that the composition $Ca_3Co_4O_9$ is better expressed in terms of the general formula $[Ca_2CoO_3][CoO_2]_{b_1/b_2}$, where b_1 refers to the periodicity along the b axis of the Ca_2CoO_3 subsystem and b_2 refers to the periodicity along the b axis of the CoO_2 subsystem. The misfit layer condition is then characterized by the ratio $b_1/b_2 \simeq 1.625 = 13/8$. Quite interestingly this ratio corresponds to a rational approximation to the golden mean irrational number

$(1 + \sqrt{5})/2$, expressed by means of two successive terms of the so-called Fibonacci series.[a] Three different forms of misfitted $[Ca_2CoO_3][CoO_2]_{b_1/b_2}$ compounds have been identified, all of them sharing similar a, b_1, and b_2 lattice values, but differing in the c parameters, which can take on the values $c_0 = 10.76$ Å (monoclinic), $c = 2c_0$ (orthorhombic), and $c = 3c_0$ (monoclinic). It is worth noticing that a *simultaneous* increase of electrical conductivity and Seebeck coefficient can be achieved above 500 K in these materials by partial substitution of cobalt atoms with Fe, Ga, or rare-earth (Pr, Dy, Er, Ho, Lu) elements (**Exercise 5.3**).

Another instance of incommensurate structure of TE interest is provided by manganese silicides of the form $MnSi_{2-x}$, which exhibit relatively good TE performance with a power factor $P = 6$ μWcm^{-1}K^{-2} at room temperature, and $ZT \simeq 0.7$ at 800 K. Due to the high natural abundance of the constituent elements Mn and Si, along with its high-temperature oxidation resistance and chemical stability, these compounds have recently received some attention as potential p-type TEMs.

Several structural formulas, for example, Mn_4Si_7, $Mn_{11}Si_{19}$, $Mn_{15}Si_{26}$, $Mn_{27}Si_{47}$, were originally proposed for these compounds, all of them belonging to the so-called Nowotny chimney–ladder phases, which consists of a "chimney"-[Si] subsystem and a "ladder"-[Mn] subsystem (Fig. 5.4a). Electron diffraction studies have revealed that both of these subsystems have a common tetragonal

---

[a] In Western civilization Fibonacci numbers were introduced by Leonardo Pisano (ca. 1170–1240) in his celebrated book *Liber Abaci* (1201) [19], although the sequence $F_n = \{1, 1, 2, 3, 5, 8, 13, 21, \ldots\}$ was not referred to as the Fibonacci series up to the nineteenth century (1877) by the French mathematician Edouard Lucas (1842–1891). The terms in this sequence are obtained from the recursive equation $F_n = F_{n-1} + F_{n-2}$, starting with $F_0 = 1$ and $F_1 = 1$. Thus, the sequence is perfectly ordered, but the rule used to generate it has nothing to do with periodicity. On the contrary, it is closely related to certain inflation properties connecting these numbers with the golden mean according to the asymptotic limit

$$\lim_{n\to\infty} \frac{F_n}{F_{n-1}} = \frac{1+\sqrt{5}}{2}, \tag{5.3}$$

which succinctly accounts for the property (discovered in 1611 by Johannes Kepler) that the ratio of two successive larger and larger Fibonacci numbers comes closer and closer to the golden mean. However, the process of convergence is very slow, since the golden mean is the "most irrational" among the irrational numbers.

$a = 5.5271$ Å axis, but the c axis lengths of the subsystems are usually different, and they are generally incommensurate to each other. Thus, introducing the incommensurability ratio $\gamma \equiv c_{\mathrm{Mn}}/c_{\mathrm{Si}}$, the compounds can be expressed as MnSi_γ with $\gamma \simeq 1.7361$, $c_{\mathrm{Mn}} = 4.3668$ Å and $c_{\mathrm{Si}} = 2.5153$ Å [229].

The calculated band structure along several high symmetry directions of the Brillouin zone for the approximant phase is shown in Fig. 5.4b. The MnSi_γ compound is a semiconductor with a gap $E_g = 0.6$ eV. Nevertheless, it is difficult to determine whether the band gap is direct or indirect because of the extremely flat bands along the $\Gamma - Z$ direction. Thus, the indirect gap between the valence band maximum at Γ point and the conduction band minimum at the Z point is $E_g = 0.595$ eV, while the direct band gap values at the Γ and Z points are $E_g = 0.596$ eV and $E_g = 0.597$ eV, respectively. In Fig. 5.4c we show the energy dependence of the DOS for an approximant phase. The sharp peak ranging from -0.7 to the Fermi energy is dominated by an Mn-3d character. We note that the presence of such a narrow peak in the DOS close to the E_F is a very favorable feature in order to obtain large Seebeck coefficient values.

5.1.3 *Quasicrystals*

Quasicrystals are metallic alloys representative of a novel condensed matter phase discovered by Shechtman in 1982,[a] which can be regarded as a natural extension of the notion of a crystal to structures with quasiperiodic, rather than periodic, long-range order. As a consequence, quasicrystals show an essentially discrete diffraction pattern (typical of well ordered systems), although they exhibit unusual arrangements of the diffraction spots, which can be accounted for in terms of (1) the existence of pentagonal, octagonal, decagonal, or dodecagonal rotational symmetry axes and (2) the presence of an inflation symmetry which gives rise to characteristic self-similar features (Fig. 5.5a and 5.6a) [21, 24].

---

[a] Daniel Shechtman was awarded the Nobel prize in Chemistry for this discovery in 2011.

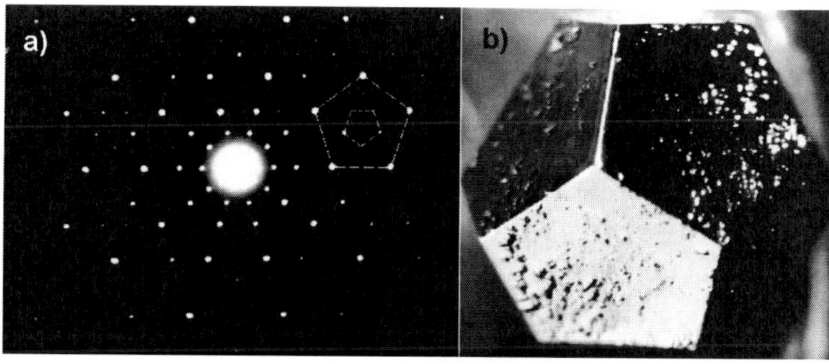

Figure 5.5 (a) Electron diffraction pattern corresponding to an AlCuFe icosahedral quasicrystal (Courtesy of J. Reyes-Gasga). A 5/10-fold symmetry axis around the origin can be clearly appreciated. If you connect all the vertices of the regular pentagons appearing in the pattern by diagonals you obtain the so-called Pythagorean pentagram shown on the upper right corner. At their intersecting points the diagonals form a smaller pentagon at the center, and the diagonals of this pentagon will form a new pentagram enclosing a yet smaller pentagon. In the ideal case, this progression can be continued ad infinitum, creating smaller and smaller pentagons and pentagrams in an endless succession exhibiting the self-similar nesting characteristic of fractal structures. (b) Due to their characteristic inflation symmetry the atomic quasiperiodic order of atoms in quasicrystals manifest itself at all scales, from microscopic to macroscopic size, as it is nicely illustrated by this picture of a HoMgZn icosahedral quasicrystal exhibiting a pentagonal dodecahedral growth habit at a millimeter scale [230].

Two main types of quasicrystals have been reported to date.[a] In the so-called decagonal quasicrystals a series of quasiperiodically arranged layers of atoms periodically stack along the main growth direction. Accordingly, decagonal quasicrystals exhibit both kinds of order, namely, periodic and quasiperiodic in just the same piece of matter. Due to this fact, these alloys exhibit highly anisotropic effects in most of their physical properties, which will be discussed in Section 5.2. On the other hand, in the so-called icosahedral quasicrystals the atoms are quasiperiodically arranged through the three directions of space. Two main growth habits have been reported for icosahedral quasicrystals, namely, pentagonal

[a]A few representatives belonging to the dodecagonal and octagonal classes have been communicated so far.

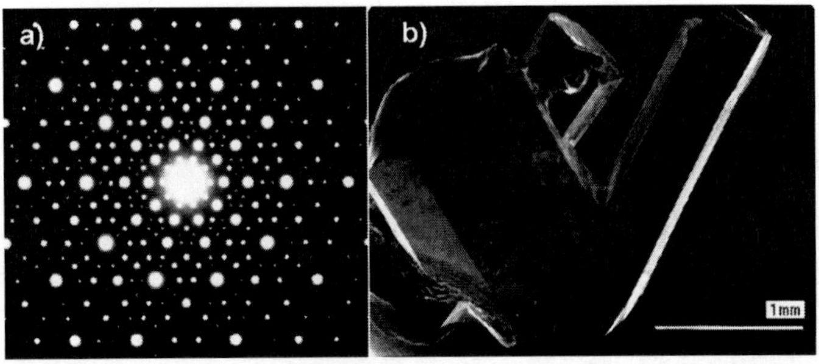

Figure 5.6 (a) Electron diffraction pattern of an $Al_{70}Ni_{15}Co_{15}$ decagonal quasicrystal showing a 10-fold symmetry axis around the origin. A progression of self-similar nested pentagons up to the third generation of the Pythagorean pentagram can be clearly seen [231]. (b) Electron scanning microscopy picture of an AlNiCo decagonal quasicrystal exhibiting a decaprismatic growth habit at a millimeter scale [230]. Reprinted with permission from I. R. Fischer et al., *Phil. Mag. B* **77**, 1601 (1998), Taylor & Francis.

dodecahedral (Fig. 5.5b) and triacontahedral, whereas decagonal quasicrystals mainly grow as elongated decaprisms (Fig. 5.6b).

In every quasicrystal-forming alloy system the quasicrystal phase is accompanied by compositionally related classical periodic crystals (having big unit cell sizes), often forming micro-twinned networks with aperiodic symmetries. These periodic crystals not only have very similar compositions, but also structures closely resembling that of the quasicrystalline counterparts, from which they can nevertheless be distinguished. Within the superspace formalism these periodic structures can be derived by projection from a parent hyperlattice in 5D or 6D onto the 3D physical space, as it occurs in the case of quasicrystals. However, whereas for quasicrystals the projection involves irrational numbers, in the case of these closely related phases the projection is expressed in terms of rational approximants to the former irrational quantities. For this reason such complex periodic crystals are called *approximant* phases.

Quasicrystals also exhibit unusual physical properties [18]. In fact, shortly after the discovery of thermodynamically stable quasicrystals of high structural quality in the AlCu(Fe,Ru,Os), AlPd(Mn,Re), ZnMg(RE), and Cd(Yb,Ca) icosahedral systems, it was progressively realized that these materials occupy an odd position

Table 5.1 Comparison between the physical properties of icosahedral quasicrystalline alloys versus typical metallic materials. I (S) stands for ionic (semiconducting) materials typical properties

Property	Metals	Quasicrystals
Mechanical	Ductility, malleability	Brittle (I)
Tribological	Moderately soft	Very hard
	Easy corrosion	Corrosion resistant
Electrical	High conductivity	Low conductivity (S)
	Resistivity increases with T	Resistivity decreases with T (S)
	Small thermpower	Large thermpower (S)
Magnetic	Paramagnetic	Diamagnetic
Thermal	High conductivity	Very low conductivity (I)
	Large specific heat	Small specific heat
Optical	Drude peak	No Drude peak
	No IR absorption	IR absorption (S)

among the well-ordered condensed matter phases. Indeed, since quasicrystals consist of metallic elements one would expect they should behave as metals. Nonetheless, it is now well established that transport properties of stable quasicrystals are quite unusual by the standard of common metallic alloys, as most of their transport properties resemble a more semiconductor-like than metallic character [21, 24].

For the sake of comparison in Table 5.1 we list a number of characteristic physical properties of both metals and icosahedral quasicrystals. By inspecting this table one realizes that quasicrystalline alloys significantly depart from metallic behavior, resembling either ionic or semiconducting materials. Thus, high-quality icosahedral quasicrystals provide an intriguing example of solids made of typical metallic atoms which do not exhibit most of the physical properties usually signaling the presence of metallic bonding.

5.1.4 *Complex Metallic Alloys*

Along with quasicrystals and their approximant phases the broad family of the so-called *complex metallic alloys* include a number

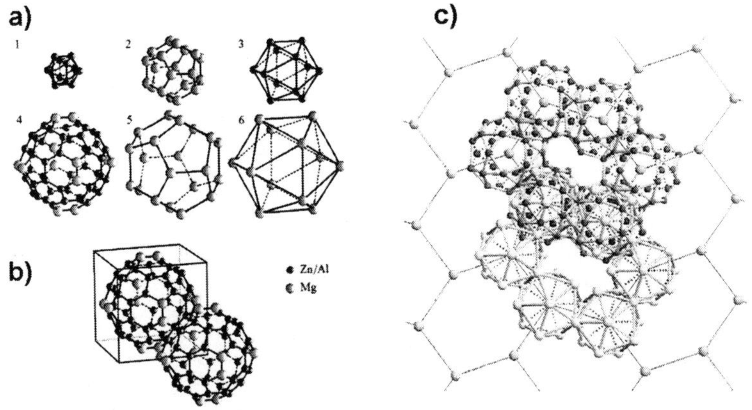

Figure 5.7 (a) Successive atomic shells of the Bergman cluster. (b) Body-centered packing of Bergman clusters (sharing a hexagonal face of the fourth shell) produce the structure of the $(Al,Zn)_{49}Mn_{32}$ Bergman phase. (c) A view of the ξ'-AlPdMn skeleton structure along the [010] direction. Mn atoms form a planar flattened-hexagon lattice and are located in the centers of pseudo-Mackay icosahedra. The two interpenetrating polyhedra (a 12-atom Pd icosahedron, white atoms) and a 30-atom Al icosidodecahedron (black atoms) that form the outer shell of the pseudo-Mackay cluster are shown (Courtesy of Janez Dolinšek).

of giant-unit-cell intermetallics, exhibiting very complex structures that contain some hundred up to several thousand atoms in the unit cell. Illustrative examples are the $Mg_{32}(Al,Zn)_{49}$ Bergman compound (Fig. 5.7a,b), with 162 atoms in the unit cell, the ξ'-$Al_{74}Pd_{22}Mn_4$ phase (Fig. 5.7c) with 258 atoms in the unit cell, λ-Al_4Mn (586 atoms in the unit cell), β-Al_3Mg_2 (1168 atoms in the unit cell), and the heavy-fermion compound $YbCu_{4.5}$, comprising as many as 7448 atoms in the cell [26, 27]. These giant unit cells contrast with elementary metals and simple intermetallics whose unit cells in general comprise from single up to a few tens atoms only. One should note, however, that these giant unit cells, with lattice parameters of several nanometers, are piled up periodically in this case. Accordingly, the translational periodicity of classical crystal lattices is recovered on the scale of many interatomic distances, whereas, on the atomic scale, the atoms are arranged in clusters with polyhedral order, where icosahedrally coordinated environments play a prominent role.

The structures of complex metallic alloys thus show duality; on the scale of several nanometers, these alloys are periodic crystals, whereas on the atomic scale, they resemble cluster aggregates. Accordingly, the notion of complex metallic alloy encompass different kinds of structurally complex materials sharing a basic property: a full description of their atomic arrangement requires the consideration of more than one spatial scale. The high structural complexity of complex metallic alloys together with the presence of two competing physical length scales—one defined by the unit-cell parameters and the other by the cluster substructure—may have a significant impact on the physical properties of these materials, such as the electronic structure and lattice dynamics. On this basis, complex metallic materials are expected to exhibit novel transport properties of interest in the field of TEs, like a combination of a relatively large Seebeck coefficient with low thermal conductivity, and electrical and thermal resistances tunable by varying the composition.

In order to gain some understanding about the role played by this multiscale feature on the physical properties of complex metallic alloys it is convenient to broadly classify them attending to two main criteria: the nature of the long-range order present in the sample and the number and local distribution of atoms in the unit cell. The extreme case corresponds to quasicrystals, whose ideal three-dimensional structure exhibits a self-similar distribution of icosahedral clusters at all scales, naturally leading to an effective unit cell of infinite size. In the case of decagonal quasicrystals, two kinds of long-range order simultaneously coexist in the same sample, namely periodic order along one direction and quasiperiodic order in the planes perpendicular to the previous one. The next step corresponds to the approximant phases exhibiting a well-defined, big unit cell which periodically repeats through the three-directions of space, though the local distribution of atoms inside this unit cell is completely isomorphous to that corresponding to closely related quasicrystals in the phase diagram. Finally, we have complex metallic alloys with giant unit cells which are neither related to quasicrystals nor to their approximant structures. According to

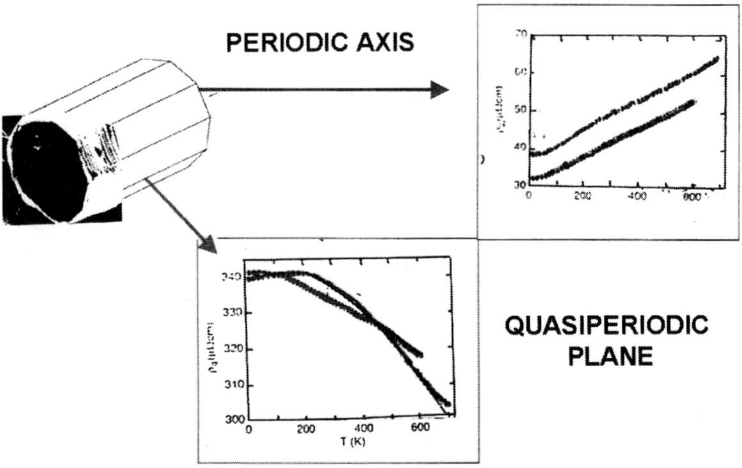

Figure 5.8 Temperature dependence of the electrical resistivity of decagonal quasicrystals of compositions $Al_{65}Cu_{20}Co_{15}$ and $Al_{70}Ni_{15}Co_{15}$ measured along the periodic and quasiperiodic directions [232]. Reprinted with permission from S. Martin, A. F. Hebard, A. R. Kortan, and F. A. Thiel, *Phys. Rev. Lett.* **67**, 719 (1991), Copyright 1991, American Physical Society.

this approximate classification scheme the role played by the local symmetry of the structural clusters progressively increases from the non–quasicrystal-related complex metallic alloys to full-fledged quasicrystals.

5.2 Decagonal Quasicrystals

As we mentioned previously, decagonal quasicrystals grow adopting the form of decagonal prisms. Cross sections of this prism show ten faceted planes containing quasiperiodically arranged clusters of atoms. These quasiperiodic planes, in turn, are periodically stacked along the prism axis, so that both kinds of order, periodic and quasiperiodic, coexist in the same sample. Accordingly, it is reasonable to expect that the interplay of periodic an aperiodic orders may bring some specific features to these materials.

Indeed, soon after the first thermodynamically stable decagonal quasicrystals were available, a remarkable anisotropy in the electrical resistivity was observed for AlCuCo, AlCuCoSi, and AlNiCo samples. This transport anisotropy is illustrated in Fig. 5.8 where we compare the temperature dependence of the electrical resistivity measured along the periodic axis and along a perpendicular direction contained in the quasiperiodic plane. One can clearly appreciate that (i) the resistivity increases (decreases) with the temperature when it is measured along the periodic (quasiperiodic) directions, respectively and (ii) the electrical resistivity in the quasicrystalline plane is about an order of magnitude higher than that along the periodic direction at low temperatures. At the same time, it must be noted that the resistivity of decagonal quasicrystals, being significantly high for alloys composed of good metallic elements, is intermediate between that of typical metals and that reported for icosahedral samples (see Fig. 5.11a).

Similar anisotropic features have been reported for other transport properties, including Hall and Seebeck coefficients as well as thermal conductivity [237]. The Seebeck coefficient measurements show small values (comparable to those observed in usual metallic systems, see Table 5.2), which linearly increase

Table 5.2 Room temperature values of the transport coefficients and FOM for decagonal quasicrystals belonging to different families

Compound	σ (Ω^{-1}cm^{-1})	S (μVK^{-1})	P (μWcm^{-1}K^{-2})	κ (Wm^{-1}K^{-1})	ZT (10^{-4})	Ref.
$Al_{70}Ni_{15}Co_{15}$						
Periodic	37037	+3.5	0.45	15.0	9	[232, 234]
Quasiperiodic	6098	−2.5	0.04	2.5	5	[232, 234]
$Al_{69.7}Ni_{20.3}Co_{10}$						
Periodic	25000	+1.9	0.09	22.0	1	[235]
Quasiperiodic	3300	+4.0	0.05	5	3	[240]
$Al_{65}Cu_{20}Co_{15}$						
Periodic	25641	+4.5	0.52			[233]
Quasiperiodic	2469	−2.0	0.01			[233]
$Al_{64}Cu_{20}Co_{15}Si_1$						
Periodic	20000	+7.5	1.12	4.0	84	[233, 236]
Quasiperiodic	2330	−4.5	0.05	1.8	8	[233, 236]

with the temperature, in agreement with Mott's relation given by Eq. (2.51). We also note that whereas in the periodic direction the Seebeck coefficient is positive, in the quasiperiodic direction it becomes negative for most measured samples. Taken together, the behavior of electrical conductivity and Seebeck coefficient suggests a more metallic-like than semiconductor-like character for the charge carriers mobility in these materials. On the other hand, the thermal conductivity takes on significantly low values, specially along the quasiperiodic direction, which progressively increase with temperature up to the samples melting point (870 K) in both the periodic and quasiperiodic directions.

In principle, decagonal quasicrystals may incorporate beneficial effects for TE performance related to the presence of two kinds of atomic arrangements in the same sample which will allow to exploit their low thermal conductivity values along quasiperiodic planes and the high electrical conductivity along the periodic stacking direction. Unfortunately, the Seebeck coefficient values are typically metallic in both possible directions, leading to very low FOM values (see Table 5.2).

It is interesting to note that the Hall coefficient depends only weakly on the temperature, and it slightly decreases with increasing the temperature in both the 10-fold axis and the quasiperiodic plane. The effective carrier density deduced from Hall measurements for decagonal quasicrystals is substantially lower than the free-electron estimations derived from low temperature-specific heat measurements (**Exercise 5.4**). This may be indicative of a strong electron–phonon mass coupling in the quasicrystalline plane. The possible existence of a pseudogap at the Fermi level in decagonal quasicrystals is still an open question, since experimental data obtained from photoemission spectroscopy, angle-resolved photoemission spectroscopy, and x-ray photoemission spectroscopy techniques remain controversial [237]. Thus, for aluminum-based decagonal quasicrystals it was shown both experimentally and numerically that a pseudogap exists only in the partial Al $3p$ states, but the high density of transition metal $3d$ states hides it in the total DOS. Some evidence for the suppression of the DOS at about 13 meV away from the Fermi level has been reported recently from low temperature (5 K) ultrafast time-resolved optical reflectivity experiments in decagonal AlNiCo quasicrystals.

A rather poorly ordered stable dodecagonal phase was reported in 1998, with composition $Ta_{62}Te_{38}$ [238]. The presence of tellurium in this compound naturally raised some expectations regarding its possible TE performance. By all indications, these tellurides seem to have electrical properties being characteristic of small band gap semiconductors. In fact, their resistivities at room temperature are similar to those of semimetals (about 1–10 mΩ cm) and they seem to exhibit a complex electronic structure near the Fermi level, so that these tellurides may have promising TE properties. At present, however, the available information on the transport properties of this family is quite scarce, particularly regarding their thermal conductivity and Seebeck coefficient.

5.3 Icosahedral Quasicrystals

5.3.1 *Transport Properties*

As its was mentioned in Section 5.1.3, it is now well established that transport properties of thermodynamically stable quasicrystals of high structural quality are quite unusual by the standard of common metallic alloys, as most of their transport properties resemble a more semiconductor-like than metallic character (see Table 5.1). On the basis of their transport properties there are three basic reasons supporting quasicrystals and their approximants as potential TEMs [240]:

(1) While most metallic alloys are good heat conductors, the thermal conductivity of quasicrystals is about two orders of magnitude lower than that of common metals, and it seems to be mainly determined by the lattice phonons (rather than the charge carriers) over a wide temperature range (Fig. 5.9). These unusually low thermal conductivity values (within the range $\kappa \sim 1$–$5\ \mathrm{Wm^{-1}K^{-1}}$ at room temperature) are comparable to those observed for thermal insulators of extensive use in aeronautical industry, such as titanium carbides and nitrides, doped zirconia, or alumina [21].

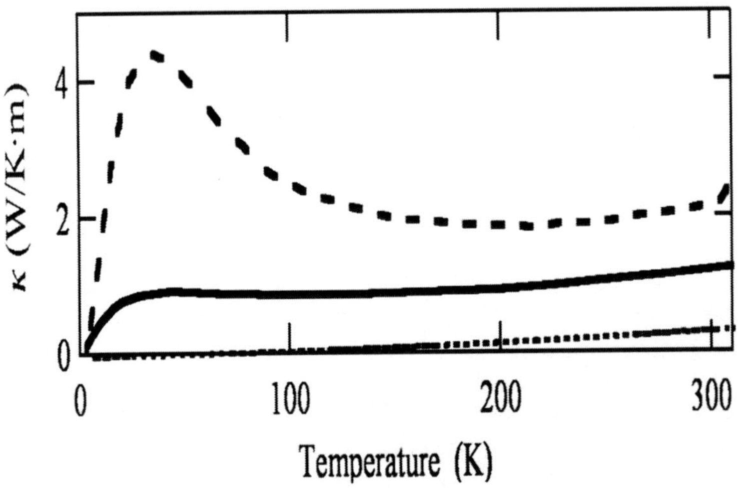

Figure 5.9 Temperature dependence of the thermal conductivity for the icosahedral $Al_{74.6}Re_{17.4}Si_8$ quasicrystal (solid line) and its 1/1-cubic approximant (dashed line). Although both compounds are supposed to have closely related structures, the quasicrystal phase exhibits a much lower thermal conductivity. The contribution due to the electrons (estimated from the Wiedemann–Franz law, dotted line) is almost negligible as compared to that due to the phonons [264] (Courtesy of Tsunehiro Takeuchi).

(2) Icosahedral quasicrystals bearing transition metals in the systems AlCu(Fe,Ru,Os) and AlPd(Mn,Re) exhibit significantly large Seebeck coefficient values (50–120 μVK^{-1}) as compared to those of typical metallic systems (1–10 μVK^{-1}) at room temperature, and the temperature dependence of the Seebeck coefficient usually deviates from the linear behavior characteristic of charge diffusion in ordinary metallic alloys (see Eq. (2.51)), exhibiting pronounced curvatures within the temperature range 100–300 K (Fig. 5.10).

(3) Although icosahedral quasicrystals exhibit unusually high electrical resistivity values (Fig. 5.11a), their electrical conductivity steadily *increases* as the temperature increases up to the melting point, as it is illustrated in Fig. 5.11b. Furthermore, in samples belonging to the icosahedral AlPdRe family, both $\sigma(T)$ and $S(T)$ increase as the temperature is increased over a broad

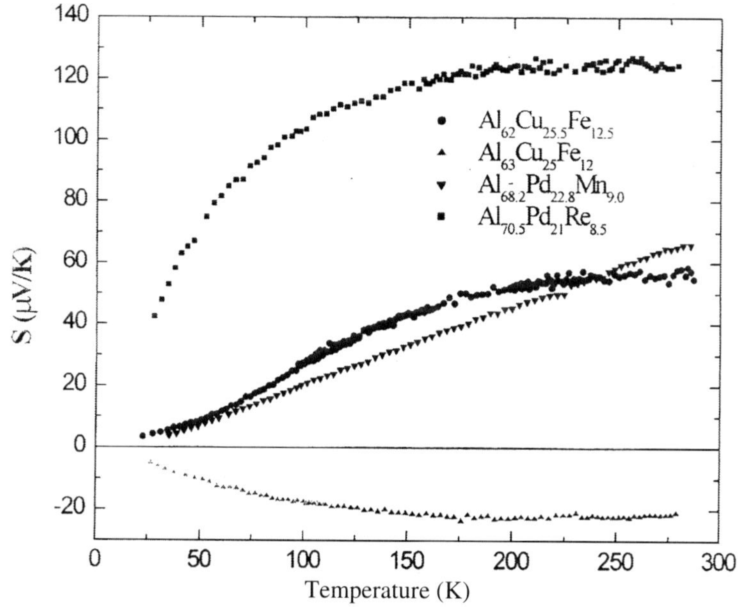

Figure 5.10 Temperature dependence of the Seebeck coefficient for different quasicrystalline samples (Courtesy of Roberto Escudero).

temperature range (~100–550 K), thus leading to a significant power factor increase [242].

These unusual electronic and thermal transport properties lead to an FOM enhancement when taken together in Eqs. (1.71) and (2.4). In addition, the transport curves are very sensitive to small variations in the chemical composition of the sample (inset, Fig. 5.11b), in close analogy to doping effects in semiconducting materials [243]. Thus, transport properties of quasicrystalline alloys exhibit unusual composition and temperature dependence, resembling more semiconductor-like than metallic character. Accordingly, quasicrystals occupy a very promising position in the quest for novel TEMs, naturally bridging the gap between semiconducting materials and metallic ones. To illustrate this point, the room temperature electrical resistivities of several classes of materials are compared in Fig. 5.12. It can be seen that quasicrystals fill the gap between metals and semiconductors, exhibiting electrical resistivity values

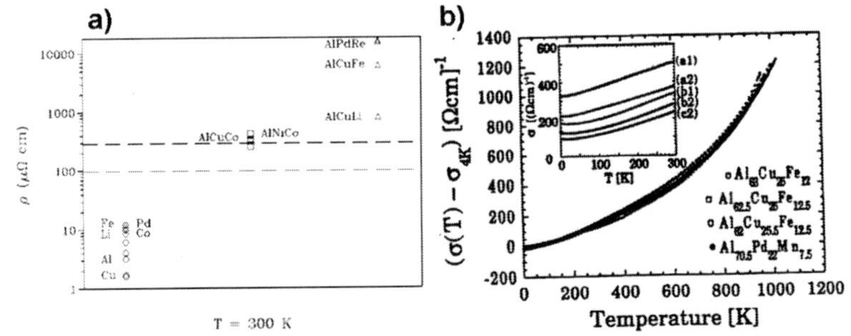

Figure 5.11 (a) The room temperature electrical resistivities of icosahedral (triangles) and decagonal (squares) quasicrystals are compared to that corresponding to their constituent elements (circles). It can be seen that the elemental solids are good metals, possessing low electrical resistivities. On the contrary, quasicrystalline alloys composed of these elements show electrical resistivities which are higher by several orders of magnitude, and much higher than typical values for conventional metallic alloys (both crystalline and amorphous), whose representative values fall in the region between the horizontal dashed lines. (b) Temperature dependence of the electrical conductivity for four different quasicrystalline samples up to 1000 K. The inset illustrates the sensitivity of the residual conductivity value to minor variations in the sample composition [241]. Reprinted with permission from D. Mayou, C. Berger, F. Cyrot-Lackmann, T. Klein, and P. Lanco, *Phys. Rev. Lett.* **70**, 3915 (1993), Copyright 1993, American Physical Society.

comparable to those reported for conducting polymers (see Section 6.2.1), metallic macrocycles, and fullerenes.

The low thermal conductivity of quasicrystals can be understood in terms of two main facts. In the first place, the charge carrier concentration close to the Fermi level is low due to the presence of a pseudogap (see Section 5.3.2), so that heat must propagate by means of atomic vibrations (phonons) and $\kappa_e \ll \kappa_l$. In the second place, as a direct consequence of the self-similar arrangement of atoms within their structure, ideal quasicrystal lattices have a fractal reciprocal space, so that the transfer of momentum to the lattice is not bounded below, leading to a significant degradation of thermal current through the sample. Thus, κ_l is greatly reduced due to a combination of a small group velocity of phonons and an enhancement of phonon–phonon scattering processes occurring

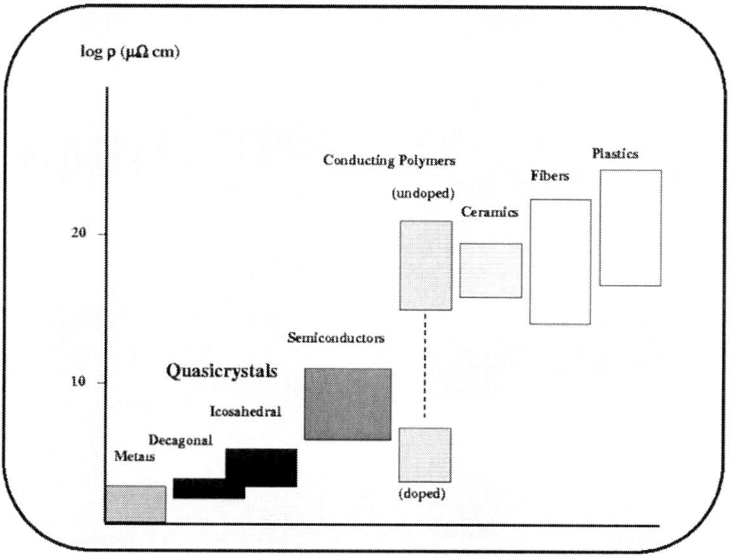

Figure 5.12 Room temperature electrical resistivity is compared for different materials of technological interest. Quasicrystals are located at the border line between metals and semiconductors [24].

at all scales in reciprocal space. In a similar way, an increased number of Umklapp electron–phonon processes in quasicrystals may enhance the phonon drag contribution to the total Seebeck coefficient as well [244].

These considerations spurred earlier experimental studies on the potential application of different quasicrystals, belonging to the icosahedral AlCuFe, AlPdMn, TiZrNi, and AlPdRe systems, as TEMs. The main goal of this systematic research was twofold: (i) to determine the most promising sample compositions among the different known quasicrystal families on the basis of their (low) thermal conductivity values, and (ii) to exploit the high sensitivity of quasicrystals transport coefficients to stoichiometric changes in order to properly enhance their power factors, without compromising their characteristic low thermal conductivity.

In Table 5.3 we list the transport coefficients for those representatives yielding the best FOM values at room temperature. By inspecting this table we see that isostructural $Al_{71}Pd_{20}Mn_9$

Table 5.3 Room temperature values of the transport coefficients and FOM for quasicrystals belonging to different families

Compound	σ (Ω^{-1}cm^{-1})	S (μVK^{-1})	P (μWcm^{-1}K^{-2})	κ (Wm^{-1}K^{-1})	ZT	Ref.
$Zn_{57}Mg_{34}Er_9$	6170	+7	0.3	4.5	0.002	[245]
$Ag_{42.5}In_{42.5}Yb_{15}$	5140	+12	0.7	4.8	0.005	[246]
$Al_{65}Cu_{20}Ru_{15}$	250	+27	0.2	1.8**	0.003	[248]
$Cd_{84}Yb_{16}$	5560	+14	1.1	9.4	0.004	[247]
$Al_{62.5}Cu_{24.5}Fe_{13}$	310	+44	0.6	1.8*	0.010	[248]
$Al_{64}Cu_{20}Ru_{15}Si_1$	390	+50	1.0	1.8**	0.020	[248]
$Al_{71}Pd_{20}Re_9$	450	+80	2.9	1.3	0.070	[249]
$Al_{71}Pd_{20}Mn_9$	714	+90	5.8	1.5	0.120	[249]

*After Ref. [61].
**Estimated upper limit.

and $Al_{71}Pd_{20}Re_9$ icosahedral samples exhibit promising ZT values, which are comparable to those reported for half-Heusler, skutterudites and clathrates in Tables 3.12, 3.13, 3.15, and 3.16, respectively. Furthermore, enhanced FOM values are obtained at higher temperatures for closely related quasicrystals (Table 5.4).

In this way, it was found that the more promising quasicrystals belong to the AlPd(Mn,Re) family. Subsequent search was then focused on refining their stoichiometries to further increase the corresponding ZT values (Table 5.5). As we can see, P and ZT values differing by more than two orders of magnitude can be attained in a single quasicrystal system by slightly changing the sample's composition by a few atomic percent (hence preserving the quasiperiodic lattice structure). Very low values of the thermal

Table 5.4 High temperature figure of merit for quasicrystals belonging to the AlPd(Mn,Re) icosahedral family

Compound	T^* (K)	ZT^*	Ref.
$Al_{71}Pd_{20}Re_9$	570	0.15	[249]
$Al_{71}Pd_{20}(Re_{0.45}Ru_{0.55})_9$	700	0.15	[250]
$Al_{71}Pd_{20}(Re_{0.35}Fe_{0.65})_9$	500	0.21	[251]
$Al_{70.8}Pd_{20.9}Mn_{8.3}$	550	0.23	[242]
$Al_{68}Ga_3Pd_{20}Mn_9$	473	0.26	[249]

Note: T^* denotes the temperature maximizing the figure of merit.

Table 5.5 Room temperature values of the transport coefficients and FOM for several AlPdRe icosahedral quasicrystals [252]

Compound	σ (Ω^{-1}cm^{-1})	S (μVK^{-1})	P (μWcm^{-1}K^{-2})	κ (Wm^{-1}K^{-1})	ZT
$Al_{69.4}Pd_{21.2}Re_{9.4}$	95	−7	0.005	1.20	0.0001
$Al_{68.5}Pd_{22.9}Re_{8.6}$	110	−10	0.01	1.16	0.0003
$Al_{67.7}Pd_{23.2}Re_{9.1}$	90	+55	0.3	0.86	0.01
$Al_{67.8}Pd_{22.2}Re_{10.0}$	180	+95	1.6	0.76	0.06

conductivity can be appreciated. We also note that both positive and negative values of the Seebeck coefficient can be obtained in this way, which allows for both the n- and p-type legs present in a typical TE cell to be fabricated from the *same* material.

In summary, it is currently agreed that, although some minor improvement may be obtained by a judicious choice of both sample composition (i.e., via a fourth element alloying) and processing conditions via trial-and-error procedures, in order to attain substantial improvement (if any) it seems convenient to adopt a more direct approach based on an electronic structure engineering strategy. To this end, a deeper understanding about the electronic structure of quasicrystals and the physical basis of their unusual transport properties becomes necessary.

5.3.2 *Electronic Structure*

It was pointed out by Hume-Rothery in 1926 that certain metallic compounds with closely related structures but apparently unrelated stoichiometries exhibit the same ratio of number of valence atoms to number of atoms (the so-called e/a ratio).[a] Hume-Rothery's rule is explained as resulting from a perturbation of the energy of the valence electrons by their diffraction by the crystal lattice. The perturbation is of such nature as to stabilize electrons with energy just equal or less than that corresponding to Bragg reflection and to destabilize electrons with a larger energy. Hence, special stability

[a]A detailed and updated account on the physical basis and the role played by Hume-Rothery mechanism in the electronic stabilization of different alloys can be found in ref. [25].

would be expected for alloys with just the right number of electrons. This number is proportional to the volume of a polyhedron in reciprocal space (the so-called Brillouin–Jones zone), corresponding to the crystallographic planes giving rise to the perturbation.

Although quasicrystals have a dense reciprocal space, only a few diffraction peaks have very strong intensities. The Hume-Rothery criterion can then be applied to quasicrystals by introducing a pseudo-Brillouin zone defined by the most intense diffraction spots. Due to the great symmetry of quasicrystals this zone is quite close to spherical shape, so that the diffraction condition can be expressed in the form $G_{hkl} = 2k_F$, where G_{hkl} is the reciprocal vector of the considered diffraction plane, $k_F = \sqrt[3]{3\pi^2 n}$ is the radius of the Fermi sphere, and n is the electron number per unit volume. This approach has been successfully used to explain the stability of quasicrystals containing elements with a full d-band, like $Zn_{43}Mg_{37}Ga_{20}$ (e/a = 2.2), $Zn_{60}Mg_{30}(RE)_{10}$ (e/a = 2.1) or $Zn_{80}Sc_{15}Mg_5$ (e/a = 2.15), by adopting the valence values Mg = +2, Zn = +2, Al = +3, Sc = +3, Ga = +3, and RE = +3 (**Exercise 5.5**).

In all these samples the redistribution of electronic states due to the Fermi sphere–pseudo Brillouin zone interaction gives rise to a significant reduction of the density of states (pseudogap) close to the Fermi energy. In addition, in the Hume-Rothery class of materials a characteristic narrow peak is present in the DOS close to the Fermi level due to the diffraction of higher energy electrons by the crystal atomic planes. As we commented on in Section 2.4.1, the presence of narrow features in the DOS generally favors the existence of large Seebeck coefficient values in the transport curves and this effect is further enhanced by the presence of a significant reduction of the DOS at the Fermi level in these compounds.

In summary, two main features can be observed in the DOS close to the Fermi energy in high-quality, thermodynamically stable icosahedral quasicrystals containing transition metal atoms: a structurally induced broad minimum (~1 eV width) due to the Fermi surface–Brillouin zone mechanism and a narrow and sharply confined dip (~0.1 eV width) due to hybridization effects involving the transition metal bands. The physical existence of this electronic pseudogap has received strong experimental support during the last two decades, as indicated by measurements of

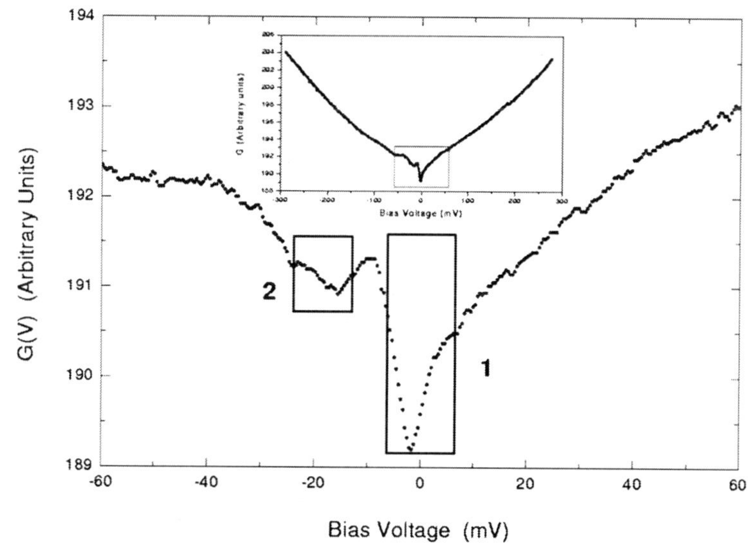

Figure 5.13 Differential conductance for an $Al_{63}Cu_{25}Fe_{12}$–Al tunnel junction at a temperature of $T = 2$ K for two different energy scales: ± 60 meV (main frame) and ± 300 meV (inset). Data file courtesy of Roberto Escudero [253]. Reprinted from Ref. [254], with permission from E. Maciá, *Phys. Rev. B* **69**, 132201 (2004), Copyright 2004, American Physical Society).

the specific heat capacity, photoemission, soft X-ray spectroscopy, magnetic susceptibility, nuclear magnetic resonance, and scanning tunneling spectroscopy probes [21]. In Fig. 5.13 we show low temperature tunneling spectroscopy measurements corresponding to the sample $Al_{63}Cu_{25}Fe_{12}$. These measurements reveal a broad pseudogap extending over an energy scale of about 0.6 eV (shown in the inset) along with some fine structure close to the Fermi level (labeled 1 and 2 in the main frame). The broad pseudogap stems from the Fermi surface–Brillouin zone interaction, while the dips may be respectively related to hybridization effects between d-Fe states and sp-states.

In order to make a meaningful comparison with experimental transport measurements one should take into account possible finite lifetime and temperature broadening effects. In so doing, it is observed that most finer details in the DOS are significantly smeared out and only the most conspicuous peaks remain in the vicinity of the Fermi level at room temperature [255, 256]. These considerations convey one to reduce the number of main spectral features necessary to capture the most relevant physics of transport processes. Accordingly, a realistic model for the electronic structure of transition metal–bearing, Al-based icosahedral quasicrystals and related phases is given by the spectral conductivity function [257],

$$\sigma(E) = \bar{\sigma} \left\{ \frac{\gamma_1}{(E - \delta_1)^2 + \gamma_1^2} + \frac{\alpha \gamma_2}{(E - \delta_2)^2 + \gamma_2^2} \right\}^{-1}, \tag{5.4}$$

characterized by the presence of two dip features modeled in terms of two Lorentzian functions. The narrowest feature (describing the *sp-d* hybridization effects) is superimposed onto a broader asymmetric pseudogap (which accounts for the Fermi surface–Brillouin zone interaction). The model includes six parameters, determining the Lorentzian's heights ($\bar{\sigma}/\gamma_i$) and widths ($\sim\gamma_i$), their positions with respect to the Fermi level, δ_i, and their relative weight in the electronic structure, $\alpha > 0$. The parameter $\bar{\sigma}$ is a scale factor measured in $(\Omega\text{cm eV})^{-1}$ units. Suitable values for the electronic model parameters $\{\gamma_i, \delta_i, \alpha\}$ can be obtained by properly combining *ab initio* calculations with experimental transport data within a systematic phenomenological approach [24]. In Fig. 5.14 the overall behavior of the $\sigma(E)$ curve is shown for a suitable choice of the model parameters. By comparing this figure with Fig. 5.13 we see that Eq. (5.4) properly captures the main spectral features of realistic samples.

5.3.3 *Band Structure Effects*

Making use of the spectral conductivity function given by Eq. (5.4) into Eqs. (2.12)–(2.14), and assuming that κ_l is negligible as compared to κ_e, one obtains [240],

$$ZT = \frac{4b\xi_1^2 T^2}{1 + \frac{420}{169} b^2 \xi_1^4 T^4}, \tag{5.5}$$

where $b \equiv e^2 L_0 = 2.44 \times 10^{-8}\ (\text{eV})^2\text{K}^{-2}$, and the parameter

$$\xi_1 = \frac{1}{2} \left(\frac{d \ln \sigma(E)}{dE} \right)_{E_F}, \tag{5.6}$$

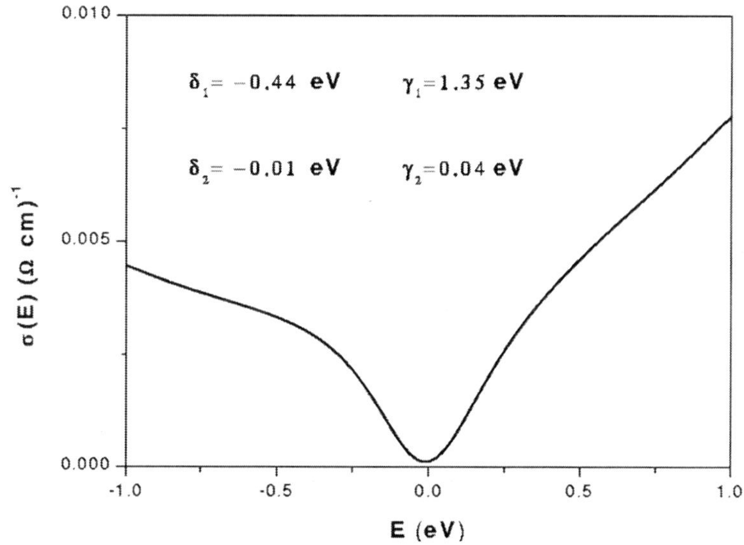

Figure 5.14 Spectral conductivity curve in the energy interval ± 1 eV around the Fermi level as obtained from Eq. (5.4) for the electronic model parameter values γ_i and δ_i indicated in the frame.

(measured in $(\text{eV})^{-1}$ units) can be experimentally determined from the Seebeck coefficient slope at low temperature after the expression [259],

$$S(T) \simeq -2|e|^{-1} b\xi_1 T. \tag{5.7}$$

According to Eq. (5.5) the FOM is an even function of ξ_1, vanishing when $\xi_1 = 0$, and attaining two well-defined peaks at both sides of the minimum, whose location depends on the temperature (Fig. 5.15 inset, **Exercise 5.6**). Quite interestingly, the overall behavior of the $ZT(\xi_1)$ curve shown in the inset of Fig. 5.15 closely correlates the composition dependence of the room temperature power factor reported for several AlPdRe samples (Fig. 5.15 main frame). Since small changes in the chemical composition of icosahedral samples give rise to a relative shift of their corresponding Fermi level positions, the reported dependence of TE power factor with e/a is clearly signaling the significant role of electronic structure effects in their TE response. In fact, it has been shown that the variation of the $S(e/a)$ curve arises from a combined effect involving the Fermi

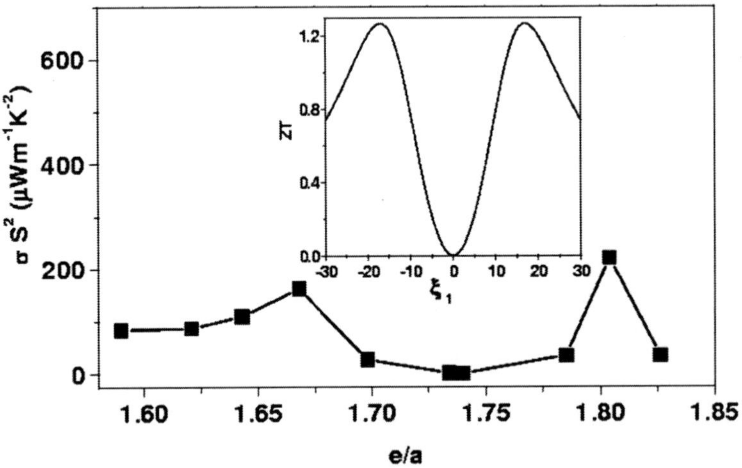

Figure 5.15 (Main frame) Composition dependence of the room temperature power factor of several icosahedral AlPdRe samples. (inset) Dependence of the FOM on the ξ_1 parameter value as determined from Eq. (5.5) evaluated at room temperature [258]. Reprinted with permission from E. Maciá, *Phys. Rev. B* **69**, 184202 (2004), Copyright 2004, American Physical Society.

level shift and the deepening of the DOS pseudogap due to covalent bond formation in certain AlPdRe samples [252]. These features can be understood in terms of the location of the Fermi level in the sample's electronic structure, as it is illustrated in Fig. 5.16. In fact, since ξ_1 depends on the slope of the $\sigma(E)$ curve, when the Fermi level is located close to the pseudogap minimum we get $\xi_1 \simeq 0$, thereby $ZT = 0$ after Eq. (5.5). On the other hand, when the Fermi level shift to the left (right) from the pseudogap's minimum, we get progressively larger ZT values as the $\sigma(E)$ slope at the Fermi level steepens (see Fig. 5.16). Therefore, the deeper the pseudogap the larger the resulting FOM at a given temperature .

The previous results have been obtained assuming that the ratio k_l/k_e is negligible in Eq. (2.41) in the considered temperature range. As soon as the contribution $k_l/k_e \neq 0$, the symmetry $ZT(-\xi_1) = ZT(\xi_1)$ is lost. For the sake of illustration, in Fig. 5.17 we plot the $ZT(\xi_1)$ curve at different temperatures for a suitable choice of the model parameters. As we can see, the inclusion of a non-

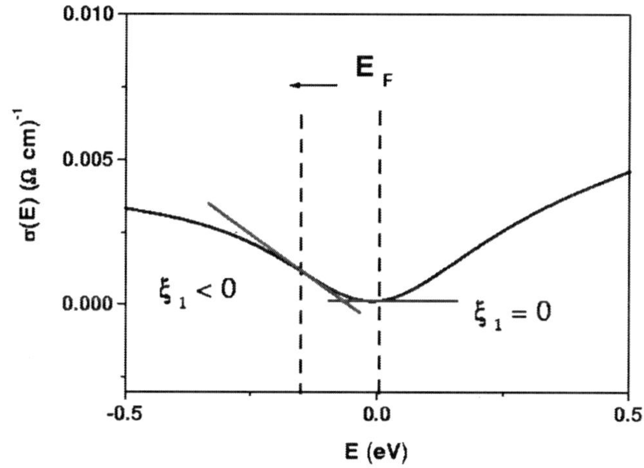

Figure 5.16 Sketch illustrating the relationship between the slope of the DOS close to the Fermi level and the ZT curve main features.

vanishing thermal conductivity lattice contribution results in a clearly asymmetric $ZT(\xi_1)$ curve, which exhibits a $ZT = 0$ minimum flanked by two maxima at about $\xi_1 \simeq -25$ (eV)$^{-1}$ and $\xi_1 \simeq +40$ (eV)$^{-1}$ at room temperature. Note that the minimum of the $ZT(\xi_1)$ curve does not coincide with $\xi_1 = 0$ in this case. As the Fermi level progressively shifts from the pseudogap's minimum, the ZT values progressively increase attaining well-defined maxima, as it occurred in the symmetric case ($k_l/k_e = 0$), but now these maxima reaches different peak values depending on the sign of ξ_1. Accordingly, we conclude that best TE performances will be expected for those stoichiometries able to locate the Fermi level *below* the minimum of the pseudogap ($S > 0$, p-type). On the other hand, we observe that by increasing the temperature the $ZT(\xi_1)$ maxima below (above) the Fermi level progressively increase (decrease) and shifts toward (away from) the value $\xi_1 = 0$. This behavior clearly indicates that, in order to obtain an optimal TE performance, the sample stoichiometry must be finely tuned, attending to the working temperature of interest. To this end, the control of the parameter ξ_1 value (due to a systematic change in the sample stoichiometry, for instance) will play a very important role in electronic structure engineering approaches.

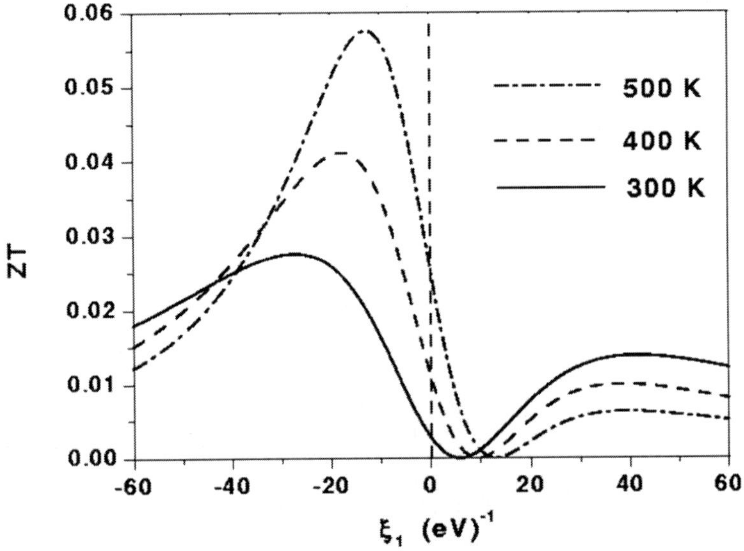

Figure 5.17 Dependence of the thermoelectric figure of merit as a function of the phenomenological coefficient ξ_1 at $T = 300$ K (solid line); $T = 400$ K (dashed line), and $T = 500$ K (dot-dashed line) for an icosahedral AlPdRe sample with $\sigma = 30$ $(\Omega\text{cm})^{-1}$ and $\kappa_l = 1.1$ $\text{Wm}^{-1}\text{K}^{-1}$ [258]. Reprinted with permission from E. Maciá, *Phys. Rev. B* **69**, 184202 (2004), Copyright 2004, American Physical Society.

To illustrate this point we shall consider the systematic variation of the $S(T)$ curve for a series of AlCuFe quasicrystals whose e/a ratio is modified by means of a controlled stoichiometric change. The influence of stoichiometric changes in the electronic structure is shown in Fig. 5.18a, where a progressive shift of the dip position in the spectral conductivity curve $\sigma(E)$ can be clearly appreciated. In Fig. 5.18b we show the temperature dependence of the Seebeck coefficient corresponding to the considered icosahedral AlCuFe samples in the temperature range 1–500 K. The overall behavior of the different $S(T)$ curves compares fairly well with the experimental measurements reported in the literature for these samples. Quite interestingly, minor changes in the Fermi level position give rise to significant differences in the TE curves which can exhibit relatively large positive ($Al_{62.5}Cu_{24.5}Fe_{13}$ and $Al_{62}Cu_{25.5}Fe_{12.5}$) or negative

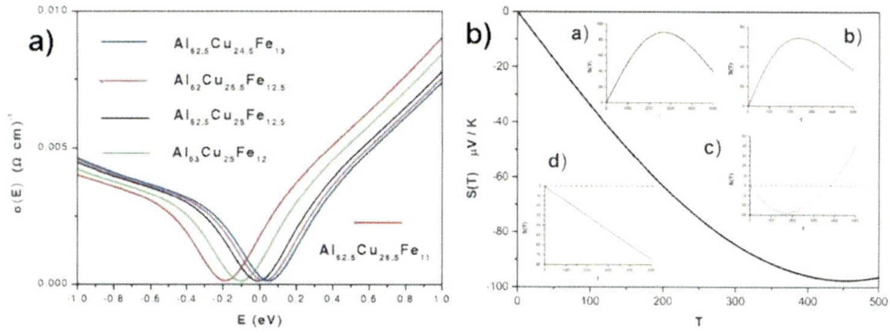

Figure 5.18 (a) Systematic variation of the spectral conductivity around the Fermi level as a function of the sample stoichiometry. (b) Temperature dependence of the Seebeck coefficient in the temperature range 1–500 K for different sample compositions: (a) $Al_{62.5}Cu_{24.5}Fe_{13}$; (b) $Al_{62}Cu_{25.5}Fe_{12.5}$; (c) $Al_{62.5}Cu_{25}Fe_{12.5}$; (d) $Al_{62.5}Cu_{26.5}Fe_{11}$; and $Al_{63}Cu_{25}Fe_{12}$ (main frame) [259]. Reprinted with permission from E. Maciá, *J. Appl. Phys.* **93**, 1014 (2003), Copyright 2003, American Institute of Physics.

($Al_{62.5}Cu_{26.5}Fe_{11}$ and $Al_{63}Cu_{25}Fe_{12}$) values or can even reverse their sign ($Al_{62.5}Cu_{25}Fe_{12.5}$). Note that the samples exhibiting larger $|S|$ values have larger ξ_1 values as well.

In Fig. 5.19a the spectral conductivity functions corresponding to different classes of aperiodic crystals are compared in an energy window close to the Fermi level. The $\sigma(E)$ curve of the quasicrystal

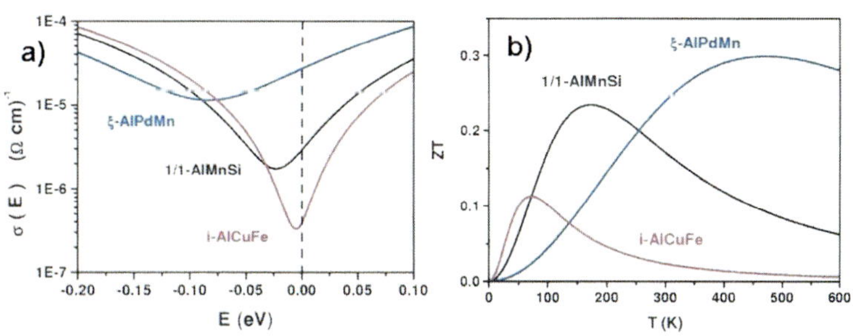

Figure 5.19 (a) Spectral conductivity function curves corresponding to suitable electronic model parameters. The dashed vertical line indicates the Fermi level position. (b) temperature dependence of the thermoelectric figure of merit for the icosahedral $Al_{63}Cu_{25}Fe_{12}$, the approximant phase 1/1-$Al_{73.6}Mn_{17.4}Si_9$, and the complex unit cell alloy ξ-$Al_{73}Pd_{22.9}Mn_{4.1}$ [260]. Reprinted with permission from E. Maciá, *Phys. Rev. B* **80**, 205103 (2009), Copyright 2009, American Physical Society.

sample is characterized by a pronounced pseudogap with steep wings, with the Fermi level located very close to the dip. Both the depth and the width of the pseudogap progressively reduce as the atomic volume of the considered alloy decreases (66. 5 at nm^{-3} for the ξ-$Al_{73}Pd_{22.9}Mn_{4.1}$ compound (see Fig. 5.7), 69.0 at nm^{-3} for the approximant phase 1/1-$Al_{73.6}Mn_{17.4}Si_9$) and the location of the pseudogap minimum progressively shifts below the Fermi level as well. The overall structure of the DOS close to the Fermi level is directly related to the TE performance of the corresponding sample, as it is illustrated in the ZT curves shown in Fig. 5.19b. In fact, since ξ_1 depends on the slope of the DOS close to the Fermi level, the steeper the slope of the $\sigma(E)$ curve at E_F, the higher the Seebeck coefficient (and consequently the ZT) curve at low temperatures. Accordingly, the quasicrystal sample exhibits a substantially larger FOM value than the other related phases in the low temperature regime. This trend is reversed at higher temperatures, so that the best TE performance is now exhibited by the ξ-$Al_{73}Pd_{22.9}Mn_{4.1}$compound.

As we saw in Section 1.4, segmented devices where different materials are joined together to improve the TE performance have been considered in some applications. In this case we must take into account the compatibility factor given by the Eq. (1.89), since materials with dissimilar s values cannot be efficiently combined in a single device. Therefore, both ZT and s must be simultaneously optimized. The TE compatibility of several materials of current technological interest has been recently reviewed, concluding that a semimetallic material with high p-type Seebeck coefficient is required for the development of segmented generators. In Table 5.6 we list the FOM and compatibility factors of different quasicrystals at room temperature as derived from data reported in the literature making use of Eq. (1.89). We observe that the largest s values are comparable to those observed in traditional TEMs, like Bi_2Te_3 or SiGe ($s \simeq 1$ V^{-1}) and PbTe ($s \simeq 1.2$ V^{-1}) [54]. The most promising quasicrystals belong to the icosahedral AlPdMn family, which exhibit an s factor larger than those corresponding to other icosahedral phases, approaching the figure reported for the TAGS compounds (see Section 3.4) at $T = 550$ K ($s \simeq 2.7$ V^{-1}) [54]. On the other hand, the s factor corresponding to AlPdRe samples is larger at

Table 5.6 Room temperature Seebeck coefficient, figure of merit, and compatibility factors for samples belonging to different quasicrystalline families

Compound	S (μVK^{-1})	ZT	s (V^{-1})	Ref.
$Al_{62.5}Cu_{24.5}Fe_{13}$	+44	0.01	0.38	[248]
$Al_{64}Cu_{20}Ru_{15}Si_1$	+50	0.02	0.66	[248]
$Cd_{84}Yb_{16}$	+16	0.01	1.04	[247]
$Al_{67.8}Pd_{22.2}Re_{10}$	+95	0.07	1.21	[252]
$Al_{71}Pd_{20}Mn_9$	+90	0.12	2.16	[249]

room temperature (where it exhibits a lower ZT value) than it is at higher temperatures (albeit it exhibits a larger ZT value). This result highlights the importance of properly balancing the ZT and s contributions in designing optimized devices.

In Fig. 5.20 we compare the room temperature compatibility factors of AlPdRe and AlPdMn icosahedral quasicrystals as a function of ξ_1. The difference between the AlPdRe and AlPdMn $s(\xi_1)$ curves is due to the significant difference between their respective residual conductivities. In the inset of Fig. 5.20 we plot the corresponding $ZT(\xi_1)$ curves, which exhibit a deep minimum, flanked by two maxima. Both ZT and s vanish at $\xi_1^0 = +5.76$ (eV)$^{-1}$. Consequently, these quasicrystals can exhibit p-type ($\xi_1 < \xi_1^0$) or n-type ($\xi_1 > \xi_1^0$) Seebeck coefficients depending on the ξ_1 value. Since the ξ_1 value can be controlled by changing the sample stoichiometry, hence shifting E_F in a scale of a few meV (see Fig. 5.18), we can confidently expect that large values of the room temperature compatibility factor, close to $s = 2.0$ V^{-1}, may be attained in AlPdMn quasicrystals with $\xi_1 \simeq -15$ (eV)$^{-1}$.

In summary, in this section we have reviewed the potential use of quasicrystals and related phases as TEMs within the context of the band structure engineering approach. As we have seen, these materials satisfy the ZT optimization requirements presented in Section 2.4.2 in a natural way, since their electronic structure exhibit a pronounced pseudogap at the Fermi level along with some fine spectral features near the Fermi level. In addition, one can efficiently exploit the high sensitivity of their transport coefficients to minor stoichiometric changes in

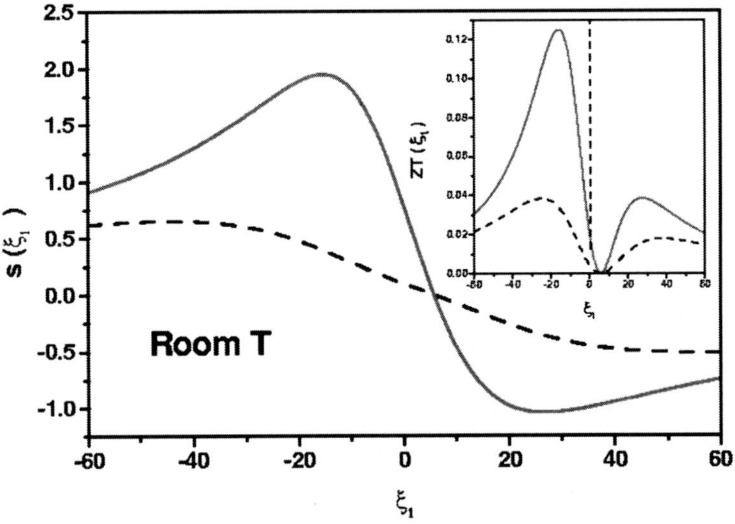

Figure 5.20 Room temperature dependence of the compatibility factor (main frame) and the thermoelectric figure of merit (inset) as a function of the phenomenological coefficient ξ_1 for AlPdRe ($\kappa = 0.7$ Wm^{-1}K^{-1} [252], $\sigma_0 = 30$ (Ωcm)$^{-1}$, dashed line), and AlPdMn ($\kappa = 1.6$ Wm^{-1}K^{-1}, $\sigma_0 = 740$ (Ωcm)$^{-1}$ [261], solid line) quasicrystals [262]. Reprinted with permission from E. Maciá, *Phys. Rev. B* **70**, 100201 (2004), Copyright 2004, American Physical Society.

order to properly enhance their power factors without sacrificing their characteristic low thermal conductivity. The most promising quasicrystals to date belong to the icosahedral AlPd(Mn,Re) family, exhibiting $ZT \simeq 0.1$ at room temperature and $ZT \simeq 0.25$ in the temperature range 475–550 K.[a] Further experimental and theoretical work is then appealing in order to fully exploit the unusual transport properties of quasicrystalline alloys in TE devices. In this regard, it is convenient to note that the monotonous increase

[a]It should be pointed out that approximant phases also exhibit very low thermal conductivities (similar to those observed in quasicrystals), along with *larger* electrical conductivity values. However, a detailed comparison between icosahedral and 1/1-AlPdMnSi and 2/1-AlPdMnSi approximant phases clearly indicates that the approximants have lower ZT values, since their Seebeck coefficient takes on typically metallic, low values [249, 263]. Accordingly, the following FOM values have been reported for their cubic approximant phases 1/1-$Al_{71.6}Re_{17.4}Si_{11}$ ($ZT = 0.10$, n-type) and 1/1-$Al_{75.6}Mn_{17.4}Si_7$ ($ZT = 0.07$, p-type) at room temperatures [264].

of κ_e with the temperature,[a] is the main restriction preventing quasicrystals to attain FOM values comparable to those reported for skutterudites or half-Heusler alloys at high temperatures. In fact, according to experimentally reported high-temperature values corresponding to the icosahedral samples $Al_{71}Pd_{20}Mn_9$ ($ZT \simeq 0.02$ at $T = 973$ K) [249], and $Al_{71}Pd_{20}Re_9$ ($ZT \simeq 0.05$ at $T = 950$ K) [251]. Therefore, quasicrystals are not suitable for high-temperature applications. Accordingly, any effort aimed at reducing the electronic contribution to the thermal conductivity must be regarded as a very promising guiding principle in order to obtain larger ZT values in quasicrystals.

5.4 Exercises

(1) Determine the lattice contribution to the thermal conductivity of calaverite at (a) room temperature and (b) $T = 600$ K.

(2) Making use of Eqs. (4.8) and (4.9) determine (a) the relaxation time of electrons in calaverite at room temperature and (b) the variation of its Seebeck coefficient if the carrier concentration is reduced by an order of magnitude.

(3) (a) The room temperature electrical resistivity and Seebeck coefficient values of $Ca_3Co_4O_{9+\delta}$ and $Ca_3Co_{3.9}Mn_{0.1}O_{9+\delta}$ cobaltites are $\rho = 0.014$ (0.016) Ωcm, and $S = +160$ (+210) μVK^{-1}, respectively [29]. Determine the relative increase of the power factor value due to the Co substitution by Mn.

(b) The electrical resistivity and Seebeck coefficient values of $Ca_3Co_{3.95}Ga_{0.05}O_{9+\delta}$ cobaltites at $T = 500$ K and $T = 1200$ K are $\rho = 10.5$ (7.5) mΩcm and $S = +152$ (+205) μVK^{-1}, respectively [29]. Determine the power factor value at each temperature.

(4) The mean value of the Hall coefficient for a number of decagonal quasicrystals is about $R_H \simeq 3 \times 10^{-4}$ cm^{3}C^{-1} (with the magnetic field perpendicular to the periodic stacking direction) and the volume of their quasi-Brillouin zone is $V \simeq 17.7$ Å^{-3} [267].

[a] According to several measurements $\kappa_e(T) \sim T^3$ above 200 K for icosahedral quasicrystals [265, 266].

Estimate: (a) the charge carriers density and (b) the carriers mobility. Comment on the obtained results.

(5) The $Al_{30}Mg_{40}Zn_{30}$ is a quasicrystal approximant phase belonging to the cubic space group Im$\bar{3}$ with a lattice constant $a = 14.355$ Å which contains $n_U = 160$ atoms in its unit cell. X-ray diffraction measurements indicate that the most intense peaks, defining the Brillouin zone, correspond to the planes (5,5,0), (5,4,3), and (7,1,0). Show that the Hume-Rothery matching rule is satisfied by this compound.

(6) Determine the value of the Seebeck coefficient leading to the maximum ZT value in Eq. (5.5).

5.5 Solutions

(1) According to the Wiedemann–Franz law we can express the lattice contribution to the thermal conductivity as $\kappa_l = \kappa - L_0 \sigma T$, so that making use of the transport data given in the text we get

$$\kappa_l = 2.6\,\mathrm{Wm^{-1}K^{-1}} - 2.44 \times 10^{-8}\mathrm{V^2K^{-2}}\,2860 \\ \times 10^2\,\Omega^{-1}\mathrm{m^{-1}}\,300\,\mathrm{K} = 0.51\,\mathrm{Wm^{-1}K^{-1}}$$

at room temperature, and

$$\kappa_l = 3.6\,\mathrm{Wm^{-1}K^{-1}} - 2.44 \times 10^{-8}\mathrm{V^2K^{-2}}\,2222 \\ \times 10^2\,\Omega^{-1}\mathrm{m^{-1}}\,600\,\mathrm{K} = 0.35\,\mathrm{Wm^{-1}K^{-1}}$$

at $T = 600$ K. Consequently, calaverite exhibits extremely low values for the lattice contribution to the thermal conductivity, which are comparable to those reported for polymers and amorphous materials (see Section 6.2.1).

(2) (a) Multiplying Eqs. (4.8) and (4.9) one can express the relaxation time in the form

$$\tau = -\left(\frac{3^{2/3}\pi^{4/3}\hbar^2}{2|e|b}\right)\frac{\sigma S}{T g_v^{2/3} n^{1/3}}$$

where $b \equiv e^2 L_0 \simeq 2.44 \times 10^{-8}\,(\mathrm{eV})^2\mathrm{K}^{-2}$ and $g_v = 1$ in this case. Plugging numerical values and making use of the transport data given in the text we have

$$\tau = +4.7854\frac{2860\ \Omega^{-1}\text{cm}^{-1}48\times10^{-6}\ \text{VK}^{-1}1.054^2\times10^{-68}\ \text{J}^2\text{s}^2}{300\ \text{K}\ \sqrt[3]{2.5}\times10^7\ \text{cm}^{-1}2.44\times10^{-8}\times1.602^2\times10^{-38}\ \text{J}^2\text{K}^{-2}|e|}$$

$$= \frac{2.8624\times10^{-33}\ \Omega^{-1}\text{Vs}^2}{|e|} = \frac{2.8624\times10^{-33}\ \text{Js}^{-1}\text{s}^2}{1.602\times10^{-19}\ \text{J}} \simeq 1.8\times10^{-14}\ \text{s}$$

where we have used the dimensional relation $[\text{W}] = \Omega^{-1}\text{V}^2$. This relaxation time figure compares well with the value reported for conduction electrons in pure aluminum metal [10].

(b) After Eq. (4.8) we get

$$\frac{S'}{S} = \left(\frac{n}{n'}\right)^{2/3} = \left(\frac{2.5\times10^{21}\ \text{cm}^{-3}}{2.5\times10^{20}\ \text{cm}^{-3}}\right)^{2/3} = 10^{2/3}$$

where the magnitudes S' and n' correspond to the sample with a reduced charge carrier concentration. Therefore, $S' = 10^{2/3}S = 48\times10^{2/3} \simeq 222.8\ \mu\text{VK}^{-1}$, and the Seebeck coefficient is significantly increased when the charge carrier concentration is reduced.

(3) (a) The relative power factor is given by

$$\frac{\Delta P}{P} = \frac{\sigma'}{\sigma}\left(\frac{S'}{S}\right)^2 - 1 = \frac{\rho}{\rho'}\left(\frac{S'}{S}\right)^2 - 1 = \frac{14}{16}\left(\frac{210}{160}\right)^2 - 1 \simeq 0.507$$

Accordingly, the power factor is increased by about 50% upon Co substitution by Mn.

(b) The power factor is given by $P = \sigma S^2$ so that,

$$P(500\ \text{K}) = \frac{10^3}{10.5}\ \Omega^{-1}\text{cm}^{-1}\ 152^2\times10^{-6}\ \mu\text{V}^2\text{K}^{-2} = 2.2\ \mu\text{Wcm}^{-1}\text{K}^{-2}$$

$$P(1200\ \text{K}) = \frac{10^3}{7.5}\ \Omega^{-1}\text{cm}^{-1}\ 205^2\times10^{-6}\ \mu\text{V}^2\text{K}^{-2} = 5.6\ \mu\text{Wcm}^{-1}\text{K}^{-2}.$$

Therefore, although the power factor is multiplied by a 2.5 factor in the high-temperature regime, this figure still falls one order of magnitude short from that reported for SiGe alloys used in current TEGs (30–60 $\mu\text{Wcm}^{-1}\text{K}^{-2}$, see Section 3.3).

(4) (a) The Hall coefficient directly measures the charge carrier density according to the expression $R_H = (en)^{-1}$, so that

$$n = \frac{1}{eR_H} = \frac{1}{1.602 \times 10^{-19}\,\mathrm{C}\,3 \times 10^{-4}\mathrm{cm}^3\mathrm{C}^{-1}}$$
$$= 2.08 \times 10^{22}\ \mathrm{cm}^{-3}.$$

On the other hand, if one assumes that the Fermi surface completely fills the available Brillouin zone volume one would have

$$V = \frac{4}{3}\pi k_F^3 = \frac{4}{3}\pi(3\pi^2 n),$$

and then

$$n = \frac{V}{4\pi^3} = \frac{17.7\ \mathrm{Å}^{-3} \times 10^{24}\ \mathrm{Å}^3\mathrm{cm}^{-3}}{4\pi^3} = 1.43 \times 10^{23}\ \mathrm{cm}^{-3}.$$

As we see, the charge carrier density values obtained from both procedures differs by a factor of about seven, and it is reasonable to conclude that the carriers concentration is not enough to allow the Fermi sphere to completely fill the Brillouin zone in this case.

(b) From the knowledge of the charge carriers density their mobility can be derived from the expression

$$\mu = \sigma R_H = \frac{\sigma}{en} = \frac{4 \times 10^3\ \Omega^{-1}\mathrm{cm}^{-1}}{1.602 \times 10^{-19}\,\mathrm{C}\,2 \times 10^{22}\ \mathrm{cm}^{-3}}$$
$$\simeq 1.25\ \mathrm{cm}^2\mathrm{V}^{-1}\mathrm{s}^{-1}$$

where the electrical conductivity value along the quasiperiodic plane has been estimated by averaging the values listed in Table 5.2, and we have made use of the dimensional relationship [V] = [ΩA] = [ΩCs^{-1}]. Thus, we conclude that the carriers mobility is significantly low in the quasiperiodic plane of decagonal quasicrystals.

(5) The Hume-Rothery matching rule is given by the relationship $2k_F = |G_{hkl}|$. For a cubic lattice we have

$$G_{hkl} = \frac{2\pi}{a}\sqrt{h^2 + k^2 + l^2}, \qquad (5.8)$$

where a is the lattice constant and $\{h, k, l\}$ denote the Miller indices. On the other hand, the radius of the Fermi sphere is given by

$$k_F = \frac{1}{a}\sqrt[3]{3\pi^2 (e/a)\, n_U}, \tag{5.9}$$

where e/a is the electron per atom ratio. Making use of Eqs. (5.8) and (5.9) the Hume-Rothery matching rule can be expressed in the form

$$h^2 + k^2 + l^2 = \left(\frac{3}{\pi}(e/a)\, n_U\right)^{2/3}$$

Adopting the valences Al $= +3$, Zn $= +2$, and Mg $= +2$ we have $e/a = 0.3 \times 3 + (0.4 + 0.3) \times 2 = 2.3$ electron per atom in the unit cell, so that

$$\left(\frac{3}{\pi}(e/a)\, n_U\right)^{2/3} = \left(\frac{3}{\pi} \times 2.3 \times 160\right)^{2/3} \simeq 49.798$$

As we see, this figure compares well with the value $h^2+k^2+l^2 = 50$ for the planes (5,5,0), (5,4,3), and (7,1,0), thereby indicating the existence of a strong overlapping between the Fermi sphere and the Brillouin zone.

(6) By imposing the extremum condition $\frac{d(ZT)}{d\xi_1} = 0$ to Eq. (5.5) one obtains $\xi_1 = 0$ (which corresponds to the minimum $ZT = 0$) and

$$\xi_1^* = \pm\frac{1}{T}\sqrt[4]{\frac{169}{420b^2}}.$$

Making use of this value of ξ_1 in Eq. (5.7) one gets

$$S^*(T) = -2|e|^{-1} b\xi_1^* T = \pm 2|e|^{-1}\sqrt[4]{\frac{169}{420}}\sqrt{b}$$

$$= \pm 2|e|^{-1}\sqrt[4]{\frac{169}{420}}\sqrt{2.44 \times 10^{-4}\,|e|\mathrm{VK}^{-1}} = \pm 248.8\ \mu\mathrm{VK}^{-1}$$

We note that this optimal Seebeck coefficient value is independent of the working temperature.

Chapter 6

Organic Semiconductors and Polymers

The inorganic TEMs we have considered in the previous chapters are hindered by issues like high cost of production, scarcity of materials, and toxicity. Because of these problems associated with inorganic compounds, organic-electronic materials have spurred a growing interest in TE community. Organic compounds provide noteworthy advantages over inorganic materials in the fact they have low-cost synthesis and solution processability over large areas, providing great material flexibility to meet the requirements of the targeted applications. Thus, the applications foreseen for large-area flexible thermogenerators and Peltier coolers are not affordable with inorganic materials, as they are both brittle and expensive.

According to Eq. (1.30), the lattice contribution to the thermal conductivity can be reduced by reducing the mean free path and the group velocity of the lattice phonons as well as by decreasing the specific heat of the solid. In the previous chapters we have focused on materials exhibiting both short-mean free paths and group velocities. In this chapter we will consider the case of materials characterized by low thermal conductivities arising from a substantial depletion of charge carriers (so that $\kappa_e \ll \kappa_l$) and a small specific heat value. In Section 6.1 we briefly describe some organic semiconducting materials. In Section 6.2.1 we will consider

Thermoelectric Materials: Advances and Applications
Enrique Maciá-Barber

ISBN 978-981-4463-52-2 (Hardcover), 978-981-4463-53-9 (eBook)
www.panstanford.com

conducting polymers, which have attracted a considerable attention since the initial discovery of doped polyacetylene in the late 1970s. Deoxyribonucleic acid (DNA) is one of the most promising nanowire materials due to the relative ease of modifications combined with the self-assembly capability which makes it possible to construct a great variety of DNA-based nanostructures. Accordingly, Section 6.2.2 will be devoted to describe the electronic structures and transport properties of this macromolecule, with a view on its possible use in TE devices. Finally, in Section 6.3 we will review some recent advances in the study of thermoelectric effects at the nanoscale.

6.1 Organic Semiconductors

Broadly speaking, organic compounds have low thermal conductivities, which according to Eq. (1.71) is an advantageous feature for application in TE devices. A drawback, however, is the poor charge transport contribution in organic materials compared with that in alloys and covalent inorganic materials. For the sake of illustration, let us consider the TE transport properties of pentacene and rubrene crystals, which are prototypical organic semiconductors (Fig. 6.1).[a] Electrical conductivity values ranging from $\sigma = 0.025$ to $1.5 \times 10^{-4}\ \Omega^{-1}\text{cm}^{-1}$ have been reported for pentacene films at temperatures between 200 and 295 K [272]. The observed Seebeck coefficients do not change significantly with the temperature and their values fall into the range 300–1000 μVK^{-1} (p-type) [268]. Finally, their room temperature thermal conductivity is about $\kappa =$ 0.4–0.5 $\text{Wm}^{-1}\text{K}^{-1}$ [271]. The significantly large Seebeck coefficient along with their low thermal conductivity values clearly indicate that these compounds may be potentially good TEMs. Nevertheless, due to their low electrical conductivity, one gets significantly low power factors, and FOM values comprised within 10^{-5} and 10^{-4} ultimately result (**Exercise 6.1**).

Enhancement of electrical conduction is therefore indispensable for these organic materials to be utilized as TE elements. Since pentacene and rubrene are intrinsic semiconductors with relatively

[a] Pentacene ($C_{22}H_{14}$) consists of five benzene rings fused forming a linear chain. Rubrene ($C_{42}H_{28}$) consists of a tetracene backbone with four phenyl side groups.

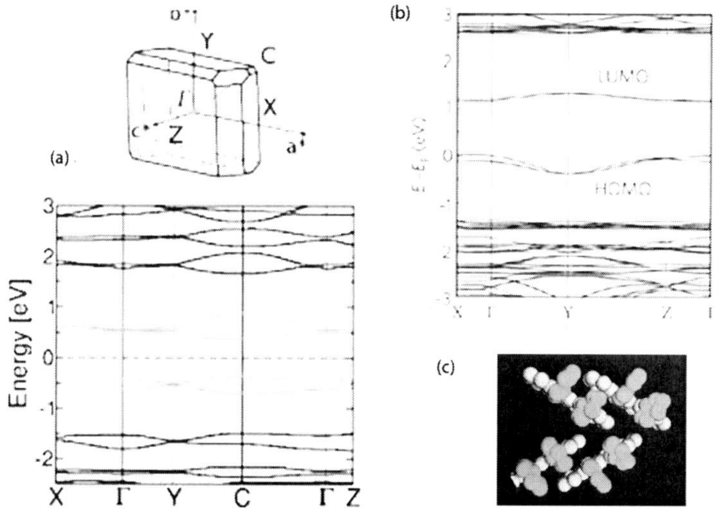

Figure 6.1 (a) Calculated band structure for the bulk pentacene. The high-symmetry points in units (2π/a,2π/b,2π/c) are Γ = (0, 0, 0), X = (0.5, 0, 0), Y = (0, 0.5, 0), C = (0.5, 0.5, 0) and Z = (0, 0, 0.5). The Fermi level is indicated by a dashed line. On top, the corresponding Brillouin zone is presented. [Reprinted with permission from C. Ambrosch-Draxl, D. Nabok, P. Puschnig and C. Meisenbichler, The role of polymorphism in organic thin films: oligoacenes investigated from first principles, *New J. Phys.* **11** (2009) 125010, doi:10.1088/1367-2630/11/12/125010]. (b) The band structure of a rubrene single crystal ; orthorombic; space group *Cmca.* [Reprinted with permission from G. Szulczewski, S. Sanvito and M. Coey, A spin of their own, *Nature Materials* **8**, 693–695 (2009) doi:10.1038/nmat2518]. (c) Packing of rubrene molecules (carbon atoms in the tertacene backbone are coloured yellow for clarity). Courtesy of O. Mitrofanov, UCL.

wide gaps ($E_g \simeq$ 0.8 eV and $E_g \simeq$ 2 eV, respectively), charge carriers must be externally provided by means of doping for their use at room temperature. Thus, several orders of magnitude-enhanced power factor values have been reported for iodine-doped pentacene thin films ($P = 0.13$ μWcm^{-1}K^{-2} at room temperature) [269]. This figure improves the values obtained by using the p-type doping molecular compound 2,3,5,6-tetrafluoro-7,7,8,8-tetracyanoquinodimethane (F_4-TCNQ), which is a strong acceptor dopant (**Exercise 6.2**).

Pentacene crystallizes in a triclinic structure, with two inequivalent molecules in a herringbone arrangement in the unit cell.

However, the growth parameters strongly affect the structural properties, giving rise to at least four polymorphs, which principally differ from each other in the tilting of the molecules with respect to the substrate. In turn, the different molecular stacking in different polymorphs can remarkably change the electronic and transport properties. The most relevant structure for thin-film electronic device applications is the triclinic system with lattice parameters $a = 5.958$ Å, $b = 7.596$ Å, $c = 15.61$ Å, $\alpha = 81.25°$, $\beta = 86.56°$, and $\gamma = 89.90°$. Rubrene crystallizes in the orthorhombic structure with lattice parameters $a = 26.86$ Å, $b = 7.19$ Å, and $c = 14.43$ Å ($V = 2786.8$ Å^3).

Their electronic band structures are shown in Fig. 6.1. Since the crystal structure of these compounds contains two molecules in a unit cell, each band in the electronic structure appears in pair. A rather dispersing lower subband (bandwidth 468 meV) and a relatively flat (bandwidth 99 meV along ΓY direction) upper subband are noted in the highest valence band of pentacene. In contrast, the band splitting in the highest valence band of rubrene is quite small. Consequently, the DOS in pentacene exhibits a sharp peak at the top of the valence band (extending from about 0.4 to 1.0 eV), whereas the DOS distribution at the top of valence band (extending from about 0.3 to 0.7 eV) is relatively smooth in rubrene. This explains the enhanced Seebeck coefficient in pentacene as compared to rubrene.

In summary, organic semiconducting materials offer low thermal conductivity as well as low cost, light weight, mechanical flexibility, yet their low power factor has so far precluded their use in TE devices. However, the large existing library of organic compounds as well as the many ways in which these compounds may be functionalized to tune electrical and thermal properties offer numerous possible routes to optimize TE properties that are as yet unexplored.

6.2 Physical Properties of Molecular Wires

Organic polymeric materials constitute a recent growing topic in thermoelectricity thanks to their easier process of preparation when compared to that of inorganic materials, their cheap price in most cases, their environmentally friendly nature, their high flexibility, and the ease to tune their properties with simple changes in their molecular structures. The organic polymers, however, present an important drawback, which is their thermal instability at high temperatures, so their possible use is limited to room temperature applications [37].

6.2.1 *Conducting Conjugated Polymers*

In contrast to metals, polymers are typically insulators, which according to Eq. (1.71) prevent them from TE applications, though polymers inherently possess a very low thermal conductivity, within the range $\kappa \simeq 0.1-0.5$ $\mathrm{Wm^{-1}K^{-1}}$ at room temperature. For this class of materials most of the phonons have mean-free paths reduced to the minimum separation between equivalent structural units, for instance the interchain distance between neighboring polymers defined by weak van der Waals interactions.

Fortunately enough, during the last two decades new classes of polymers have been synthesized that are capable of carrying unusually high electrical currents. In general, conjugated polymers are semiconductors that provide the electronic conductivity after doping with suitable dopant agents. There are a variety of p-dopants or oxidizing substances that have been used to dope different polymers for improved electrical conductivity: iodine, iron chloride, camphor sulphonic acid, arsenic pentachloride, and so forth. For instance, a doped form of the polyacetylene molecule (as a consequence of an acidic treatment) was reported to have an electrical conductivity $\sim 6 \times 10^4$ $\Omega^{-1}\mathrm{cm}^{-1}$, comparable to that of certain semiconducting materials. Other polymers such as polyaniline, polypyrrole, or polythiophene were found to have as well high electrical conductivities when chemically doped in a proper way (Table 6.1).

Along with the electrical conductivity, Seebeck coefficient values have also been reported in the literature for polyacetylene, polyanilines, polypyrroles, and polythiophenes. The best conjugated polymer up to now is polyacetylene ($ZT \simeq 0.6$ at room temperature) due to its good electrical conductivity and relatively high Seebeck coefficient. Nevertheless, it has poor stability in the (required) doped state, even in an inert atmosphere and it is also insoluble. The

Table 6.1 Room temperature values of the transport coefficients, power factor and FOM for different polymer materials (After Ref. [34]). The FOM value is estimated by assuming $\kappa \simeq 0.5$ Wm^{-1}K^{-1} as a representative upper bond for these materials. Positive values of the Seebeck coefficient are obtained for all considered (doped) samples, indicating that the charge carriers are holes

Sample	σ (Ω^{-1}cm^{-1})	S (μVK^{-1})	P (μWcm^{-1}K^{-2})	ZT (lower bond)
polyacetylene	40,000	+16	10	0.6
polyaniline	7000	+7	0.3	0.02
polythiophene	100	+22	0.05	1×10^{-3}
polypyrrole	200	+7	10^{-4}	6×10^{-6}

other considered polymers show better stability in the doped state, but lower Seebeck coefficient and electrical conductivity, which leads to small estimated FOM values (Table 6.1).

On the basis of theoretical band structure calculations, indicating the presence of both flat and dispersive bands close to the Fermi level,[a] some conducting polymers, like polythiophene-, polyaminosquarine-, and polycarbazole-doped derivatives have been recently considered as potential TEMs (Fig. 6.2). The experimentally measured TE parameters are listed in Table 6.2. By comparing the data listed in Tables 6.1 and 6.2 we see that, though polycarbazole derivatives exhibit significantly larger S values, their electrical conductivity is far too small to yield ZT figures of practical interest.

In summary, albeit significant progresses have been made in order to improve the charge carrier transport in doped-conjugated polymers, the obtained electrical conductivity values are not yet high enough to make them feasible candidates for TE applications. To solve this problem several attempts have recently been made to prepare suitable polymer-based composite materials. For instance, poly(vinyl-acetate)-carbon nanotube composites, with a carbon nanotube concentration of 20 wt %, exhibit $\sigma = 48\ \Omega^{-1}$cm^{-1}, $\kappa =$

[a]A band structure similar to that previously discussed for organic semiconductors (see Fig. 6.1). The presence of flat bands close to the Fermi level is a convenient feature for TEMs (see Section 2.4.1).

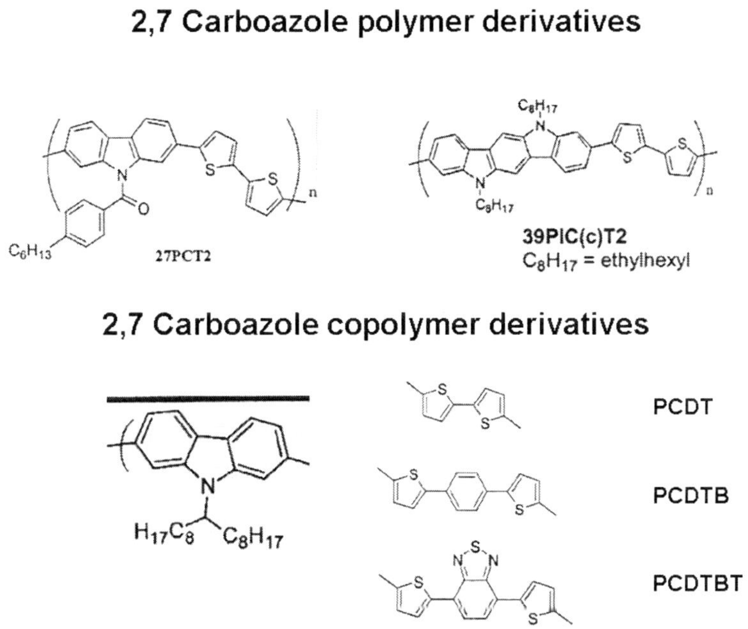

Figure 6.2 Structure of some polycarbazole derivatives proposed as potential TEMs. Reprinted from Ref. [274], with permission from L'evesque I., et al., 2007 *Chem. Mater.* **19**, 2128, Copyright 2007, American Chemical Society, and Ref. [275], with permission from Aïch R D, et al., 2009 *Chem. Mater.* **21**, 751, Copyright 2009, American Chemical Society.

0.34 $\mathrm{Wm^{-1}K^{-1}}$, and $ZT = 0.006$ at room temperature [276]. The TE properties of carbon nanotube-filled polymer composites can be further enhanced by modifying the junctions between the carbon nanotubes, yielding higher electrical conductivities (up to $\sigma = 400$ $\Omega^{-1}\mathrm{cm}^{-1}$) without significantly altering the Seebeck coefficient and eventually attaining $ZT \simeq 0.02$ at room temperature [277]. In a similar way, appreciable electrical conductivity values ($\sigma = 440$ $\Omega^{-1}\mathrm{cm}^{-1}$ at room temperature), yielding a power factor $P = 0.26$ $\mu\mathrm{Wcm^{-1}K^{-2}}$, have been reported for composites based on carbon nanotubes and polycarbazoles [34].

As a new strategy along the same line, composites formed by two molecular networks have gained much attention due to the fact that such materials possess unusual physical properties resulting from

Table 6.2 Room temperature values of the transport coefficients, power factor, and FOM for different polycarbazole derivatives whose chemical structure is shown in Fig. 6.2. The 27PCT2 compound was prepared in form of a pellet whereas, the remaining samples were prepared as thin films. The FOM value is estimated by assuming $\kappa \simeq 0.5$ Wm^{-1}K^{-1} as a representative upper limit for these materials

Sample	σ (Ω^{-1}cm^{-1})	S (μVK^{-1})	P (μWcm^{-1}K^{-2})	ZT (lower limit)
27PCT2	0.3	+71	10^{-3}	10^{-4}
PCDT	23	+53	0.07	0.004
PCDBT	87	+40	0.15	0.008
PCDTBT	160	+34	0.20	0.011

the combination of two different materials. In particular, composite materials consisting of both organic and inorganic compounds can be tuned for achieving high electrical conductivity (contributed by the inorganic material) and a poor thermal conductivity (arising from the organic compound). Thus, composites based on bismuth telluride (Bi_2Te_3) nanorods and polyaniline nanoparticles have recently been synthesized. The composite consists of a highly ordered chain structure of polyaniline nanoparticles on Bi_2Te_3 nanorods, and it exhibits a higher Seebeck coefficient than either pure polymer or bismuth telluride. At room temperature the thermal conductivity of the composite is lower than that of its pure constituents, due to selective phonon scattering by the interface. Nevertheless, the resulting FOM ($ZT = 0.004$ at room temperature) is not significantly improved as compared to that previously reported for bulk polymer materials (**Exercise 6.3**).

A thermoelectric generator entirely based on polymer materials has been recently tested. The all-organic TEG consists of 54 thermocouples with lengths of approximately 40 μm. The p-type leg is made of poly(3,4-ethylenedioxythiophene) with $S \simeq 70$ μVK^{-1} at optimal doping, whereas the n-type is based on tetrathiafulvalene-TCNQ with $S \simeq -28$ μVK^{-1}. The maximum power output at optimal external conditions (see Section 1.4.1) was $W \simeq 24$ nWcm^{-2} working at a temperature difference of $\Delta T = 10$ K [279].

6.2.2 *Transport Properties of DNA*

The structural complexity of macromolecules of biological interest is certainly larger than that of the conjugated molecules we have considered in the previous section, and one may expect these macromolecules to exhibit relatively unusual transport properties. To check such a possibility, a number of experiments measuring the electrical conductivity of dry proteins and nucleic acids were performed in the 1960s yielding results that were interpreted in terms of the classical expression for semiconductor conduction

$$\sigma = \sigma_0 \exp\left(-\frac{E_g}{k_B T}\right) \quad (\Omega\ \mathrm{cm})^{-1}, \tag{6.1}$$

with room temperature values $\sigma_0 = 8 \times 10^{-8}$ $(\Omega\ \mathrm{cm})^{-1}$ for DNA $(E_g = 0.9\ \mathrm{eV})$, $\sigma_0 = 1 \times 10^{-10}$ $(\Omega\ \mathrm{cm})^{-1}$ for collagen $(E_g = 0.9\ \mathrm{eV})$, and $\sigma_0 = 8 \times 10^{-13}$ $(\Omega\ \mathrm{cm})^{-1}$ for myosin $(E_g = 0.5\ \mathrm{eV})$. As we see, the electrical conductivity values for macromolecules are remarkably lower than those measured in doped conducting polymers.

In those early measurements, however, some specimens were in crystalline state and others in the powdered and fibrous state. In the latter, the results are certainly influenced by charge transport between polymer strands in close proximity to each other, which cast some doubts about the reliability of these experiments in order to extract information about the intrinsic conduction properties of the polymers themselves. Accordingly, first measurements able to infer charge transfer from fluorescence measurements on double-stranded DNA spurred a great interest. In these studies a donor and an acceptor moieties are attached to the DNA at a given distance, and upon photo-excitation a single charge carrier (usually a hole) is injected into the chain, travels the distance and finally, recombines at the acceptor site. Thus, the DNA conductivity was assessed from charge transfer rates as a function of the distance between the donor and acceptor sites, confirming that duplex DNA is more effective than proteins as a medium for charge transfer but it does not function as a molecular wire [280].

A number of experimental measurements aimed to directly probe the electric current as a function of the potential applied across the DNA molecules were subsequently reported during the last decade [281]. These experiments were performed in a variety of conditions, where important factors including DNA–substrate interaction, contact effects with the electrodes, relative humidity, the spatial distribution of counterions, and the nucleotide sequence nature (i.e., periodic or aperiodic one), are not kept constant. This state of affairs is considerably difficult for a proper comparison among different experimental reports, which range from completely insulating ($\sigma < 10^{-6}\ \Omega^{-1}\text{cm}^{-1}$) to semiconducting ($\sigma = 800\ \Omega^{-1}\text{cm}^{-1}$) and even superconducting behaviors. From the collected data three main conclusions can be drawn:

- First, long DNA samples of biological origin are typically more insulating than short synthetic oligomers, generally exhibiting a semiconducting behavior.
- Second, by all indications the structure of the DNA helix when deposited on dry surfaces may be very different from that found by crystallization of DNA in solution, so that the DNA–substrate interactions are critical in determining the conductivity of an immobilized molecule, generally leading to poor conductivity.
- Third, the role of contacts deserves a particular attention. In many measurements, contact with metal electrodes was achieved by laying down the molecules directly on the electrodes. In this case, it is rather difficult to prove that the DNA molecule is in direct contact with the electrodes. More reliable results were obtained from measurements performed in DNA molecules whose extremes are previously functionalized in order to attach them to the contacts via chemical bonding.

Two representative experimental layouts are shown in Fig. 6.3. In a setup, a nanoelectronic platform based on single-walled carbon nanotubes was fabricated for measuring electrical transport in single-molecule single-strand DNA and double-strand DNA samples of a 80 base pair long DNA fragment. To enhance the contact efficiency a covalent bonding between an amine-terminated DNA molecule and a carboxyl-functionalized carbon nanotube was established and the DNA molecule was suspended over a nanotrench in order to mitigate the problem of compression-induced perturbation on the charge transport. A nonlinear I–V characteristic curve was observed indicating a semiconducting behavior (gap width $\sim$1 eV,

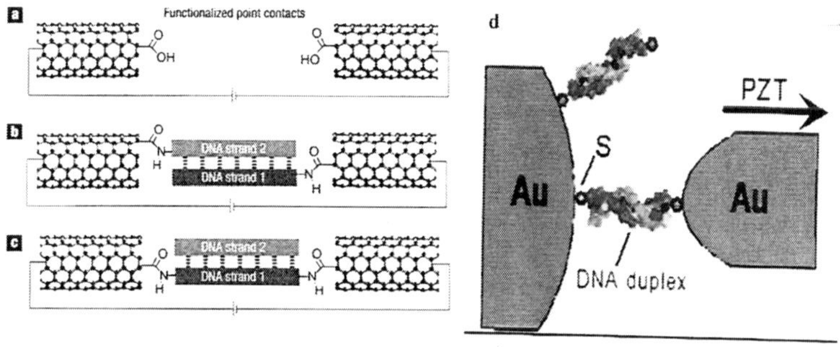

Figure 6.3 Schematics illustrating a method to chemically attach duplex DNA strands with molecular nanocontacts. (a) Functionalized point contacts made through the oxidative cutting of a single-walled nanotube wired into a device, (b) bridging by functionalization of both strands with amine functionality, and (c) bridging by functionalization of one strand with amines on either end [282]. Reprinted with permission from Macmillan Publishers Ltd.: *Nat. Nanotechnol.* **3** 163, Copyright 2008. (d) The contact is formed through a thiolated chemical bond between the electrode (Au) and the DNA molecule, whose 3' end has been modified with a C_3H_6SH linker. In the same buffer solution a gold-scanning tunnel microscope tip, which is covered with an insulating layer over most of the tip surface except for its end, is brought into contact. Once contact is formed the tip is pulled backwards and the resulting current is monitored with a piezoelectric transducer [283]. Reprinted with permission from B. Xu, P. Zhang, X. Li, N. Tao, 2004 *Nano Lett.* **4**, 1105, Copyright 2004, American Chemical Society.

p-type conduction) in both aqueous (sodium acetate buffer) and vacuum (10^{-5} torr) conditions. From basic principles it is expected that a single-strand DNA molecule will carry only a feeble current due to lack of structural integrity. Indeed, a current of about a 25–40 pA (0.5–1.5 pA) at 1 V bias was measured for double-strand DNA (single-strand DNA) duplexes, respectively, at ambient conditions. Accordingly, the conductance of the double-stranded structures is about an order of magnitude higher than that of single-stranded ones with similar number of bases (**Exercise 6.4**).[a] This observation clearly demonstrates that the interactions between the base pairs and stacking effects play a vital role in charge transport through DNA.

[a]Such a conductivity difference is significantly greater for oligo-C, oligo-T, and oligo-A chains, where C, T, and A stand for cytosine, thymine, and adenine, respectively [284].

On the other hand, making use of the experimental setup shown in Fig. 6.3d measurements were performed on DNA duplexes of the form 5'-CGCG(AT)$_m$CGCG-3', where some guanine-cytosine (GC) base pairs are replaced by adenine-thymine (AT) ones, in order to analyze sequence effects on the transport properties. The conductance data can be described by an expression of the form $G = Ae^{-\beta L}$, where L is the length of the AT bridge, with $A = (1.3 \pm 0.1) \times 10^{-3}\ G_0$ and $\beta = 0.43 \pm 0.01$ Å^{-1}, where $G_0 \equiv e^2/h = 1/12906\ \Omega^{-1}$ is the so-called conductance quantum [283]. These findings are consistent with a tunneling process across AT regions between the GC domains, in good agreement with the idea that the guanine highest occupied molecular orbitals (HOMO) favor charge migration, whereas short AT sequences create a tunneling barrier for charge hopping through guanines along the DNA stack.

In fact, the electronic structure of different DNA oligomers has been studied by means of ab initio calculations based on the density functional theory [281]. The cases of the homopolymers polyG-polyC and polyA-polyT have been extensively considered, along with some related structures like poly(GC)-poly(CG). In order to reduce the computational effort earlier calculations did not explicitly take into account either the water shell or the cations around the sugar-phosphate backbone. Accordingly, these preliminary works focused on the dry A-DNA electronic structure. Close to the Fermi level it shows well defined, narrow bands separated by a broad gap (2–3 eV). The valence bands in A-form polyG-polyC and A-form polyA-polyT consist of 11 states, that is, one per base pair in the unit cell. In the case of polyG-polyC the top-most valence band has a very small bandwidth (Fig. 6.4a). This band is associated with the π-like HOMO of the guanine. The charge density of the states associated with this band appears almost exclusively on the guanines, with negligible weight either in the backbones or in the cytosines (Fig. 6.4c). The lowest conduction band is significantly broader and it is made of the LUMO of the cytosines. Similar results are obtained for A-form polyA-polyT chains, where the charge density appears concentrated on the HOMO orbitals of the adenines and exhibit a broader valence band width (~ 0.25 eV).

When the presence of a water shell and sodium ions distributed along the sugar–phosphate backbone is taken into account, a

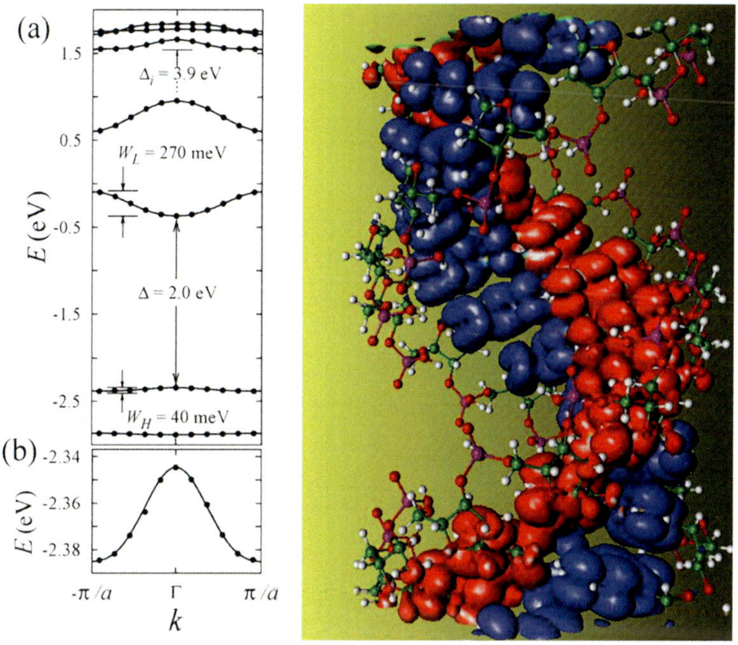

Figure 6.4 (a,b) Energy bands close to the Fermi level as a function of the wave vector k of a polyG-polyC molecule in dry conditions. In the plot results obtained from ab initio calculations (dots) are compared to those derived from a one-dimensional tight-binding model with one orbital per unit cell (solid line). Δ indicates the HOMO–lowest unoccupied molecular orbital (LUMO) gap, Δ_i the gap between closest orbitals in the guanine system (relevant to optical transitions), and $W_{H(L)}$ are the HOMO (LUMO) bandwidths, respectively. (c) Surfaces of constant charge density for the states corresponding to the lowest unoccupied band (light gray) and highest occupied band (dark gray) of a polyG-polyC molecule in the A-form in dry conditions [285]. Reprinted with permission from E. Artacho, M. Machado, D. Sánchez-Portal, P. Ordejón, and J. M. Soler, 2003 *Molecular Phys.* **101,** 1587, Copyright 2003, Taylor & Francis (Courtesy of E. Artacho).

number of localized states appear in the HOMO-LUMO gap due to the presence of Na-water ions around phosphate groups. In fact, the phosphate groups of the DNA molecule are negatively charged. Hence, positive protons or metal cations (usually referred to as counterions) are necessary to neutralize and stabilize DNA in physiological conditions. Water also plays a crucial role to this end. Hydrophobic forces compel DNA to adopt the B-form, and the polarity of the water molecules helps to screen DNA's charges. Thus, the inclusion of Na^+ cations evenly distributed through the backbone gives rise to the presence of a band related to the Na-phosphate groups between the π-electron bands of the base molecules, so that the LUMO moves from cytosines to the phosphate-cations system when in presence of Na^+ for both A-form polyA-polyT and A-form polyG-polyC [286]. Accordingly, the water shell and the counterions can lead to the presence of a number of states in the main π-π^* energy gap (which can be regarded as impurity states), hence effectively doping the DNA molecule. Nevertheless, the mobility of the charge carriers, proceeding through the overlapping of the $\pi-\pi$ orbitals of consecutive base pairs along the helical axis, is not appreciably affected by the presence of these states. Accordingly, the presence of flat valence and conduction bands could be fully exploited in principle (see Section 2.4.1) to improve the Seebeck coefficient in DNA.

The main features of the electronic structure obtained from numerical results have been experimentally confirmed by means of some spectroscopic techniques. In particular, it has been confirmed that the HOMO originates in the DNA bases, in agreement with numerical calculations, for both polyG-polyC and polyA-polyT duplexes forming a mixture of A- and B-DNA forms. It has been also demonstrated that when holes are injected in polyG-polyC by chemical oxidation the hole charge is localized on G, but not on cytosine, deoxyribose, or phosphates [287].

In summary, the reported experiments demonstrate the high sensitivity of DNA electrical conductivity to several factors. Firstly, we have the structural complexity of nucleic acids, which is significantly influenced by its close surrounding chemical environment (humidity degree, counterions distribution) affecting the integrity of the base-pair stack, as well as by the unavoidable presence of thermal fluctuations. Secondly, the kind of order present in the DNA macromolecule plays an important role in determining its transport characteristics: periodically ordered polyG-polyC chains exhibit semiconducting behavior, whereas biological samples are more insulating. Finally, measuring charge transport in a DNA chain is strongly biased by the invasive role of contacts, the charge injection mechanism, the quality of the DNA–electrode interface, and the possible interaction with some inorganic substrate, or other components of the experimental layout.

In comparison to the large number of studies on the electrical conductivity of DNA molecules only a scarce number of works have been devoted to measuring their thermal conductivity. Since it is expected that the heat transport will be mainly determined by phonons rather than charge carriers in this molecule, a relatively low value can be anticipated for its thermal conductivity.

A suitable estimation of the thermal conductivity of a molecular wire in contact with two thermal reservoirs at the ends relies on the quantum of thermal conductance $g_0^T = \pi^2 k_B^2 T/(3h) = 9.46 \times 10^{-13} T$ WK^{-1},[a] which represents the maximum possible value of energy transported per phonon mode (assuming ideal coupling conditions and ballistic heat transport) [288]. In the regime of low temperatures four main modes, arising from dilatational, torsional, and flexural degrees of freedom are expected for a quantum wire [289]. Therefore, the thermal conductivity of a DNA oligomer of length $L_N = 0.34N$ nm and cross-section $A = \pi R^2$ (where N is the number of base pairs and $R \simeq 1$ nm is the helix radius) will be given by

$$\kappa_N \simeq 4 g_0^T \frac{L_N}{A} = 4\left(\frac{\pi^2 k_B^2}{3h}\right)\frac{0.34N}{\pi R^2} T.$$

For a DNA oligomer with $N = 5$ base pairs one gets $\kappa = 0.02$ $\mathrm{Wm^{-1}K^{-1}}$ (at $T = 10$ K) and $\kappa \simeq 0.6$ $\mathrm{Wm^{-1}K^{-1}}$ (at room temperature) under optimal conditions. This figure compares well with the estimations based on experimental measurements of the DNA specific heat at low temperatures (**Exercise 6.5**), as well as more recent direct measurements of the thermal conductivity of different DNA samples reporting values ranging between $\kappa = 0.60$ and $\kappa = 0.82$ $\mathrm{Wm^{-1}K^{-1}}$ at room temperature [28].

---

[a]We note that the relation between the electrical conductance quantum and the thermal conductance quantum obeys the Wiedemann–Franz law. In fact, $\frac{g_0^T}{G_0} = \frac{\pi^2 k_B^2 T}{3h}\frac{h}{e^2} = L_0 T$.

6.3 Thermoelectricity at the Nanoscale

As an alternative to bulk materials the study of the TE properties of single molecules may underpin novel thermal devices such as molecular-scale Peltier coolers, and provide new insight into mechanisms for molecular-scale transport. Indeed, from the study of TE voltage over a molecule attached to two metallic leads one can gain valuable information regarding the location of the Fermi energy relative to the molecular levels. In particular, from the sign of the Seebeck coefficient it is possible to deduce the conduction mechanism, with a positive sign indicating p-type conduction (the Fermi level is closer to the HOMO level), whereas a negative sign indicates n-type conduction (the Fermi level is closer to the LUMO level).

In fact, the extreme sensitivity of Seebeck coefficient to finer details in the electronic structure suggests that one could optimize the device's TE performance by properly engineering its electronic structure. For instance, by shifting the Fermi level position in order to optimize the TE performance of a given molecular arrangement.

In order to explore such a possibility the TE properties of molecular junctions created by trapping aromatic molecules between gold electrodes with thiol end groups have been systematically investigated with a suitably modified scanning tunneling microscope (Fig. 6.5). As the scanning tunneling microscope is attached to a molecule, a thermal gradient is applied and the Seebeck coefficient is measured by applying a voltage, so that no electrical current passes through the junction (Fig. 6.5a). This procedure is repeated many times and a histogram of the voltage required to achieve a vanishing current is obtained for different temperature gradients, $\Delta T = 10\text{–}30$ K. Then, the peak (i.e., most probable) voltage in the histograms is plotted as a function of ΔT, and the Seebeck coefficient is obtained from a linear fitting.

The experimentally measured transport coefficients (in air at ambient conditions) are listed in Table 6.3. As we see, the Seebeck coefficient values measured for multiple molecule junctions are in very good agreement with the measurements corresponding to single-molecule junctions (when available). This correlation indicates that the intermolecular interactions in the monolayer

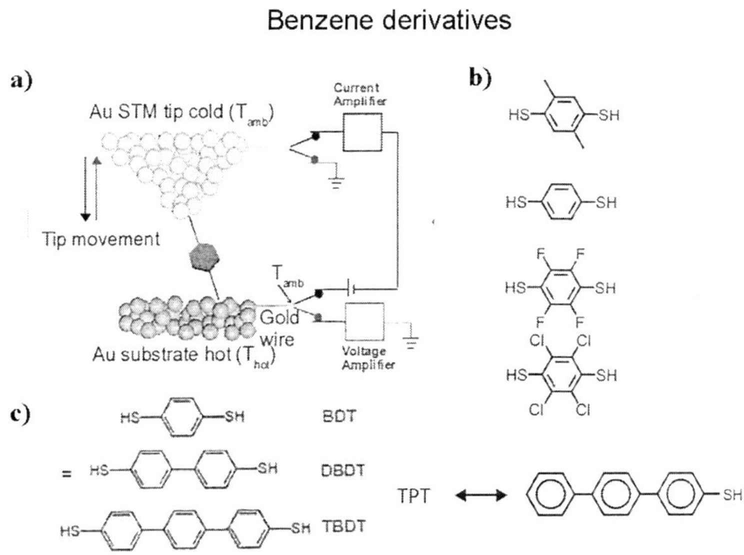

Figure 6.5 (a) Single-molecule molecular junction setup, (b) structure of simple benzene dithiol derivatives, and (c) structure of the polybenzodithiol and TPT derivatives considered in the experiments [291–293]. Reprinted with permission from O. Reddy, et al., *Science* **315**, 1568 (2007), copyright 2007; K. Baheti, et al., *Nano Lett.* **8**, 715 (2008) Copyright (2008) American Chemical Society; A. Tan, et al., *Appl. Phys. Lett.* **96**, 013110 (2010), Copyright 2010, American Institute of Physics.

arrangement are weak. The end groups of the aromatic molecules (thiol, –SH, cyan, –CN, isocyan, –NC, or amine $-NH_2$) where systematically varied to study the effect of contact coupling strength in the transport properties. The thiol-terminated aromatic molecular junctions reveal a positive Seebeck coefficient that increases linearly with the molecule length, in contrast with the measurements performed in polymer films, where no significant dependence was appreciated among polycarbazole derivatives [274]. Positive values of the Seebeck coefficient are obtained for all considered molecules when contacted through thiol groups, indicating that the charge transport is primarily associated with the HOMO level in this case. On the contrary, a negative value is obtained for a benzene molecule contacted to gold electrodes with cyanide or isocyanide end-groups, due to charge transport primarily occurring through the LUMO.

Table 6.3 Room temperature ($T = 293$ K) Seebeck coefficient and resitance for different molecules. The number in their labels indicates the number of monomers, B stands for benzene derivatives, MT and DT indicate monothiol and dithiol end-groups, respectively. The third column list single-molecule measurements, while the next columns on the right list measurements for monolayers containing about 100 molecules attached to the scanning microscope tip

Sample	Ref.	S (μVK^{-1})	S (μVK^{-1})	R (kΩ)
1BMT	[294]		$+8.1 \pm 0.8$	1000
1BDT	[291, 294]	$+7.0 \pm 0.2$	$+9.8 \pm 0.6$	80
2,5 dimethyl-BDT	[292]	$+8.3 \pm 0.3$		
4F-BDT	[292]	$+5.4 \pm 0.4$		
4Cl-BDT	[292]	$+4.0 \pm 0.6$		
B-CN	[292]	$-(1.3 \pm 0.5)$		
3B-NC	[294]	$-(1.0 \pm 0.4)$		
2BMT	[294]		$+13.6 \pm 1.2$	3000
2BDT	[292, 294]	$+12.9 \pm 2.2$	$+11.7 \pm 1.3$	90
3BMT	[294]		$+17.0 \pm 1.0$	8000
3BDT	[292, 294]	$+14.2 \pm 3.2$	$+15.4 \pm 1.0$	150
3BD(NH_2)	[295]		6.4 ± 0.4	
TPT	[293]	$+16.9 \pm 1.4$		
4BMT	[294]		$+21.0 \pm 1.3$	

Thus, end-groups are key to controlling the very nature of charge carriers.

Therefore, by properly varying end-groups and molecular junction constituents one can engineer metal–molecule heterostructures with targeted TE properties. This appealing possibility has been further analyzed by performing the simultaneous measurement of the conductance and Seebeck coefficient of single molecule junctions in order to estimate their power factor in a straightforward way. Conductance values were obtained by measuring the current across the gold-molecule-gold junction at an applied bias voltage of 10 mV. The Seebeck coefficient values are determined on the same junction from the measured thermoelectric current through the junction held under a temperature gradient, while maintaining a zero bias voltage across the junction. The experimentally obtained power factors range from $GS^2 = 3.7 \times 10^{-20}$ WK^{-2} for 1,5-bis-

(diphenylphosphenyl)acetylene to $GS^2 = 8.3 \times 10^{-18}$ WK^{-2} for 4,4'-diaminostilbene [296]. In order to compare with power factor values reported for bulk compounds we can adopt a cross-section of $\sim$1 Å^2 and a length of $\sim$15 Å for the 4,4'-diaminostilbene molecule to get $P \simeq 1.2 \times 10^{-2}$ μW·cm^{-1}K^{-2}, a figure comparable to that obtained for longer polymers.

In summary, the experiments indicated that molecular junctions have favorable TE properties, hence suggesting that devices incorporating molecular junctions may be good candidates for a next generation of nanodevices for TE applications. Accordingly, it is convenient to pay some attention on the theoretical aspects related to this kinds of TE devices.

6.3.1 *Transport Coefficients for Molecular Junctions*

The configuration we have in mind is a junction comprised of two leads separated by a nanoscale element: a quantum dot, a nanotube, a molecule, etc. Consider such a junction, where the leads on the left and on the right are held at different temperatures, T_L and T_R, respectively. The corresponding temperature difference, $\Delta T = T_R - T_L$, gives rise to both a heat current and charge current. If the circuit is open, after a transient time charges accumulate on one side of the junction and deplete on the other, so that a zero charge current is achieved and a voltage drop across the junction is formed. By analogy with bulk macroscopic systems the Seebeck coefficient is defined as (minus) the amount of voltage generated in the nanoelements under an applied temperature difference between the leads at the state of vanishing electrical current, namely[a]

$$S = - \lim_{\Delta T \to 0} \left| \frac{\Delta V}{\Delta T} \right|_{I=0}. \tag{6.2}$$

The starting point for calculating the Seebeck coefficient within a single particle picture is the Landauer expression for the electrical current

$$I = \frac{2e}{h} \int_{-\infty}^{+\infty} \mathcal{T}_N(E, V)[f_L(E, V) - f_R(E, V)]dE, \tag{6.3}$$

[a]Most materials presented below are borrowed from the excellent review by Dubi and Di Ventra [35].

where $\mathcal{T}_N(E, V)$ is the transmission coefficient for a molecule composed of N monomers and $f_{L,R}$ are the Fermi distributions of the left and right leads, respectively. In the limit of a small external bias V and small temperature gradient (i.e., $|e\Delta V| \ll E_F$ and $|\Delta T| \ll T$, where T is the background temperature), the Fermi distribution functions are given by

$$f_{L,R}(E, V) \simeq f_0(E, V) + \frac{\partial f}{\partial E}\left(E_F - E_F^{L,R}\right) + \frac{\partial f}{\partial E}(E - E_F)\frac{T_{L,R} - T}{T}, \tag{6.4}$$

where $f_0(E, V)$ is the equilibrium distribution function and $E_F^{L,R}$ denote the Fermi energy levels at each lead. Inserting Eq. (6.4) into Eq. (6.3) and equating the electrical current to zero, one obtains

$$0 = \Delta V \int_{-\infty}^{+\infty} \mathcal{T}_N(E)\left(-\frac{\partial f}{\partial E}\right) dE + \frac{\Delta T}{eT}\int_{-\infty}^{+\infty} \mathcal{T}_N(E)\left(-\frac{\partial f}{\partial E}\right)(E - E_F)dE, \tag{6.5}$$

where $\Delta V \equiv (E_F^L - E_F^R)/e$. Finally, Eq. (6.5) can be arranged in the form

$$-\left|\frac{\Delta V}{\Delta T}\right|_{I=0} = \frac{1}{eT}\frac{\int_{-\infty}^{+\infty} \mathcal{T}_N(E)\left(-\frac{\partial f}{\partial E}\right)(E - E_F)dE}{\int_{-\infty}^{+\infty} \mathcal{T}_N(E)\left(-\frac{\partial f}{\partial E}\right)dE} = S_N, \tag{6.6}$$

where we have made use of Eq. (6.2). If there are not resonances (i.e., peak features) in the transmission coefficient close to the equilibrium Fermi level one can further simplify this expression for the Seebeck coefficient using a Taylor expansion of the transmission coefficient at the Fermi level to obtain

$$S_N(T) = -|e|L_0\left(\frac{\partial \ln \mathcal{T}_N(E)}{\partial E}\right)_{E_F} T, \tag{6.7}$$

where $T = (T_R + T_L)/2$ is the average temperature of the leads. Thus, in the coherent tunneling limit at zero external bias the Seebeck coefficient only depends on the temperature difference between the two leads and the slope of the transmission coefficient at the Fermi energy, in close analogy with the Mott's expression for bulk materials (see Eq. (2.51)).

Therefore, the Landauer formalism provides a simple interpretation of the Seebeck coefficient at the molecular scale in terms of the transmission coefficient $\mathcal{T}_N(E)$. On the other hand, within the same formalism, the electrical conductance is given by

$$G_N(E_F) = G_0 \mathcal{T}_N(E_F). \tag{6.8}$$

By comparing Eqs. (6.7) and (6.8) we see that the conductance strongly depends on the transmission coefficient value, whereas the Seebeck coefficient remains essentially invariant upon a change in the $\mathcal{T}_N(E)$ intensity. Since the transmission coefficient is determined by the coupling between the molecule and the leads the applicability of these expressions can be tested by systematically changing the chemical nature of end-groups in molecular junctions. In fact, by inspecting Table 6.3 we see that when the coupling contact is reduced by removing one of the thiol groups, the resistance changes by almost two orders of magnitude, while the Seebeck coefficient values remain essentially the same, as expected. Consequently, we can confidently assume that the above expressions properly account for the experimental data presented in the previous section.

Thus, in order to determine both the conductance and the Seebeck coefficient we must calculate the transmission coefficient. To this end, as a first approximation we shall consider that the charge carrier dynamics can be decoupled from vibrational atomic motions in the molecule and that the coupling between the contacts and the molecule τ is weak enough, so that the lead-molecule lead junction can be described in terms of three non-interacting subsystems according to the molecular junction model shown in Fig. 6.6. The molecule is described in terms of a linear chain with an orbital per site (on-site energies ε_α, ε_β, and ε_γ), where each lattice site represents a monomer (squares) and t and ηt are the hopping terms between them. The molecule is connected to leads modeled as semi-infinite one-dimensional chains of atoms (circles) with one orbital per site, with an on-site energy ε_M, and t_M ($>\tau$) is the lead hopping term.

Within the transfer matrix framework, and considering nearest–neighbors interactions only, the Schrödinger equation corresponding to the molecular junction model shown in Fig. 6.6 can be expressed in the form

$$\begin{pmatrix} \psi_{N+1} \\ \psi_N \end{pmatrix} = \mathbf{T}_{N+1}\mathbf{T}_N \ldots \mathbf{T}_1\mathbf{T}_0 \begin{pmatrix} \psi_0 \\ \psi_{-1} \end{pmatrix}, \tag{6.9}$$

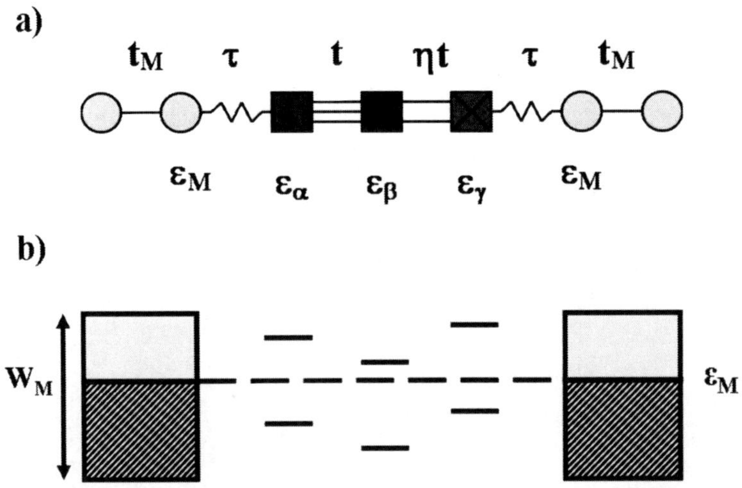

Figure 6.6 (a) Tight-binding molecular junction model. (b) Energy band structure of the molecular junction model sketched in (a). The bandwidth of the contacts is $W_M = 4t_M$. The dashed horizontal line indicates the location of the contacts Fermi level. The segments below (over) the dashed line correspond to the HOMO (LUMO) orbitals of each monomer, respectively [297].

where ψ_n is the wavefunction amplitude for the energy E at site n and

$$\mathbf{T}_n(E) = \begin{pmatrix} \dfrac{E - \varepsilon_n}{t_{n,n+1}} & -\dfrac{t_{n,n-1}}{t_{n,n+1}} \\ 1 & 0 \end{pmatrix}, \tag{6.10}$$

is the local transfer matrix. The lead-molecule-lead zero bias transmission coefficient, $\mathcal{T}_N(E)$, describing the fraction of charge carriers transmitted through a chain of length N in the absence of any applied voltage, can then be obtained from the knowledge of the leads dispersion relation, $E(k) = \varepsilon_M + 2t_M \cos k$, and the matrix elements of the molecular junction global transfer matrix $\mathbf{M}(E) \equiv \prod_{n=N+1}^{0} \mathbf{T}_n(E)$, by means of the relationship [24]

$$\mathcal{T}_N(E) = \frac{4\sin^2 k}{[M_{12} - M_{21} + (M_{11} - M_{22})\cos k]^2 + (M_{11} + M_{22})^2 \sin^2 k}. \tag{6.11}$$

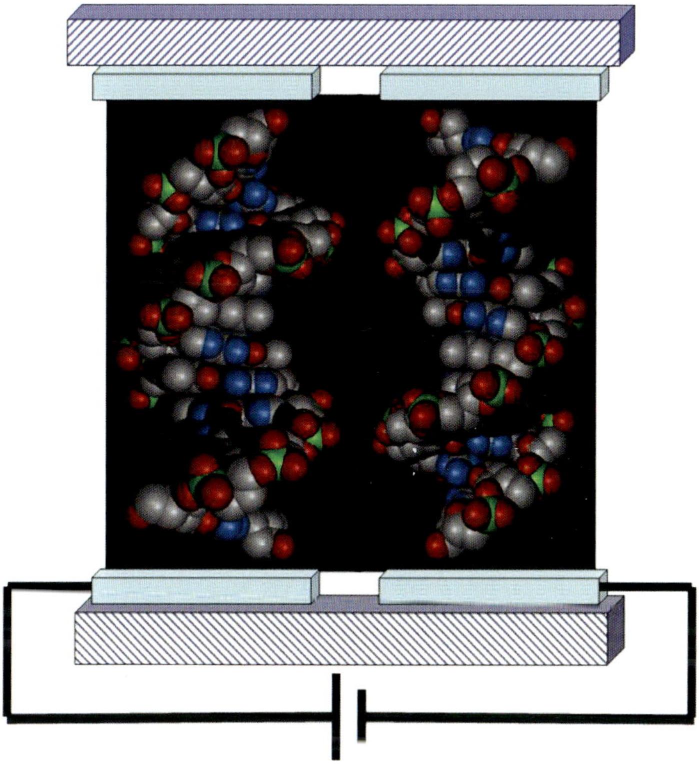

Figure 6.7 Sketch illustrating the basic features of a nanoscale DNA-based Peltier cell. A polyA-polyT (polyG-polyC) oligonucleotide, playing the role of n-type, left (p-type, right) semiconductor legs, are connected to organic wires (light boxes) deposited onto ceramic heat sinks (dark boxes) [298]. Reprinted with permission from E. Maciá, 2007 *Phys. Rev. B* **75**, 035130, Copyright 2007, American Physical Society).

6.3.2 *DNA-Based Thermoelectric Devices*

For the sake of illustration, in this section we shall consider the nano-Peltier cell sketched in Fig. 6.7. The physical motivations inspiring this device are basically two-fold, and they are based on experimental current–voltage curves showing that:

- periodically ordered, synthetic DNA chains like polyG-polyC or polyA-polyT exhibit a semiconducting behavior,

- double-stranded polyA-polyT chains behave as n-type semiconductors, whereas polyG-polyC ones behave as p-type semiconductors [299]. Thus, charge transfer mainly proceeds via hole (electron) propagation through the purine (pyrimidine) bases, where the HOMO (LUMO) carriers are respectively located in polyG-polyC (polyA-polyT) chains.

Accordingly, these synthetic DNAs may provide the basic building blocks necessary to construct a nanoscale TE cell, where the DNA chains will play the role of semiconducting legs in standard Peltier cells.

From an experimental viewpoint the possible use of DNA-related molecules in the design of nanoscale TE devices was opened up by the measurement of an appreciable TE power ($+18$ μVK^{-1} at room temperature) over guanine molecules adsorbed on a graphite substrate using a STM tip [300]. This figure is larger than those reported for benzene-dithiol derivatives in Table 6.3, albeit guanine molecules were deposited onto a substrate (physorption) rather than being chemically connected to it as in the case of molecular junction measurements. In any event there certainly exists a very long way from TE measurements performed at the single nucleotide scale to the full-fledged helicoidal structure of duplex DNA chains we are interested in (Fig. 6.7). In order to estimate the expected TE performance of short DNA chains let us consider an experimental layout similar to that shown in Fig. 6.5a, where the benzene derivatives are replaced by properly functionalized (e.g., thioled) nucleobases adenine, guanine, cytosine, thymine, and uracil or short oligonucleotides made from different combinations of these bases. As a first approximation, the resulting contact-molecule-contact arrangement can be described within the molecular model introduced in Section 6.3.1, and the corresponding transport coefficients calculated in terms of Eqs. (6.7) and (6.8).

The room temperature Seebeck coefficient curves, as a function of the Fermi energy position, for molecular junctions containing a single nucleobase are shown in Fig. 6.8. To gain some physical insight into these $S(E_F)$ curves, let us first consider the energy dependence of the corresponding transmission coefficients, which are shown in the inset of Fig. 6.8. As we see, the $\mathcal{T}(E_F)$ curves are very similar

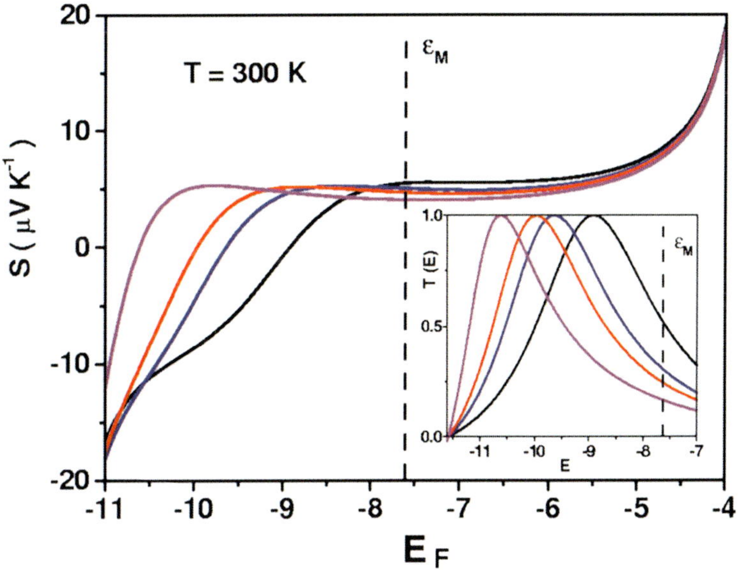

Figure 6.8 Room temperature Seebeck coefficient as a function of the Fermi energy for the nucleobases G, A, C, and T (from right to left) with $\tau = 1.0$ eV, $t_M = 2.0$ eV, and $\varepsilon_M = -7.58$ eV. (inset) The transmission coefficient as a function of the Fermi energy for the nucleobases G, A, C, and T (from right to left). The vertical dashed lines indicate the position of the contact's band center ε_M.

in shape for the different nucleobases, exhibiting a well-defined maximum defining the resonance energy

$$E_\upsilon = \varepsilon_M + \frac{4\Gamma\tau^2}{\varepsilon_\upsilon - \varepsilon_M}, \tag{6.12}$$

where $\upsilon = \{G,A,C,T\}$ labels the considered nucleobase, and the factor

$$\Gamma \equiv \left(\frac{t_M}{\tau}\right)^2 - 1,$$

depends on the coupling strength between the nucleobase and the leads (since $t_M \geq \tau$ we have $\Gamma \geq 0$). As expected, when $\tau = t_M$ ($\Gamma = 0$) the resonance energy coincides with the leads on-site energy ($E_\upsilon = \varepsilon_M$). If the Fermi energy just occurs at the transmission peak position one has $S_\upsilon = 0$ in the Seebeck coefficient curve, as prescribed by Eq. (6.7). Otherwise, as the Fermi level approaches the resonance energy from above we get positive values of the Seebeck coefficient, which vanishes at $E_F = E_\nu$, and changes its sign as the Fermi level further shifts toward the spectral window bottom edge. The location of the contact's band center ε_M marks two distinct behaviors. For $E_F > \varepsilon_M$, all the considered bases have very similar values of Seebeck coefficient ($S_\nu \simeq 5$ μVK^{-1}) over a wide energy range up to $E_F \simeq -5$ eV, where the Seebeck coefficient curve suddenly rises as E_F progressively approaches the spectral window upper edge, $E_+ = \varepsilon_M + 2t_M$. On the contrary, due to resonance effects involving the molecular orbitals of the bases, in the energy range $-11 \lesssim E \lesssim -9$, the Seebeck coefficient curves corresponding to each nucleobase are clearly discernible from each other, and relatively large (negative) values of Seebeck coefficient can be reached when E_F approaches the spectral window lower edge, $E_- = \varepsilon_M - 2t_M$.

In the case of chains composed of two nucleobases of energies ε_n and ε_m, respectively coupled with a hopping term t_{nm}, one can appreciate the existence of characteristic dimerization effects on the transport properties. For the sake of illustration, in the main frame of Fig. 6.9 we compare the monomer T and dimer TT Seebeck coefficient curves for $\tau = 1.0$ eV. Similar curves are obtained when comparing the dimers GG, AA, and CC curves with the corresponding single nucleobase $S(E_F)$ curves. An overall increase of the dimer Seebeck coefficient as compared to the single base one is clearly appreciated over a wide energy range, in analogy with the trend experimentally observed for the polybenzene thiol series (see Table 6.3), reaching a well-defined peak ($S \simeq 22$ μVK^{-1}) at $E_F \simeq -9.2$ eV. Similar results are obtained for the other bases as well. In the inset of Fig. 6.9, we compare the transmission coefficients for T and TT bases, respectively. As we can see, orbital overlapping effects in the thymine dimer result in the presence of two resonance energies and the overall narrowing of the transmission coefficient, which is centered around $E_F \simeq -10.2$ eV. As a consequence of this narrowing effect the $\mathcal{T}(E_F)$ gets steeper, which according to Eq. (6.7) ultimately leads to a Seebeck coefficient enhancement close to the transmission resonant peaks. In addition, the formation of a dimer nucleobase results in an electrical conductance reduction through dimers as compared to that observed in monomers, and this

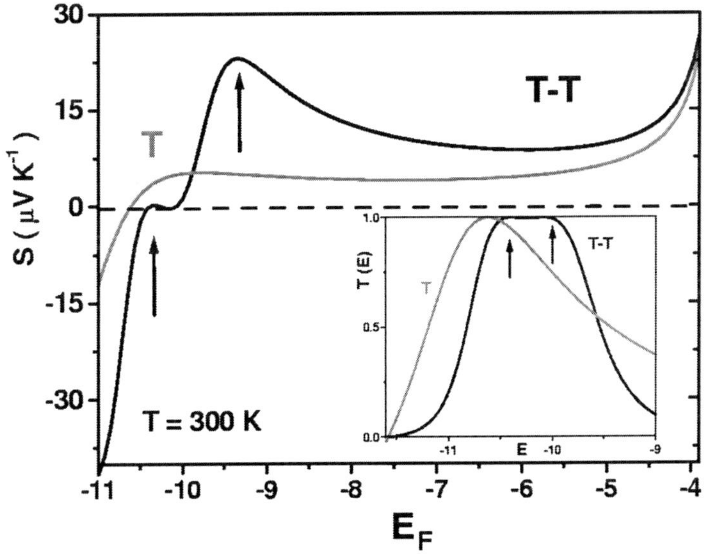

Figure 6.9 Room temperature Seebeck coefficient as a function of the Fermi energy for TT and T nucleobases with $\tau = 1.0$ eV, $t_m = 2.0$ eV, $t_{TT} = 0.4$ eV, and $\varepsilon_m = -7.58$ eV. (inset) Transmission coefficient as a function of the Fermi energy for T and TT bases. The vertical arrows indicate the location of the resonance energy values (inset) and the related peaks in the Seebeck coefficient curve (main frame).

effect is significantly enhanced as the orbital overlapping decreases leading to smaller interbase hopping terms.

As one may expect from general considerations, the presence of narrow, multiple peaked features in the transmission coefficients further increases as one goes from dimer molecules to chains composed of three nucleobases, and beyond. In Fig. 6.10 we show the transmission and Seebeck coefficient curves for a suitable choice of the model parameters. The $S(E_F)$ curves corresponding to the GGG, AAA, and CCC (not shown) trimers are characterized by two peaks and a crossing point which defines two different regimes exhibiting p-type or n-type Seebeck coefficient, respectively. The Seebeck coefficient value attained at the peaks is significantly large (200–300 μVK^{-1} in absolute value),[a] and compare well with values

[a]The peak values for the CCC trimer are about one order of magnitude larger [297].

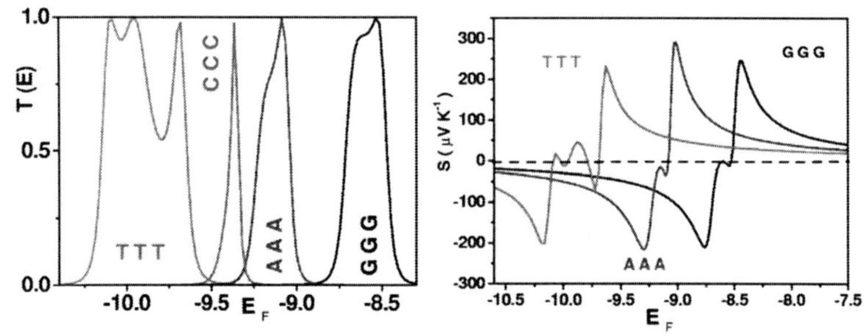

Figure 6.10 (a) Transmission coefficient as a function of the Fermi energy for TTT, CCC, AAA, and GGG trimers with $\tau = 0.5$ eV, $t_M = 2.0$ eV, $t_{TT} = 0.16$ eV, $t_{AA} = 0.07$ eV, $t_{CC} = 0.04$ eV, $t_{GG} = 0.08$ eV, and $\varepsilon_M = -7.58$ eV. (b) Room temperature Seebeck coefficient as a function of the Fermi level energy for TTT, AAA, and GGG trimers with $\tau = 0.5$ eV, $t_M = 2.0$ eV, and $\varepsilon_M = -7.58$ eV.

reported for benchmark TEMs. The Seebeck coefficient of the TTT trimer exhibits two additional minor peaks between the larger ones. These additional features stem from a stronger coupling between neighboring T–T bases in the trimer which gives rise to a characteristic triplet structure in the transmission coefficient shown in Fig. 6.10a. In fact, such a triplet structure is not well resolved for the other trimers due to the significantly weaker value of their molecular orbital overlapping. By inspecting Fig. 6.10b two main conclusions can be drawn. First, the Seebeck coefficient curves for different homomer chains peak at characteristic energy values, which are different enough to allow for a reliable energy resolution among different XXX moieties. Second, the Seebeck coefficient curves corresponding to certain trimers display complementary responses among them. Thus, the AAA Seebeck coefficient curve peaks within the energy range $-9.1 \lesssim E_F \lesssim -8.6$ showing a p-type behavior, whereas the GGG curve takes negative values in the same energy interval instead. The same occurs for CCC and AAA (not shown), or TTT and AAA in the energy windows $-9.3 \lesssim E_F \lesssim -9.1$, and $-9.7 \lesssim E_F \lesssim -9.1$, respectively. In addition the $S_{TTT}(E_F)$ curve exhibits two subsidiary (positive) peaks located at an energy interval where the other three trimers take on negative values. The presence of these characteristic peaks provide an additional criterion highlighting the presence of TTT trimers. By properly combining all these characteristic features one could confidently discriminate among different homomers present in a test sample from Seebeck coefficient measurements. In fact, the presence of resonance effects among electronic levels in oligonucleotides composed of three nucleobases leads to a significant enhancement of the TE signal. By comparing the transport curves corresponding to different types of trimers we see that a characteristic TE signature can be used to identify the XYX type codons from XXX homonucleotide ones on the basis of their different TE responses.

This result suggests we may consider the possibility of sequencing short DNA fragments connected between metallic contacts at different temperatures by employing TE measurements, allowing for a direct sensing of triplet nucleobases associations (including codons in codifying regions) via their *TE signature* [301]. Indeed, there currently exists a growing interest in the search for new sequencing methods entirely based on physical principles able to allow for non-invasive analysis of a huge number of nucleotides along the DNA strands. In this regard, scanning tunnel spectroscopy, which directly detects the molecular levels of single DNA bases, has been exploited during the last few years. In fact, nucleobase-modified tip STM measurements demonstrate the ability to identify the different DNA nucleobases due to selective chemical interactions, although it remains a chemically based rather than a purely physically based technique [31].

Up to now we have considered single-stranded DNA molecules. Also, we have neglected the possible influence of environmental effects. However, more realistic treatments should take into account the presence of a number of counterions located along the DNA sugar-phosphate backbone (mainly in the vicinity of negatively charged phosphates) as well as the grooves of the DNA helix (mainly near the nitrogen electronegative atoms of guanine and adenine). Interestingly enough, in the case of more realistic double-stranded chains completely analogous results are obtained for the length dependence of the transport coefficients as well as for their sensitivity to minor changes in their sequence composition. For the sake of illustration, in Fig. 6.11 we compare the Seebeck coefficient as a function of the energy for different values of the sugar-phosphate on-site energy, γ, for a polyG-polyC chain with

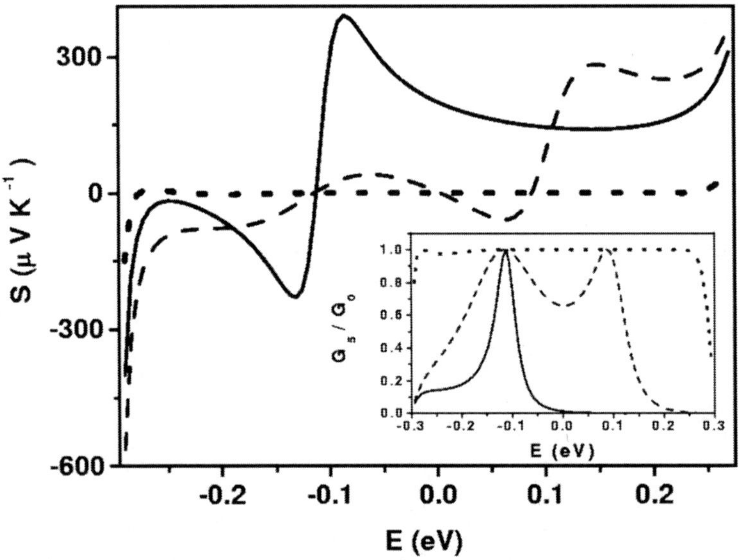

Figure 6.11 Room temperature Seebeck coefficient as a function of the Fermi energy for a polyG-polyC oligomer with $N = 5$ base pairs and $\gamma =$ 4.5 eV (solid curve), $\gamma = 4.0$ eV (dashed curve), and $\gamma = 3.0$ eV (dotted curve) with $\tau = t_M = 0.15$ eV, and $\varepsilon_M = 0$ eV. Inset: Landauer conductance as a function of the Fermi level energy for the same samples shown in the main frame [298]. Reprinted with permission from E. Maciá 2007 *Phys. Rev. B* **75** 035130, Copyright 2007, American Physical Society.

$N = 5$ Watson–Crick base pairs. By inspecting this plot we realize the remarkable role played by environmental effects on the Seebeck coefficient value. In fact, by systematically varying the on-site energy parameter from $\gamma = 4.5$ eV (no environmental effects) to $\gamma = 3.0$ eV, the TE response of the DNA chain can be modulated from typically semiconducting values to typically metallic ones. As expected from Eqs. (6.7) and (6.8), this Seebeck coefficient reduction is related to a progressive enhancement of the DNA conductance. This result is shown in the inset of Fig. 6.11, where we plot the systematic variation of the polyG-polyC oligomer conductance as γ is progressively decreased.

In summary we conclude that the TE response of short double-strand DNA chains strongly depends on the relative position between the contacts Fermi level and the DNA molecular levels.

Thus, while the TE power of polyA-polyT oligomers is quite insensitive to the number of base pairs composing the chain, polyG-polyC oligomers exhibit a strong dependence on the chain length. Accordingly, we can efficiently optimize the power factor of DNA chains by properly shifting the Fermi level position close to the resonance energy, which plays the role of a tuning parameter. On the other hand, depending on the E_F position, n-type and p-type TE responses can be simultaneously obtained for polyA-polyT and polyG-polyC DNA chains, respectively. This is a very convenient feature in order to design DNA-based TE devices, where both oligomers would play the role that semiconducting materials legs usually play in standard Peltier cells (Fig. 6.7). To this end, the relatively low value of the polyA-polyT chain Seebeck coefficient could be significantly improved by connecting it to adenine wires, rather than guanine ones, in order to get a proper alignment between the contacts Fermi level and the DNA molecular levels.

By properly combining the previous results, making use of the typical values $L_N = 0.34 \times N$ nm for the length, and $R = 1$ nm for the radius of B-form DNA, we can determine the magnitude of the TE power factor $P_N = \sigma_N S_N^2 = G_N L_N S_N^2/(\pi R^2)$ for the considered samples. In the case $N = 1$ the power factor takes on relatively small values over a broad range of energies located around the conductance peak, but it significantly increases as the Fermi level approaches the band edges, as it was previously discussed. In the case $N = 5$, in addition to this general behavior the power factor also attains significantly large values close to the resonance energy of the polyG-polyC chain due to the presence of the above mentioned Seebeck coefficient peaks. The values of the power factor maxima attained in this case ($P_5 = 15$–30 μWcm^{-1}K^{-2}) nicely fit with those reported for benchmark TEMs. The potential of DNA oligomers as TE materials will ultimately depend on their thermal transport properties, which according to the data presented in Section 6.2.2 will fall within the range $\kappa = 0.6$–0.8 Wm^{-1}K^{-1} at room temperature. By taking the value $\kappa \simeq 0.8$ Wm^{-1}K^{-1} as a conservative estimation, from the power factors values previously obtained, we get $ZT \simeq 0.6$–1.5 for polyG-polyC chains with five base pairs at room temperature. These remarkably high figure of merit values must be properly balanced with the significant role played by unavoidable environmental effects, stemming from the presence of a cation–water molecules atmosphere around the DNA chain, on the actual TE efficiency of DNA-based nano-cells.

6.4 Exercises

(1) (a) Room temperature transport coefficients for pentacene measured at different charge carrier concentration values read $\sigma_1 = 0.01$ ($\sigma_2 = 1.5 \times 10^{-4}$) $\Omega^{-1}\text{cm}^{-1}$ and $S_1 = +300$ ($S_2 = +600$) μVK^{-1}, respectively [272]. Obtain the corresponding power factors.

(b) At room temperature pentacene films obey the relationship given by Eq. (2.1). Their optimal TE performance was found for a carrier concentration $n = 2.5 \times 10^{20}$ cm^{-3} where a mobility $\mu = 0.01$ cm^{2}V^{-1}s^{-1} was obtained [272]. Determine the corresponding power factor and the FOM value ($\kappa = 0.51$ Wm^{-1}K^{-1} at room temperature for this material [271]).

(2) In order to improve its TE performance a thin film of pentacene was doped by co-evaporation with the acceptor organic molecule F_4-TCNQ. The transport coefficients at best doping concentrations read $\sigma_1 = 2.5 \times 10^{-2}$ $\Omega^{-1}\text{cm}^{-1}$ and $S_1 = +270$ μVK^{-1}at room temperature. Subsequently, samples with a bilayer structure, composed of undoped pentacene capped with a 2 nm thick F_4-TCNQ layer were grown. At the optimal pentacene thickness (6 nm) the transport coefficients read $\sigma_2 = 0.43$ $\Omega^{-1}\text{cm}^{-1}$ and $S_2 = +200$ μVK^{-1} at room temperature. The thermal conductivity of the pentacene-undoped layer is $\kappa = 0.48$ Wm^{-1}K^{-1} at room temperature [273]. Determine: (a) the power factor and (b) the FOM in both cases.

(3) The room temperature transport coefficient values of a composite formed by Bi_2Te_3 nanorods and polyaniline nanoparticles read $\sigma = 12$ $\Omega^{-1}\text{cm}^{-1}$, $S = +38$ μVK^{-1} and $\kappa = 0.11$ Wm^{-1}K^{-1}. The charge carrier density derived from Hall coefficient measurements is $n = 8.9 \times 10^{17}$ cm^{-3} [278]. Determine the carriers mobility, the power factor and the FOM of the composite.

(4) The radius of the duplex DNA is $r = 1$ nm and the distance between neighboring base pairs along the helix axis is $d = 0.34$ nm (hydrated B-DNA form). Making use of these data determine the electrical conductivity of the DNA samples indicated in the text.

(5) The specific heat of a calf thymus DNA sample ($d = 1.7$ gcm^{-3}) is $C_p/T^3 = 0.012$ Jkg^{-1}K^{-4} within the temperature range 0.5–5 K [290]. The mean sound speed of DNA is $v = 2000$ ms^{-1}. Estimate the thermal conductivity of this sample at $T = 5$ K.

6.5 Solutions

(1) (a) The power factor is given by $P = \sigma S^2$ so that we get,

$$P_1 = 10^{-2}\ \Omega^{-1}\text{cm}^{-1}\ 300^2 \times 10^{-6}\ \mu\text{V}^2\text{K}^{-2}$$
$$= 9 \times 10^{-4}\ \mu\text{Wcm}^{-1}\text{K}^{-2}$$
$$P_2 = 1.5 \times 10^{-4}\ \Omega^{-1}\text{cm}^{-1}\ 600^2 \times 10^{-6}\ \mu\text{V}^2\text{K}^{-2}$$
$$= 5.4 \times 10^{-5}\ \mu\text{Wcm}^{-1}\text{K}^{-2}.$$

Therefore, we obtain very low power factor values albeit the Seebeck coefficient values are quite large.

(b) The electrical conductivity at the optimal concentration is given by

$$\sigma = en\mu = 1.602 \times 10^{-19}\ \text{C}\ 2.5 \times 10^{20}\ \text{cm}^{-3}\ 10^{-2}\ \text{cm}^2\text{V}^{-1}\text{s}^{-1}$$
$$= 0.4\ \Omega^{-1}\text{cm}^{-1}. \tag{6.13}$$

On the other hand, making use of Eq. (2.1) we have

$$\frac{S_1}{S_2} = \frac{A - \ln\sigma_1}{A - \ln\sigma_2} \Rightarrow A = \frac{S_1 \ln\sigma_2 - S_2 \ln\sigma_1}{S_2 - S_1} = 0.405.$$

Making use of this value along with the electrical conductivity value given by Eq. (6.13) into Eq. (2.1), we obtain

$$S = 87\ \mu\text{VK}^{-1}\ (0.405 - \ln 0.4) = 115\ \mu\text{VK}^{-1}$$

Finally, the power factor and the FOM are respectively given by

$$P = 0.4\ \Omega^{-1}\text{cm}^{-1}\ 115^2 \times 10^{-6}\ \mu\text{V}^2\text{K}^{-2}$$
$$= 5.3 \times 10^{-3}\ \mu\text{Wcm}^{-1}\text{K}^{-2}$$

which substantially improves the figures previously obtained in (a), and

$$ZT = 0.4\ \Omega^{-1}\text{cm}^{-1}\ 10^2\ \text{m}^{-1}\text{cm}\ \frac{115^2 \times 10^{-12}\ \text{V}^2\text{K}^{-2} \times 300\ \text{K}}{0.51\ \text{Wm}^{-1}\text{K}^{-1}}$$
$$= 3.1 \times 10^{-4}$$

This value is comparable to that obtained for decagonal quasicrystals in Chapter 5.

(2) (a) The power factor is given by $P = \sigma S^2$ so that we get,

$$P_1 = 0.025\ \Omega^{-1}\text{cm}^{-1}\ 270^2 \times 10^{-6}\ \mu\text{V}^2\text{K}^{-2}$$
$$= 1.8 \times 10^{-3}\ \mu\text{Wcm}^{-1}\text{K}^{-2}$$
$$P_2 = 0.43\ \Omega^{-1}\text{cm}^{-1}\ 200^2 \times 10^{-6}\ \mu\text{V}^2\text{K}^{-2}$$
$$= 1.7 \times 10^{-2}\ \mu\text{Wcm}^{-1}\text{K}^{-2}$$

for the doped layer and the bilayer structure, respectively.

(b) The corresponding FOM are given by

$$(ZT)_1 = \frac{P_1 T}{\kappa} = \frac{1.8 \times 10^{-3}\ \mu\text{Wcm}^{-1}\text{K}^{-2}\ 10^2\ \text{m}^{-1}\text{cm}\ 300\ \text{K}}{0.48\ \text{Wm}^{-1}\text{K}^{-1}}$$
$$= 1.13 \times 10^{-4}$$
$$(ZT)_2 = \frac{P_2 T}{\kappa} = \frac{1.7 \times 10^{-2}\ \mu\text{Wcm}^{-1}\text{K}^{-2}\ 10^2\ \text{m}^{-1}\text{cm}\ 300\ \text{K}}{0.48\ \text{Wm}^{-1}\text{K}^{-1}}$$
$$= 1.06 \times 10^{-3}$$

By comparing these figures with those obtained in the previous exercise we realize that the bilayer structure significantly improves both the power factor and ZT values in pentacene-based systems [273].

(3) From the knowledge of the charge carriers density their mobility can be derived from the expression

$$\mu = \sigma R_H = \frac{\sigma}{en} = \frac{12\ \Omega^{-1}\text{cm}^{-1}}{1.602 \times 10^{-19}\ \text{C}\ 8.9 \times 10^{17}\ \text{cm}^{-3}}$$
$$\simeq 84\ \text{cm}^2\text{V}^{-1}\text{s}^{-1}$$

where we have made use of the dimensional relationship $[\text{V}] = [\Omega\text{A}] = [\Omega\text{Cs}^{-1}]$. Thus, we see that the very low charge carriers density is compensated by a relatively large mobility in order to yield a moderate electrical conductivity. The power factor is given by

$$P = 12\ \Omega^{-1}\text{cm}^{-1}\ 38^2 \times 10^{-6}\ \mu\text{V}^2\text{K}^{-2} = 0.017\ \mu\text{Wcm}^{-1}\text{K}^{-2}$$

which coincides with that obtained in the Exercise 6.2b for a sample with a lower conductivity and a larger Seebeck coefficient. Finally, the FOM is

$$ZT = \frac{PT}{\kappa} = \frac{0.017\ \mu\text{Wcm}^{-1}\text{K}^{-2}\ 10^2\ \text{m}^{-1}\text{cm}\ 300\ \text{K}}{0.11\ \text{Wm}^{-1}\text{K}^{-1}} = 4.6 \times 10^{-3}$$

which is four times larger than the ZT value obtained for the sample considered in the Exercise 6.2b. As we see, this improvement entirely results from the four-fold lower thermal conductivity measured in the composite structure.

(4) The Ohm's law can be expressed in the form

$$\Delta V - IR = I\rho\frac{L}{A},$$

where $A = \pi r^2$ is the cross-section of the DNA molecule, and $L = Nd$ is its length, where N is the number of base pairs. Thus,

$$\sigma = \frac{Nd}{\pi r^2}\frac{I}{\Delta V}.$$

The considered DNA samples contained $N = 80$ base pairs and the current was measured under a potential drop of $\Delta V = 1$ V, so that one obtains

$$\sigma = \frac{80 \times 0.34 \times 10^{-7}\ \text{cm}}{\pi \times 10^{-14}\ \text{cm}^2}\frac{25 \times 10^{-12}\ \text{A}}{1\ \text{V}} = 2.2 \times 10^{-3}\ (\Omega\text{cm})^{-1}$$

$$\sigma = \frac{80 \times 0.34 \times 10^{-7}\ \text{cm}}{\pi \times 10^{-14}\ \text{cm}^2}\frac{40 \times 10^{-12}\ \text{A}}{1\ \text{V}} = 3.5 \times 10^{-3}\ (\Omega\text{cm})^{-1},$$

for the double-stranded DNA molecules, and

$$\sigma = \frac{80 \times 0.34 \times 10^{-7}\ \text{cm}}{\pi \times 10^{-14}\ \text{cm}^2}\frac{0.5 \times 10^{-12}\ \text{A}}{1\ \text{V}} = 4.3 \times 10^{-5}\ (\Omega\text{cm})^{-1}$$

$$\sigma = \frac{80 \times 0.34 \times 10^{-7}\ \text{cm}}{\pi \times 10^{-14}\ \text{cm}^2}\frac{1.5 \times 10^{-12}\ \text{A}}{1\ \text{V}} = 1.3 \times 10^{-4}\ (\Omega\text{cm})^{-1},$$

for the single-stranded DNA molecules.

(5) Making use of Eq. (1.31)

$\kappa_l = \frac{d}{3} C_p \bar{v} l$, we get

$$\kappa_l = \frac{1.7}{3} \times 10^3 \text{ kgm}^{-3} \times 1.5 \text{ Jkg}^{-1}\text{K}^{-1} \times 2 \times 10^3 \text{ ms}^{-1} \times 3.4 \times 10^{-8} \text{ m}$$

$$= 0.058 \text{ Wm}^{-1}\text{K}^{-1}$$

where we have assumed that the phonons-mean free path is of about 100 base pairs (34 nm) at $T = 5$ K.

Bibliography

1. T. J. Seebeck, Magnetische Polarisation der Metalle und Erze durch Temperatur-Differenz, *Abh. K. Akad. Wiss.* 265–373 (1823).
2. J. C. Peltier, Nouvelles espériences sur la caloricité des courants électriques, *Ann. Chim. Phys.* **56**, 371–385 (1834).
3. P. Tunbridge, A letter by William Thomson on the "Thomson effect", *Notes and Records of the Royal Society London*, Vol. **26**, No. 2, December 1971.
4. L. Onsager, Reciprocal relations in irreversible processes, *Phys. Rev.* **37**, 405 (1931).
5. H. Bohr, *Collected Mathematical Works. II Almost periodic functions*, Dansk Matematisk Forening, Copenhagen (1952).
6. IUCr Report of the Executive Committee for 1991 *Acta Cryst. A* **48**, 922 (1992). See also http://www.iucr.ac.uk/iucr-top/comm/capd/terms.html. 1971.
7. A. F. Ioffe, *Semiconductor Thermoelements and Thermoelectric Cooling*, Infosearch Ltd., London (1957).
8. J. M. Ziman, *Electrons and Phonons*, Clarendon, Oxford (1960).
9. N. W. Ashcroft and N. D. Mermin, *Solid State Physics*; Saunders College Publ., Cornell (1976).
10. U. Mizutani, *Introduction to the Electron Theory of Metals*, Cambridge University Press, Cambridge (2001).
11. G. S. Nolas, J. Sharp, and H. J. Goldsmid, *Thermoelectrics Basic Principles and New Materials Developments*, Springer-Verlag, Berlin (2001).
12. A. D. Avery and B. L. Zink Peltier cooling and Onsager reciprocity in ferromagnetic thin films *Phys. Rev. Lett.* **111**, 126602 (2013).
13. T. C. Harman and J. M. Honig, *Thermoelectric and Thermomagnetic Effects and Applications* , McGraw-Hill, New York (1967).
14. N. F. Mott and F. A. Davis, *Electronic Processes in Non-crystalline Materials*, Clarendon Press, Oxford (1979).

15. G. D. Mahan, *Solid State Physics vol. 51: Solid State Physics-Advances in Research and Applications*, Academic, New York (1988), pp. 81–157.
16. G. D. Mahan, *Many Particle Physics*, Plenum, New York (1990).
17. A. G. Slack, In *CRC Handbook of Thermoelectrics*, D. M. Rowe, Ed., CRC Press, Boca Raton, FL (1995).
18. A. P. Tsai, In *Physical Properties of Quasicrystals*, Z. M. Stadnik, Ed.; Springer Series in Solid-State Physics Vol. **126**, Springer, Berlin (1999), p. 5.
19. L. E. Sigler, *Fibonacci's Liber Abaci: A Translation into Modern English of Leonardo Pisano's Book of Calculation Sources and Studies in the History of Mathematics and Physical Sciences*, Springer, New York (2002).
20. M. Baake, In *Quasicrystals: An introduction to Structure, Physical Properties, and Applications*, Suck J. B., Schreiber M., and H.äussler P, Ed., Springer, Berlin (2002), p. 17.
21. J. M. Dubois, *Useful Quasicrystals*, World Scientific, Singapore, (2005).
22. W. Steurer and S. Deloudi, *Crystallography of Quasicrystals—Concepts, Methods and Structures*, Springer Series in Materials Science **126**, Springer-Verlag, Berlin (2009).
23. T. Janssen, G. Chapuis, and M. de Boissieu, *Aperiodic Crystals: From Modulated Phases to Quasicrystals*, IUCr Monographs on Crystallography **20,** Oxford University Press, Oxford (2009).
24. E. Maciá, *Aperiodic Structures in Condensed Matter: Fundamentals and Applications*, CRC Press, Boca Raton, FL (2009).
25. U. Mizutani, *Hume-Rothery Rules for Structurally Complex Alloy Phases*, CRC Press, Boca Raton, FL (2011).
26. J. M. Dubois and E. Belin-Ferré, *Introduction to the science of complex metallic alloys* in *Complex Metallic Alloys. Fundamentals and Applications*, J. M. Dubois and E. Belin-Ferré, Eds., Wiley-VCH, Weinheim (2011), p. 1.
27. E. Maciá and M. de Boissieu, *Properties of CMAs: theory and experiments* in *Complex Metallic Alloys. Fundamentals and Applications*, J. M. Dubois and E. Belin-Ferré, Eds; Wiley-VCH, Weinheim (2011), p. 41.
28. J. J. Jin and J. Gerard (eds.), *Materials Science of DNA*, CRC Press, Boca Raton, FL (2012).
29. C. J. Liu, Thermopower of misfit-layered calcium cobaltites, in *Thermoelectric Power*, W. P. Dempsey, Ed., Nova Science Pub., New York (2012).

30. C. Wood, Materials for thermoelectric energy conversion, *Rep. Prog. Phys.* **51**, 459 (1988).
31. R. G. Endres, D. L. Cox, and R. R. P. Singh, *Colloquium*: The quest for high-conductance DNA, *Rev. Mod. Phys.* **76**, 195 (2004).
32. E. Maciá, The role of aperiodic order in science and technology, *Rep. Prog. Phys.* **69**, 397 (2006).
33. L. Li, Z. Chen, M. Zhou, and R. Huang, Developments in semiconductor thermoelectric materials, *Front. Energy* **5**, 125 (2011).
34. N. Dubey and M. Leclerc, Conducting polymers: Efficient thermoelectric materials, *Polymer Phys.* **49**, 467 (2011).
35. Y. Dubi and M. Di Ventra, *Colloquium*: Heat flow and thermoelectricity in atomic and molecular junctions, *Rev. Mod. Phys.* **83**, 131 (2011).
36. M. Zebarjadi, K. Esfarjani, M. S. Dresselhaus, Z. F. Ren, and G. Chen, Perspectives on thermoelectrics: from fundamentals to device applications, *Energy Environ. Sci.* **5**, 5147 (2012).
37. M. Martín-González, O. Caballero-Calero, and P. Díaz-Chao, Nanoengineering thermoelectrics for 21st century: Energy harvesting and other trends in the field, *Renew. Sustainable Energy Rev.* **24**, 288 (2013).
38. Y. Pei, H. Wang, and G. J. Snyder, Band engineering of thermoelectric materials, *Adv. Mater.* **24**, 6125 (2012).
39. K. Schmalzl, D. Strauch, and H. Schiber, Lattice-dynamical and ground state properties of CaF_2 studied by inelastic neutron scattering and density-functional methods, *Phys. Rev. B* **68**, 144301 (2003).
40. P. A. Kalugin, M. A. Chernikov, A. Bianchi, and H. R. Ott, Structural scattering of phonons in quasicrystals, *Phys. Rev. B* **53**, 14 145 (1996).
41. L. Qiu, I. P. Swainson, G. S. Nolas, and M. A. White, Structure, thermal and transport properties of the clathrates $Sr_8Zn_8Ge_{38}$, $Sr_8Ga_{16}Ge_{30}$, and $Ba_8Ga_{16}Si_{30}$, *Phys. Rev. B* **70**, 035208 (2004).
42. J. Sun, M. L. Yeh, B. J. Jung, B. Zhang, J. Feser, A. Majumdar, and H. E. Katz, Simultaneous increase in Seebeck coefficient and conductivity in a doped poly(alkylthiophene) blend with defined density of states, *Macromolecules* **43**, 2897 (2010).
43. K. Kuroki, and R. Arita, "Pudding mold" band drives large thermopower in Na_xCoO_2, *J. Phys. Soc. Jpn.* **76**, 083707 (2007).
44. M. V. Vedernikov and E. K. Iordanishvili, A. F. Iofee and the origin of modern semiconductor thermoelectric energy conversion, *17th International Conference on Thermoelectrics* (1998), pp. 37–42.

45. D. Kraemer, B. Poudel, H. P. Feng, J. C. Caylor, B. Yu, X. Yan, Y. Ma, X. Wang, D. Wang, A. Muto, K. McEnaney, M. Chiesa, Z. Ren, and G. Chen, High-performance flat-panel solar thermoeletric generators with high thermal concentration, *Nature Mater.* **10**, 532 (2011).
46. H. J. Goldsmid and R. W. Douglas, The use of semiconductors in thermoelectric refrigeration, *Brit. J. Appl. Phys.* **5**, 386 (1954).
47. R. V. Littleton IV, Terry M. Tritt, J. W. Kolis, and D. R. Ketchum, Transition-metal pentatellurides as potential low-temperature thermoelectric refrigeration materials, *Phys. Rev. B* **60**, 13453 (1999).
48. L. E. Bell , Cooling, heating generating power and recovering waste heat with thermoelectric systems, *Science* **321**, 1457 (2008).
49. E. J. Sandoz-Rosado, S. J. Weinstein, and R. J. Stevens, On the Thomson effect in thermoelectric power devices, *Int. J. Thermal Sci.* **66**, 1 (2013).
50. C. B. Vining, An inconvenient truth about thermoelectrics, *Nature Mater.* **8**, 83 (2009).
51. D. Nemir and J. Beck, On the significance of the thermoelectric figure of merit Z, *J. Electron. Mats.* **39,** 1897 (2011).
52. D. Narducci, Do we really need high thermoelectric figures of merit? A critical appraisal to the power conversion efficiency of thermoelectric materials, *Appl. Phys. Lett.* **99**, 102104 (2011).
53. G. Jeffrey Snyder, E. S. Toberer, R. Khanna, and W. Seifert, Improved thermoelectric cooling based on Thomson effect, *Phys. Rev. B* **86**, 045202 (2012).
54. G. Jeffrey Snyder, Application of the compatibility factor to the design of segmented and cascaded thermoelectric generators, *Appl. Phys. Lett.* **84**, 2436 (2004).
55. G. Jeffrey Snyder and Ursell T. S., Thermoelectric efficiency and compatibility, *Phys. Rev. Lett.* **91**, 148301 (2003).
56. C. Goupil, Thermodynamics of the thermoelectric potential, *J. Appl. Phys.* **106**, 104907 (2009).
57. H. Wang, Y. Oei, A. D. LaLonde, and G. J. Snyder, Weak electron-phonon coupling contributing to high thermoelectric performance in *n*-type PbSe, *Proc. Matl. Acad. Sci. USA* **109**, 9705 (2012).
58. T. Harman, J. H. Cahn, and M. Logan, Measurement of thermal conductivity by utilization of the Peltier effect, *J. Appl. Phys.* **30**, 1351 (1959).
59. S. Y. Kwon, Y. G. Kim, S. Lee, and J. C. Kim, Evaluation system for figure of merit of thermoelectric devices, *Jpn. J. Appl. Phys.* **50**, 11RE02 (2011).

60. J. de Boor and V. Schnidt, Efficient thermoelectric van der Pauw measurements, *Appl. Phys. Lett.* **99**, 022102 (2011).

61. J. M. Dubois, S. S. Kang, P. Archembault, and B. Colleret, Thermal diffusivity of quasicrystalline and related crystalline alloys, *J. Mat. Res.* **8**, 38 (1993).

62. E. Maciá, Thermal conductivity of complex metallic alloys: Beyond Wiedemann-Franz law, *Phys. Rev. B* **79**, 245119 (2009).

63. K. C. Lukas, W. S. Liu, G. Joshi, M. Zebarjadi, M. S. Dresselhaus, Z. F. Ren, G. Chen, and C. P. Opeil, Experimental determination of the Lorenz number in $Cu_{0.01}BiTe_{2.7}Se_{0.3}$ and $Bi_{0.88}Sb_{0.12}$, *Phys. Rev. B* **85**, 205410 (2012).

64. W. H. Butler and R. K. Williams, Electron-phonon interaction and thermal conductivity, *Phys. Rev. B* **18**, 6483 (1978).

65. E. S. Toberer, M. Christensen, B. B. Iversen, and G. Jeffrey Snyder, High temperature thermoelectric efficiency in $Ba_8Ga_{16}Ge_{30}$, *Phys. Rev. B* **77**, 075203 (2008).

66. M. P. Dieter, H. B. Mattheus, R. A. Jeffcoat, and R. F. Museman, Comparison of lead bioavailability in F344 rats fed lead acetate, lead oxide, lead sulfide, or lead ore concentrate from Skagway, Alaska, *J. Toxicol. Environ. Health* **39**, 79 (1993).

67. R. W. Christy, Electrical conductivity and thermoelectric power in ionic crystals, *Am. J. Phys.* **28**, 457 (1960).

68. A. F. May, E. S. Toberer, A. Saramat, and G. Jeffrey Snyder, Characterization and analysis of thermoelectric transport in *n*-type $Ba_8Ga_{16-x}Ge_{30+x}$, *Phys. Rev. B* **80**, 125205 (2009).

69. G. Rogl, A. Grytsiv, P. Rogl, E. Bauer, and M. Zehetbauer, A new generation of p-type didymium skutterudites with high ZT, *Intermetallics* **18**, 546 (2011).

70. S. B. Inayat, K. R. Rader, and M. M. Hussain, Nano-materials enabled thermoelectricity from window glasses, *Scientific Reports* **2**, 1 (2012).

71. G. S. Nolas and H. J. Goldsmid, The figure of merit in amorphous thermoelectrics, *Phys. Status Solidi A* **194**, 271 (2002).

72. A. F. May, J. P. Fleurial, and G. Jeffrey Snyder, Thermoelectric performance of lanthanum telluride produced via mechanical alloying, *Phys. Rev. B* **78**, 125205 (2008).

73. G. Jeffrey Snyder, M. Christensen, E. J. Nishibori, T. Caillat, and B. B. Iversen, Disordered zinc in Zn_4Sb_3 with phonon- glass and electron-crystal thermoelectric properties, *Nature Mater.* **3**, 458 (2004).

74. J. S. Rhyee, K. H. Lee, S. M. Lee, E. Cho, S. I. Kim, E. Lee, Y. S. Kwon, J. H. Shin, and G. Kotliar, Peierls distortion as a route to high thermoelectric performance in $In_4Se_{3-\delta}$ crystals, *Nature*, **459**, 965 (2009).

75. A. F. Lalonde, Y. Pei, and G. Jeffrey Snyder, Reevaluation of $PbTe_{1-x}I_x$ as high performance n-type thermoelectric material, *Energy Environ. Sci.* **4**, 2090 (2011).

76. Y. Pei, X. Shi, A. LaLonde, H. Wang, L. Chen, and G. Jeffrey Snyder, Convergence of electronic bands for high performance bulk thermoelectrics, *Nature*, **473**, 66 (2011).

77. K. F. Hsu, S. Loo, F. Guo, W. Chen, J. D. Dyck, C. Uher, T. Hogan, E. K. Polykroniadis, and M. G. Kanatzidis, Cubic AgPbSbTe: Bulk thermoelectric materials with high figure of merit, *Science*, **303**, 818 (2004).

78. R. Venkatasubramanian, E. Silvola, T. Colpitts, and B. O'Quinn, Thin-film thermoelectric devices with high room-temperature figures of merit, *Nature*, **413**, 597 (2001).

79. T. C. Harman, P. J. Taylor, M. P. Walsh, and B. E. LaForge, Quantum dot superlattice thermoelectric materials and devices, *Science*, **297**, 2229 (2002).

80. H. Littman and B. Davidson, Theoretical bound on the thermoelectric figure of merit from irreversible thermodynamics, *J. Appl. Phys.* **32**, 217 (1961).

81. A. F. May, E. Falge-Larsen, and G. J. Snyder, Electron and phonon scattering in the high-temperature thermoelectric $La_3Te_{4-z}M_z$ (M = Sb, Bi), *Phys. Rev. B* **81**, 125205 (2010).

82. G. Min and D. M. Rowe, Thermoelectric figure of merit barrier at minimum lattice thermal conductivity, *Appl. Phys. Lett.* **77**, 860 (2000).

83. D. M. Rowe and G. Min, α-In σ plot as a thermoelectric material indicator, *J. Mater. Lett.* **14**, 617 (1995).

84. E. J. Skoug and D. T. Morelli, Role of lone-pair electrons in producing minimum thermal conductivity in nitrogen-group chalcogenide compounds, *Phys. Rev. Lett.* **107**, 235901 (2011).

85. Y. Zhang, E. Skoug, J. Cain, V. Ozolinš, D. Morelli, and C. Wolverton, First-principles description of anomalously low lattice thermal conductivity in thermoelectric Cu-Sb-Te ternary semiconductors, *Phys. Rev. B* **85**, 054306 (2012).

86. J. S. Rhyee, K. H. Lee, S. M. Lee, E. Cho, S. I. Kim, E. Lee, Y. S. Kwon, J. H. Shim, and G. Kotliar, Peierls distortion as a route to high thermoelectric performance in $In_4Se_{3-\delta}$ crystals, *Nature (London)* **459**, 965 (2009).

87. D. G. Cahill, S. K. Watson, and R. O. Pohl, Lower limit to the thermal conductivity of disordered crystals, *Phys. Rev. B* **46**, 6131 (1992).
88. R. O. Pohl, X. Liu, and E. Thompson, Low-temperature thermal conductivity and acoustic attenuation in amorphous solids, *Rev. Mod. Phys.* **74**, 991 (2002).
89. W. P. Hsieh, M. D. Losego, P. V. Braun, S. Shenogin, P. Keblinski, and D. G. Cahill, Testing the minimum conductivity model for amorphous polymers using high pressure, *Phys. Rev. B* **83**, 174205 (2011).
90. L. D. Hicks and M. S. Dresselhaus, Effect of quantum-well structures on the thermoelectric figure of merit, *Phys. Rev. B* **47**, 12727 (1993).
91. L. D. Hicks and M. S. Dresselhaus, Thermoelectric figure of merit of a one-dimensional conductor, *Phys. Rev. B* **47**, 16631 (1993).
92. H. Beyer, J. Nurnus, H. Böttner, A. Lambrecht, T. Roch, and G. Bauer, PbTe based superlattice structures with high thermoelectric efficiency, *Appl. Phys. Lett.* **80**, 1216 (2002).
93. D. A. Broido and T. L. Reinecke, Thermoelectric transport in quantum well superlattices, *Appl. Phys. Lett.* **70**, 2835 (1997).
94. D. A. Broido and T. L. Reinecke, Theory of thermoelectric power factor in quantum well and quantum wire superlattices, *Phys. Rev. B* **64**, 045324 (2001).
95. J. E. Cornett and O. Rabin, Universal scaling relations for the thermoelectric power factor of semiconducting nanostructures, *Phys. Rev. B* **84**, 205410 (2011).
96. P. Pichaunusakorn and P. R. Bandaru, The optimal Seebeck coefficient for obtaining the maximum power factor in thermoelectrics, *Appl. Phys. Lett.* **94**, 223108 (2009).
97. K. Durczewski and A. Krzywicki, Enhanced thermoelectric power and cooling ability near the Anderson transition, *Phys. Rev. B* **58**, 10302 (1998).
98. T. Takeuchi, Conditions of electronic structure to obtain large dimensionless figure of merit for developing practical thermoelectric materials, *Mat. Trans.* **50**, 2359 (2009).
99. E. Maciá, May quasicrystals be good thermoelectric materials?, *Appl. Phys. Lett.* **77**, 3045 (2000).
100. Z. Fan, H. Q. Wang, and J. C. Zheng, Searching for the best thermoelectrics through the optimization of transport distribution function, *J. Appl. Phys.* **109**, 073713 (2011).
101. J. E. Enderby and A. C. Barnes, Electron transport at the Anderson transition, *Phys. Rev. B* **49,** 5062 (1994).

102. E. Maciá, Compatibility factor of segmented thermoelectric generators based on quasicrystalline alloys, *Phys. Rev. B* **70**, 100201(R) (2004).

103. D. Nozaki, H. Sevinçli, W. Li, R. Gutiérrez, and G. Cuniberti, Engineering the figure of merit and thermopower in single molecule devices connected to semiconducting electrodes, *Phys. Rev. B* **81,** 235406 (2010).

104. G. D. Mahan and J. O. Sofo, The best thermoelectric, *Proc. Natl. Acad. Sci. USA* **93**, 7436 (1996).

105. E. Maciá, Theoretical prospective of quasicrystals as thermoelectric materials, *Phys. Rev. B* **64**, 094206 (2001).

106. E. Maciá, Quasicrystals as thermoelectric materials: A theoretical prospective, *J. Alloys Compoun.* **342**, 460 (2002).

107. G. S. Nolas, J. Yang, and R. W. Ertenberg, Transport properties of $CoGe_{1.5}Se_{1.5}$, *Phys. Rev. B* **68**, 193206 (2003).

108. J. L. Cohn, G. S. Nolas, V. Fessantidis, T. H. Metcalf, and G. A. Slack, Glassline heat conduction in high mobility crystalline semiconductors, *Phys. Rev. Lett.* **82**, 779 (1999).

109. G. A. Slack and V. G. Tsoukala, Some properties of semiconducting $IrSb_3$, *J. Appl. Phys.* **76**, 1665 (1994).

110. M. Christensen, A. B. Abrahamsen, N. B. Christensen, D. Juranyi, N. H. Andersen, K. Lefmann, J. Andreasson, C. R. H. Bahl, and B. B. Iversen, Avoided crossing of rattler modes in thermoelectric materials, *Nature Mats.* **7**, 811 (2008).

111. C. G. Gallo, B. S. Chandrasekhar, and P. H. Sutter, Transport properties of bismuth single crystals, *J. Appl. Phys.* **34**, 144 (1963).

112. Y. Liu and E. Allen, Electronic structure of semimetals Bi and Sb, *Phys. Rev. B* **52**, 1566 (1995).

113. G. A. Slack, Thermal conductivity of elements with complex lattices; B, P, S, *Phys. Rev.* **139**, A 507 (1965).

114. G. A. Slack, D. W. Oliver, and F. H. Horn, Thermal conductivity of boron and some boron compounds, *Phys. Rev. B* **4**, 1714 (1971).

115. K. C. Lukas, G. Joshi, K. Modic, Z. F. Ren, and C. P. Opeil, Thermoelectric properties of Ho-doped $Bi_{0.88}Sb_{0.12}$, *J. Mater. Sci.* **47**, 5729 (2012).

116. B. Lenoir, A. Dauscher, M. Cassart, Yu. I. Ravich and H. Scherrer, Effect of antimony content on the thermoelectric figure of merit of $Bi_{1-x}Sb_x$ alloys, *J. Phys. Chem. Solids*, **59**(1), 129–134 (1998).

117. J. J. Shen, T. J. Zhu, X. B. Zhao, S. N. Zhang, S. H. Yang, Z. Z. Yin, *Energy Environ. Sci.* **3**, 1519 (2010).

118. F. D. Rosi, B. Abeles, and R. V. Jensen, Materials for thermoelectric refrigeration, *J. Phys. Chem. Solids* **10**, 191 (1959).

119. S. K. Mishra, S. Satpathy, and O. Jepsen, Electronic structure and thermoelectric properties of bismuth telluride and bismuth selenide, *J. Phys. Condens. Matter* **9**, 461 (1997).

120. S. Wang, W. Hie, H. Li, and X. Tang, Enhanced performances of melt spun $Bi_2(Te,Se)_3$ for n-type thermoelectric legs, *Intermetallics* **19**, 1024 (2011).

121. J. F. Li, W. S. Liu, L. D. Zhao, and M. Zhou, High-performance nanostructured thermoelectric materials, *NPG Asia Mater.* **2**, 152 (2010).

122. P. Jund, R. Viennois, X. Tao, K. Niedziolka, and J. C. Tédenac, Physical properties of thermoelectric zinc antimonide using first-principles calculations, *Phys. Rev. B* **85,** 224105 (2012).

123. C. Okamura, T. Useda, and K. Hasezaki, Preparation of single-phase ZnSb thermoelectric materials using a mechanical grinding process, *Mater. Trans.* **51**, 860 (2010).

124. W. Chen and J. Li, Origin of the low thermal conductivity of the thermoelectric material β-Zn_4Sb_3: An ab initio theoretical study, *Appl. Phys. Lett.* **98,** 241901 (2011).

125. Y. L. Pei and Y. Liu, Electrical and thermal transport properties of Pb-based chalcogenides: PbTe, PbSe, and PbS, *J. Alloys Compounds* **514**, 40 (2012).

126. P. K. Rawat, B. Paul, and P. Banerji, Thermoelectric properties of $PbSe_{0.5}Te_{0.5}$: x (PbI_2) with endotaxial nanostructure: A promising n-type material, *Nanotechnology* **24**, 215401 (2013).

127. J. P. Heremans, V. Jokovic, E. S. Toberer, A. Saramat, K. Kurosaki, A. Charoenphakdee, S. Yamanaka, and G. Jeffrey Snyder, Enhancement of thermoelectric efficiency in PbTe by distortion of the electronic density of states, *Science* **321**, 554 (2008).

128. G. W. Crabtree and N. S. Lewis, Solar energy conversion, *Phys. Today* **60** (3), 37 (2007).

129. H. Wang, J. F. Li, M. Zou, and T. Sui, Synthesis and transport properties of $AgSbTe_2$ as a promising thermoelectric compound, *Appl. Phys. Lett.* **93**, 202106 (2008).

130. E. Skrabeck and D. Trimmer, U. S. Patent no. 3945855 (1976).

131. K. F. Hsu, S. Loo, F. Guo, W. Chen, J. S. Dyck, C. Uher, T. Hogan, E. K. Polychroniadis, and M. G. Kanatzidis, Cubic AgPbSbTe: Bulk thermoelectric materials with high figure of merit, *Science* **303**, 919 (2004).

132. A. Kosuga, M. Uno, K. Kurosami, and S. Yamanaka, Thermoelectric properties of stoichiometric $Ag_{1-x}Pb_{18}SbTe_{20}$, *J. Alloys Compounds* **391**, 288 (2005).

133. A. Kosuga, M. Uno, K. Kurosami, and S. Yamanaka, Thermoelectric properties of stochiometric $Ag_{1-x}Pb_{18}SbTe_{20}$ ($x = 0$, 0.1, 0.3), *J. Alloys Compounds* **387**, 52 (2005).

134. N. Chen, F. Gascoin, G. Jeffrey Snyder, E. Müller, G. Karpisnski, and C. Stiewe, Macroscopic thermoelectric inhomogeneities in $(AgSbTe_2)_x(PbTe)_{1-x}$, *Appl. Phys. Lett.* **87**, 171903 (2005).

135. J. Androulakis, K. F. Hsu, R. Pcionek, H. Kong, C. Uher, J. J. D'Angelo, A. Downey, T. Hogan, and M. G. Kanatzidis, Nanostructuring and high thermoelectric efficiency in p-type $Ag(Pb_{1-x}Sn_x)_mSbTe_{2+m}$, *Adv. Mater.* **18**, 1170 (2006).

136. P. F. P. Poudeu, A. Guéguen, T. Hogan, and M. G. Kanatzidis, High figure of merit in nanostructured n-type KPb_mSbTe_{m+2} thermoelectric materials, *Chem. Mater.* **22**, 1046 (2010).

137. B. Wölfing, C. Kloc, J. Teubner, and E. Bucher, High performance thermoelectric Tl_9BiTe_6 with an extremely low thermal conductivity, *Phys. Rev. Lett.* **86**, 4350 (2001).

138. S. Yamanaka, A. Kosuga, and K. Kurosaki, Thermoelectric properties of Tl_9BiTe_6, *J. Alloys Compound.* **352**, 275 (2003).

139. J. W. Sharp, B. C. Sales, D. G. Mandrus, and B. C. Chakoumakos, Thermoelectric properties of Tl_2SnTe_5 and Tl_2GeTe_5, *Appl. Phys. Lett.* **74**, 3794 (1999).

140. J. W. Sharp, US Patent no. 6.169.245.

141. M. G. Kanatzidis et al., New and old concepts in thermoelectric materials, *Angew. Chem. Int. Ed.* **48**, 8616 (2009).

142. K. Kurosaki, A. Kosuga, H. Muta, M. Uno, and S. Yamanaka, Ag_9TlTe_5: A high-performance thermoelectric bulk material with extremely low thermal conductivity, *Appl. Phys. Lett.* **87**, 061919 (2005).

143. K. Kurosaki, H. Uneda, H. Muta, and S. Yamanaka, Extremely low thermal conductivity of AgTlTe, *J. Alloys Compound.* **395**, 304 (2005).

144. D.-Y. Chung, T. Hogan, P. Brazis, M. Rocci-Lane, C. Kannewurf, M. Bastea, C. Uher, and M. G. Kanatzidis, $CsBi_4Te_6$: A high-performance

thermoelectric material for low-temperature applications, *Science* **287**, 1024 (2000).

145. H. Hohl, A. P. Ramirez, C. Goldmann, G. Ernst, B. Wölfing, and E. Bucher, Efficient dopants for ZrNiSb-based thermoelectric materials, *J. Phys. Condens. Matter* **11**, 1697 (1999).
146. C. Uher, J. Yang, and G. P. Meisner, Thermoelectric properties of Bi-doped half-Heusler alloys, 18th International Conference on Thermoelectrics (1999), p. 56.
147. D.-Y. Chung, T. P. Hogan, M. Rocci-Lane, P. Brazis, J. R. Ireland, C. R. Kannewurf, M. Bastea, C. Uher, and M. G. Kanatzidis, A new thermoelectric material: $CsBi_4Te_6$, *J. Am. Chem. Soc.* **126**(20), 6414–6428 (2004).
148. S. Sakurada and N. Shutoh, Effect of Ti substitution on the thermoelectric properties of (Zr,Hf)NiSn half-Heusler compounds, *Appl. Phys. Lett.* **86**, 082105 (2005).
149. K. Mastronardi, D. Young, C. C. Wang, P. Khalifah, R. J. Cava, and A. P. Ramirez, Antimonides with the half-Heusler structure: New thermoelectric materials, *Appl. Phys. Lett.* **74**, 1415 (1999).
150. D. P. Young, P. Khalifah, R. J. Cava, and A. P. Ramirez, Thermoelectric properties of pure and doped MFeSb (M=V, Nb), *J. Appl. Phys.* **87**, 317 (2000).
151. Y. Xia, S. Bhattacharya, V. Ponnambalam, A. L. Pope, S. J. Poon, and T. M. Tritt, Thermoelectric properties of semimetallic (Zr,Hf)CoSb half-Heusler phases, *J. Appl. Phys.* **88**, 1952 (2000).
152. T. Sekimoto, K. Kurosaki, H. Muta, and S. Yamanaka, Thermoelectric and thermophysical properties of TiCoSb, ZrCoSb, HfCoSb prepared by SPS, *International Conference on Thermoelectrics* 2005, p. 335
153. C. Yu, T. J. Zhu, R. Z. Shi, Y. Zhang, X. B. Zhao, and J. He, High performance half-Heusler thermoelectric materials HfZrNiSnSb prepared by levitation melting spark plasma sintering, *Acta Mater.***57**, 2757 (2009).
154. Q. Shen, L. Chen, T. Goto, T. Hirai, J. Yang, G. P. Meisner, and C. Uher, Effects of partial substitution of Ni and Pd on the thermoelectric properties of ZrNiSn-based half-Heusler compounds, *J. Appl. Phys.* **79**, 4165 (2001).
155. S. R. Culp, S. J. Poon, N. Hickman, T. M. Tritt, and J. Blumm, Effect of substitutions on the thermoelectric figure of merit of half-Heusler phases at 800°C, *J. Appl. Phys.* **88,** 042106 (2006).

156. H. Hazama, M. Matsubara, and R. Asahi, Thermoelectric properties of off stoichiometric TiNiSn half-Heusler systems, *J. Electronic Mats.* DOI: 10.1007/s11664-012-2848-z.

157. H. H. Xie, C. Yu, B. He, T. J. Zhu, and X. B. Zhao, Thermoelectric properties and n- to p-type conversion of Co-doped ZrNiSn based half-Heusler alloys, *J. Electronic Mats.* DOI: 10.1007/s11664-012-2077-7.

158. M. M. Koza, L. Capogna, A. Leithe-Jasper, H. Rosner, W. Schnelle, H. Mutka, M. R. Johnson, C. Ritter, and Y. Grin, Vibrational dynamics of filled skutterudites MFeSb (M = Ca, Sr, Ba, and Yb), *Phys. Rev. B* **81**, 174302 (2010).

159. W. Jeitschko and D. Braun, $LaFe_4P_{12}$ with filled $CoAs_3$-type structure and isotypic lanthanoid-transition metal polyphosphides, *Acta Crystallogr.* **B33**, 3401 (1977).

160. B. Chen, J. H. Xu, C. Uher, D. T. Morelli, G. P. Meisner, J. P. Fleurial, T. Caillat and A. Borshchevsky, Low-temperature transport properties of the filled skutterudites $CeFe_{4-x}Co_xSb_{12}$, *Phys. Rev. B* **55**, 1476 (1995).

161. V. L. Kuznetsov, L. A. Kuznetsova, and D. M. Rowe, Effect of partial void filling on the transport properties of $Nd_xCo_4Sb_{12}$ skutterudites, *J. Phys. Condens. Matter* **15**, 5035 (2003).

162. T. He, J. Chen, D. Rosenfeld, and M. A. Subramanian, Thermoelectric properties of indium-filled skutterudites, *Chem. Mater.* **18,** 759 (2006).

163. G. A. Lamberton Jr., S. Bhattacharya, R. T. Littleton IV, M. Kaeser, R. H. Tedstrom, T. M. Tritt, J. Yamg, and G. S. Nolas, High figure of merit in Eu-filled $CoSb_3$-based skutterudites, *Appl. Phys. Lett.* **80**, 598 (2002).

164. B. C. Sales, B. C. Chakoumakos, and D. Mandrus, Thermoelectric properties of thallium-filled skutterudites, *Phys. Rev. B* **61**, 2475 (2000).

165. G. S. Nolas, J. L. Cohn, and G. A. Slack, Effect of partial void filling on the lattice thermal conductivity of skutterudites, *Phys. Rev. B* **58**, 164 (1998).

166. N. R. Dilley, E. D. Bauer, M. B. Maple, and B. C. Sales, Thermoelectric properties of chemically substituted skutterudites $Yb_yCo_4Sn_xSb_{12-x}$, *J. Appl. Phys.* **88**, 1948 (2000).

167. J. F. Dyck, W. Chen, C. Uher, L. Chen, X. Tang, and T. Hirai, Thermoelectric properties of the n-type filled skutterudite $Ba_{0.3}Co_4Sb_{12}$ doped with Ni, *J. Appl. Phys.* **91**, 3698 (2002).

168. P. F. Quiu, J. Yang, R. H. Liu, X. Shi, X. Y. Huang, G. J. Snyder, W. Zhang, and L. D. Chen, High-temperature electrical and thermal transport

properties of fully filled skutterudites RFeSb (R=Ca, Sr, Ba, La,Ce, Pr, Nd, Eu, and Yb), *J. Appl. Phys.* **109**, 063713 (2011).

169. B. C. Sales, D. Mandrus, and R. K. Williams, Filled skutterudite antimonides: A new class of thermoelectric materials, *Science* **272**, 1325 (1996).
170. G. S. Nolas, M. Kaeser, R. T. Littleton IV, and T. M. Tritt, High figure of merit in partially filled ytterbium skutterudite materials, *Appl. Phys. Lett.* **77**, 1855 (2000).
171. M. Puyet, A. Dauscher, B. Lenoi, M. Dehmas, C. Stiewe, E. Müller, and J. Hejtmanek, Beneficial effect of Ni substitution on the thermoelectric properties in partially filled $Ca_yCo_{4-x}Ni_xSb_{12}$ skutterudites, *Appl. Phys.* **97**, 083712 (2005).
172. Y. Z. Pei, L. D. Chen, W. Zhang, X. Shi, S. Q. Bai, X. Y. Zhao, Z. G. Mei, and X. Y. Li, Synthesis and thermoelectric properties of KCoSb, *Appl. Phys. Lett.* **89,** 221107 (2006).
173. A. Watcharapasorn, R. S. Feigelson, T. Caillat, A. Vorxhcjevsky, G. Jeffrey Snyder, and J. P. Fleurial, Preparation and thermoelectric properties of $CeFe_4As_{12}$, *J. Appl. Phys.* **91**, 1344 (2002).
174. D. T. Morelli and G. P. Meisner, Low temperature properties of the filled skutterudite $CeFe_4Sb_{12}$, *J. Appl. Phys.* **77**, 3777 (1995).
175. B. C. Sales, D. Mandrus, B. C. Chakoumakos, V. Keppens, and J. R. Thompson, Filled skutterudites antimonides: Electron crystals and phonon glasses, *Phys. Rev. B* **56**, 15081 (1997).
176. M. M. Koza, A. Leithe-Jasper, H. Rosner, W. Schnelle, H. Mutka, M. R. Johnson, M. Krisch, L. Capogna, and Y. Grin, Vibrational dynamics of the filled skutterudites MFeSb (M=Ca, Sr, Ba, and Yb): Temperature response, dispersion relation, and material properties, *Phys. Rev. B* **84**, 014306 (2011).
177. R. P. Hermann, F. Grandjean, and G. J. Long, Einstein oscillators that impede thermal transport, *Am. J. Phys.* **73**, 110 (2005).
178. C. B. Vining, Half-full glasses, *Nature Mats.* **7**, 765 (2008).
179. X. Shi, J. Yang, J. R. Salvador, M. F. Chi, J. Y. Cho, H. Wang, S. Q. Bai, J. H. Yang, W. Q. Zhang, and L. D. Chen, Multiple-filled skutterudites: High thermoelectric figure of merit through separately optimizing electrical and thermal transports, *J. Am. Chem. Soc.* **133**, 7837 (2011).
180. G. Rogl, A. Grytsiv, P. Rogl, E. Bauer, M. B. Kerber, M. Zehetbauer, and S. Puchegger, Multifilled nanocrystalline p-type didymium - Skutterudites with ZT $>$ 1.2, *Intermetallics* **18**, 2435 (2010).

181. M. M. Koza, M. R. Johnson, R. Viennois, H. Mutka, L. Girard, and D. Ravot, Breakdown of phonon glass paradigm in La- and Ce-filled Fe_4Sb_{12} skutterudites, *Nature Mats.* **7**, 805 (2008).

182. D. Wee, B. Kozinsky, and M. Fornari, Frequency of filler vibrations in CoSb skutterudites: A mechanical interpretation, *J. Phys. Soc. Jpn.* **82**, 014602 (2013).

183. V. L. Kuznetsov, L. A. Kuznetsova, A. E. Kaliazin, and D. M. Rowe, Preparation and thermoelectric properties of ABB clathrate compounds, *J. Appl. Phys.* **87**, 7871 (2000).

184. G. S. Nolas, T. J. R Weakly, J. L. Cohn, and R. Sharma, Structural properties and thermal conductivity of crystalline Ge clathrates, *Phys. Rev. B* **61,** 3845 (2000).

185. A. Saramat, G. Svensson, A. E. C. Palmqvist, C. Stiewe, E. Mueller, D. Platzek, S. G. K. Williams, D. M. Rowe, J. D. Bryan, and G. D. Stucky, Large thermoelectric figure of merit at high temperature in Czochralski-grown clathrate $Ba_8Ga_{16}Ge_{30}$, *J. Appl. Phys.* **99**, 023708 (2006).

186. S. Paschen, W. Carrillo-Cabrera, A. Bentien, V. H. Tran, M. Baenitz, Yu. Grin, and F. Steglich, Structural, transport, magnetic, and thermal properties of $Eu_8Ga_{16}Ge_{30}$, *Phys. Rev. B* **64,** 214404 (2001).

187. N. A. Avila, K. Suekuni, K. Umeo, H. Fukuoka, S. Yamanaka, and T. Takabatake, $Ba_8Ga_{16}Si_{30}$ with type-I clathrate structure: Drastic suppression of heat conduction, *Appl. Phys. Lett.* **92**, 041901 (2008).

188. H. Euchner, S. Pailhès, L. T. K. Nguyen, W. Assmus, F. Ritter, A. Haghighirad, Y. Grin, S. Paschen, and M. de Boissieu, Phononic filter of rattling phonons in the thermoelectric clathrate $Ba_8Ga_{40+x}Ni_{16-x}$, *Phys. Rev B* **86**, 224303 (2012).

189. D. S. Dalafave, Thermoelectric properties of $Re_6Ga_xSe_yTe_{15-y}$ ($0 \leq x \leq 2$; $0 \leq y \leq 7.5$), *Mats. Chem. Phys.* **119**, 195 (2010).

190. Y. Wu, J. Nylén, C. Naseyowma, N. Newman, F. J. Garcia-Garcia, and U. Häussermann, Comparative study of the thermoelectric properties of amorphous $Zn_{41}Sb_{59}$ and crystalline Zn_4Sb_3, *Chem. Mater.* **21**, 151 (2009).

191. C. Roche, P. Percher, G. Toussaint, A. Jenny, H. Scherrer, and S. Scherrer, Study of Chevrel phases for thermoelectric applications: band structure calculations on MMo_6Se_8 compounds (M = metal), *J. Phys. Condens. Matter* **10,** L333 (1998).

192. G. S. Nolas, H. Takizawa, T. Endo, H. Sellinschegg, and D. C. Johnson, Thermoelectric properties of Sn-filled skutterudites, *Appl. Phys. Lett.* **77**, 52 (2000).

193. G. S. Nolas, G. A. Slack, D. T: Morelli, T. M. Tritt, and A. C. Ehrlich, The effect of rare-earth filling on the thermal conductivity of skutterudites, *J. Appl. Phys.* **79**, 4002 (1996).

194. R. D. Shannon, Revised effective ionic radii and systematic studies of interatomic distances in halides and chalcogenides, *Acta Crystallogr.* **A32**, 751 (1976).

195. A. B. Shick, J. B. Ketterson, D. L. Novikov, and A. J. Freeman, Electronic structure, phase stability, and semimetal-semiconductor transitions in Bi, *Phys. Rev. B* **60,** 15484 (1999).

196. P. Larson, Effect of $p_{1/2}$ corrections in the electronic structure of Bi_2Te_3 compounds, *Phys. Rev. B* **68**, 155121 (2003)

197. M. Kim, A. J. Freeman, and C. B. Geller, Screened exchange LDA determination of the ground and excited state properties of thermoelectrics: Bi_2Te_3, *Phys. Rev. B* **72**, 035205(R) (2005).

198. O. V. Misochko, Nonequilibrium phase transition in V-group semimetals induced by ultrashort laser pulses, *Bull. Rus. Acad. Sci.* **74**, 589 (2010).

199. A. Decker, G. A. Landrum, and R. Dronskowski, Structural and electronic Peierls transition in the elements: The crystal structure of tellurium, *Z. Anorg. Allg. Chem.* **628**, 295 (2002).

200. O. Rubel and D. Laughton, Lone-pair states as a key to understanding impact ionization in chalcogenide semiconductors, *J. Phys.: Condens. Matter* **22,** 355803 (2010).

201. A. Coke, T. Lee, and T. P. Das, Investigation of the electronic properties of tellurium energy band structure, *Phys. Rev. B* 22, 2968 (1980).

202. G. Kresse, J. Furthmüller, and J. Hafner, Theory of the crystal structure of selenium and tellurium: The effect of generalized-gradient corrections to the local-density approximations, *Phys. Rev. B* **50**, 13181 (1994).

203. A. N. Quiu, L. T. Zhang, and J. S. Wu, Crystal structure, electronic structure, and thermoelectric properties of β-Zn_4Sb_3 from first principles, *Phys. Rev. B* **81**, 035203 (2010).

204. M. S. Lee and S. D. Mahanti, Validity of the rigid band approximation in the study of the thermopower of narrow band gap semiconductors, *Phys. Rev. B* **85**, 165149 (2012).

205. H. Kim and M. Kaviany, Effect of thermal disorder on high figure of merit in PbTe, *Phys. Rev. B* **85**, 045213 (2012).

206. C. M. Jaworski, M. S. Nielsen, H. Wang, S. N. Girard, W. Cai, W. D. Porter, M. G. Kanatzidis, and J. P. Heremans, Valence-band structure of

highly efficient p-type thermoelectric PbTe-PbS alloys, *Phys. Rev. B* **87**, 045203 (2013).

207. M. W. Oh, B. S. Kim, S. D. Park, D. M. Wee, and H. W. Lee, First-principle calculation of the electronic structure of $HfTe_5$, *Solid State Comm.* **146**, 454 (2008).
208. N. D. Lowhorn, T. M. Tritt, E. E. Abbott, and J. W. Kolis, Enhancement of the power factor of the transition metal pentatelluride $HfTe_5$ by rare-earth doping, *Appl. Phys. Lett.* **88**, 022101 (2006).
209. G. N. Kamm, D. J. Gillespe, A. C. Ehrlich, and D. L. Peebles, Fermi surface, effective masses, and energy bands of $HfTe_5$ as derived from the Shubnikov-de Haas effect, *Phys. Rev. B* **35**, 1223 (1985).
210. B. M. Zawilski, R. V. Littleton IV, and Terry M. Tritt, Investigation of the thermal conductivity of the mixed pentatellurides $Hf_xZr_{1-x}Te_5$, *Appl. Phys. Lett.* **77**, 2319 (2000).
211. A. F. May, D. J. Singh, and G. J. Snyder, Influence of band structure on the large thermoelectric performance of lanthanum telluride, *Phys. Rev. B* **79**, 153101 (2009).
212. X. Wang, R. Yang, Y. Zhang, P. Zhang, and Y. Xue, Rare earth chalcogenide Ce_3Te_4 as high efficiency high temperature thermoelectric material, *Appl. Phys. Lett.* **98**, 222110 (2011).
213. A. F. May, M. McGuire, C. Cantoni, and B. C. Sales, Physical properties of $Ce_{3-x}Te_4$ below room temperature, *Phys. Rev. B* **86**, 035135 (2012).
214. D. P. Young, C. L. Brown, P. Khalifah, R. J. Cava, and A. P. Ramirez, Thermoelectric properties of Ag_3AuTe_2, *J. Appl. Phys.* **88**, 5221 (2000).
215. R. P. Chasmar and R. Stratton, The thermoelectric figure of merit and its relation to thermoelectric generators, *J. Electron. Control* **7**, 52 (1959).
216. W. Lu, X. Tan, K. Yin, H. Liu, X. Tang, J. Shi, Q. Zhang, and C. Uher, Convergence of conduction bands as a means of enhancing thermoelectric performance of *n*-type $Mg_2Si_{1-x}Sn_x$ solid solutions, *Phys. Rev. Lett.* ***108***, 166601 (2012).
217. V. K. Zaitsev, M. I. Federov, E. A. Gurieva, I. S. Eremin, P. P. Konstantinov, A. Yu. Samubib, and M. V. Vedernikov, Highly effective $Mg_2Si_{1-x}Sn_x$ thermoelectrics, *Phys. Rev. B* **74**, 045207 (2006).
218. N. Okinaka and T. Akiyama, Latent property of defect-controlled metal oxide: Nonstoichiometric titanium oxides as prospective materials for high-temperature thermoelectric conversion, *Jpn. J. Appl. Phys.* **45**, 7009 (2006).

219. I. Terasaki, Y. Sasago, K. Uchinokura, Large thermoelectric power in $NaCo_2O_4$ single crystals, *Phys. Rev. B* **56**, R12685 (1997).

220. P. M. de Wolff, The pseudo-symmetry of modulated crystals structures, *Acta Cryst. A* **30,** 777 (1974).

221. S. L. Penfiedl and W. E. Ford, Ueber den Calaverite, *Z. Kristallogr. Mineral.* **35**, 430 (1902).

222. H. Smith, Ueber das bemerkenswerthe Problem der Entwickelung der Kristllformen des Calaverite, *Z. Kristallogr. Mineral.* **37**, 209 (1903).

223. V. Goldschmidt, Ch. Palache, and M. Peacock, Ueber Calaverit, *N. Jahrbuch f. Mineralogie, Beil. Abt. A* **63**, 1 (1931).

224. B. Dam, A. Janner, and J. D. H. Donnay, Incommensurate morphology of calaverite $AuTe_2$ crystals, *Phys. Rev. Lett.* **55,** 2301 (1985).

225. B. C. H. Krutzen and J. E. Inglesfield, First-principles electronic structure calculations for incommensurately modulated calaverite, *J. Phys.: Condens. Matter* **2**, 4829 (1990).

226. G. van Tendeloo, P. Gregoriades, and S. Amelinckx, Electron-microscopy studies of modulated structures in (Au,Ag)Te_2. 1 Calaverite $AuTe_2$, *J. Sol. State Chem.* **50**, 321 (1983).

227. A. van Triest, W. Folkerts, and C. Haas, Electronic structure and photoelectron spectra of calaverite $AuTe_2$, *J. Phys.: Condens. Matter* **2**, 8733 (1990).

228. A. Charoenphakdee, K. Kurosaki, A. Harnwunggmoung, H. Muta, and S. Yamanaka, Thermoelectric properties of gold telluride: $AuTe_2$, *J. Alloys Compound.* **496**, 53 (2010).

229. Y. Miyazaki, D. Igarashi, K. Hayashi, T. Kajitani, and K. Yubuta, Modulated crystal structure of chimney-ladder higher manganese silicides $MnSi_\gamma$ ($\gamma \sim 1.74$), *Phys. Rev. B* **78**, 214104 (2008).

230. I. R. Fischer, Z. Islam, A. F. Panchula, K. O. Cheon, M. J. Cramer, P. C. Canfield, and A. L. Goldman, Growth of large-grain of R-Mg-Zn quasicrsytals from the ternary melt (R =Y, Er, Ho, Dy and Tb), *Phil. Mag. B* **77**, 1601 (1998).

231. K. Hiraga, F. J. Lincoln, and W. Sun, Structure and structural change of AlNiCo decagonal quasicrystals by high-resolution electron microscopy, *Materials Trans. JIM* **32**, 308 (1991).

232. S. Martin, A. F. Hebard, A. R. Kortan, and F. A. Thiel, Transport properties of $Al_{65}Cu_{20}Co_{15}$ and $Al_{70}Ni_{15}Co_{15}$ decagonal quasicrystals, *Phys. Rev. Lett.* **67,** 719 (1991).

233. L. Shu-yuan, W. Xue-mei, Z. Dian-lin, X. L. He, and K. X. Kuo, Anisotropic transport properties of a stable two-dimensional quasicrystal: $Al_{62}Si_3Cu_{20}Co_{15}$, *Phys. Rev. B* **41,** 9625 (1990).

234. L. Shuyuan, L. Guohong, and Z. Dian-lin, Thermopower of decagonal $Al_{73}Ni_{17}Co_{10}$ single quasicrystals: Evidence for a strongly enhanced electron-phonon coupling in the quasicrystalline plane, *Phys. Rev. Lett.* **77**, 1998 (1996).

235. J. Dolinšek, Anisotropy effects in complex metallic alloys, *Chem. Soc. Rev.* **41**, (2012).

236. Z. Dian-lin, C. Shao-chun, W. Yun-ping, and W. Xue-mei, Anisotropic thermal conductivity of the 2D single quasicrystals: $Al_{65}Ni_{20}Co_{15}$ and $Al_{62}Si_3Cu_{20}Co_{15}$, *Phys. Rev. Lett.* **66,** (2006).

237. D. Zhang, Electronic properties of stable decagonal quasicrystals, *Phys. Status Solidi A* **207**, 2666 (2010).

238. M. Conrad, F. Krumeich, and B. Harbrecht, A dodecagonal quasicrystalline chalcogenide, *Angew. Chem. Int. Ed.* **37**, 1383 (1998).

239. E. Maciá, T. Takeuchi, and T. Otagiri, Modeling the spectral conductivity of Al-Mn-Si quasicrystalline approximants: A phenomenological approach, *Phys. Rev. B* **72**, 174208 (2005).

240. E. Maciá, Quasicrystals and the quest for next generation thermoelectric materials, *Rev. Sol. State Mats. Sci.* **37**, 215 (2012).

241. D. Mayou, C. Berger, F. Cyrot-Lackmann, T. Klein, and P. Lanco, Evidence for unconventional electronic transport in quasicrystals, *Phys. Rev. Lett.* **70,** 3915 (1993).

242. A. L. Pope, T. M. Tritt, M. A. Chernikov, and M. Feuerbacher, Thermal and electrical transport properties of the single-phase quasicrystalline material $Al_{70.8}Pd_{20.9}Mn_{8.3}$, *Appl. Phys. Lett.* **75**, 1854 (1999).

243. A. Carlsson, And now quasi-semiconductors? *Nature* **353**, 15 (1991).

244. F. Cyrot-Lackmann, Quasicrystals as potential candidates for thermoelectric materials, *Mater Sci. Eng. A*, **294–296**, 611 (2000).

245. K. Giannò, A. V. Sologubenko, M. A. Chernikov, H. R. Ott, I. R. Fisher, and P. C. Canfield, Electrical resistivity, thermopower, and thermal conductivity of single grained (Y,Tb,Ho,Er)-Mg-Zn icosahedral quasicrystals, *Mater. Sci. Eng. A* **294–296**, 715 (2000).

246. Y. K. Kuo, K. M. Sivakumar, H. H. Lai, C. N. Ku, S. T. Lin, and A. B. Kaiser, Thermal and electrical transport properties of Ag-In-Yb quasicrystals: An experimental study, *Phys. Rev. B* **72,** 054202 (2005).

247. Y. K. Kuo, H. H. Lai, C. H. Huang, W. C. Ku, C. S. Lue, and S. T. Lin, Thermoelectric properties of binary Cd-Yb quasicrystal and Cd_6Yb, *J. Appl. Phys.* **95**, 1900 (2004).

248. F. S. Pierce, P. A. Bancel, B. D. Biggs, Q. Guo, and S. J. Poon, Composition dependence of the electronic properties of Al-Cu-Fe and Al-Cu-Ru-Si semimetallic quasicrystals, *Phys. Rev. B* **47**, 5670 (1993).

249. Y. Takagiwa, T. Kamimura, S. Hosoi, J. T. Okada, and K. Kimura, Thermoelectric properties of polygrained icosahedral $Al_{71-x}Ga_xPd_{20}Mn_9$ ($x = 0, 2, 3, 4$) quasicrystals, *J. Appl. Phys.* **104**, 073721 (2008).

250. T. Nagata, K. Kirihara, and K. Kimura, Effect of Ru substitution for Re on the thermoelectric properties of AlPdRe icosahedral quasicrystals, *J. Appl. Phys.* **94,** 6560 (2003).

251. J. T. Okada, T. Hamamatsu, S. Hosoi, T. Nagata, and K. Kimura, Improvement of thermoelectric properties of icosahedral AlPdRe quasicrystals by Fe substitution for Re, *J. Appl. Phys.* **101**, 103702 (2007).

252. K. Kirihara and K. Kimura, Composition dependence of thermoelectric properties of AlPdRe icosahedral quasicrystals, *J. Appl. Phys.* **92**, 979-986 (2002).

253. R. Escudero, J. C. Lasjaunias, Y. Calvayrac, and M. Boudard, Tunnelling and point contact spectroscopy of the density of states in quasicrystalline alloys, *J. Phys. Condens. Matter* **11**, 383 (1999)

254. E. Maciá, Modeling the thermopower of icosahedral $Al_{63}Cu_{25}Fe_{12}$ quasicrystals: Spectral fine structure, *Phys. Rev. B* **69**, 132201 (2004).

255. Z. M. Stadnik, D. Pordie, M. Garnier, Y. Baer, A. P. Tsai, A. Inoue, K. Edagawa, and S. Takeuchi, Electronic structure of icosahedral alloys studied by ultrahigh energy resolution photoemission spectroscopy, *Phys. Rev. Lett.* **77,** 1777 (1996).

256. Z. M. Stadnik, D. Pordie, M. Garnier, Y. Baer, A. P. Tsai, A. Inoue, K. Edagawa, S. Takeuchi, and K. H. J. Buschow, Electronic structure of quasicrystals studied by ultrahigh-energy-resolution photoemission spectroscopy, *Phys. Rev. B* **55,** 10 938 (1997).

257. C. V. Landauro, E. Maciá, and H. Solbrig, Analytical expressions for the transport coefficients of icosahedral quasicrystals, *Phys. Rev. B* **67**, 184206 (2003).

258. E. Maciá, Thermoelectric figure of merit of AlPdRe icosahedral quasicrystals: Composition-dependent effects, *Phys. Rev. B* **69** 184202 (2004).

259. E. Maciá, Thermoelectric properties of icosahedral quasicrystals: A phenomenological approach, *J. Appl. Phys.* **93**, 1014 (2003).

260. E. Maciá, Optimizing the thermoelectric efficiency of icosahedral quasicrystals and related complex alloys, *Phys. Rev. B* **80**, 205103 (2009).

261. P. Popčević, I. Batistić, E. Tutiš, K. Velebit, M. Heggen, and M. Feuerbacher, The generalization of the kinetic equations and the spectral conductivity function to anisotropic systems: Case T-$Al_{72.5}Mn_{21.5}Fe_6$ complex metallic alloy, *Croat. Chem. Acta* **83**, 95 (2010).

262. E. Maciá, Compatibility factor of segmented thermoelectric generators based on quasicrystalline alloys, *Phys. Rev. B* **70**, 100201 (2004).

263. Y. Takagiwa, T. Kamimura, S. Hosoi, J. T. Okada, and K. Kimura, Thermoelectric performance of Al-Pd-Mn quasicrystals: comparison with (1/1,2/1-)AlPdMnSi approximants and improvement by Ga substitution for Al, *Z. Kristallogr.* **224**, 21 (2009).

264. T. Takeuchi, T. Otagiri, H. Sakagami, T. Kondo, and U. Mizutani, Thermoelectric properties of the Al-TM-Si (TM=Mn,Re) 1/1-cubic approximant, in *Mater. Res. Soc. Symp. Proc.* **805**; Materials Research Society, Boston (2004), p. 105.

265. T. Takeuchi, Thermal conductivity of the Al-based quasicrystals and approximants, *Z. Kristallogr.* **224**, 35 (2009).

266. O. Rapp and S. J. Poon, Conductivity of icosahedral quasicrystals, *Phys. Rev. B* **84**, 174206 (2011).

267. W. Xue-mei, Z. Dian-lin, and L. F. Chen, Universality of the Hall-effect anisotropy in decagonal quasicrystals, *Phys. Rev. B* **48,** 10542 (1993).

268. K. P. Perstich, B. Rossner, and B. Batlogg, Field-effect-modulated Seebeck coefficient in organic semiconductors, *Nature Mater.* **7**, 321 (2008).

269. K. Hayashi, T. Shinano, Y. Miyazaki, and T. Kajitani, Fabrication of iodine-doped pentacene thin films for organic thermoelectric devices, *Appl .Phys. Lett.* **109**, 023712 (2011).

270. S. Kitagawa, H. Kotegawa, H. Tou, H. Ishii, K. Kudo, M. Nohara and H Harima, Pressure-induced superconductivity in mineral calaverite $AuTe_2$, *J. Phys. Soc. Jpn.* **82**, 113704, 248–249 (2013).

271. N. Kim, B. Domercq, S. Yoo, A. Christensen, B. Kippelen, and S. Grahm, Thermal transport properties of thin films of small molecule organic semiconductors, *Appl. Phys. Lett.* **87**, 241908 (2005).

272. G. H. Kim, M. Shtein, and K. P. Pipe, Thermoelectric and bulk mobility measurements in pentacene thin films, *Appl. Phys. Lett.* **98**, 093303 (2011).

273. K. Harada, M. Sumino, C. Adachi, S. Tanaka, and K. Miyazaki, Improved thermoelectric performance of organic thin-film elements utilizing a bilayer structure of pentacene and 2,3,5,6-tetrafluoro-7,7,8,8-tetracyanoquinodimethane (F_4-TCNQ), *Appl. Phys. Lett.* **96**, 253304 (2010).

274. I. Lévesque, P. O. Bertrand, N. Blouin, M. Leclerc, S. Zecchin, G. Zotti, C. I. Ratcliffe, D. D. Klug, X. Gao, F. Gao, and J. S. Tse, Synthesis and thermoelectric properties of polycarbazole, polyindolocarbazole, and polydiindolocarbazole derivatives, *Chem. Mater.* **19**, 2128 (2007).

275. R. B. Aïch, N. Blouin, A. Bouchard, and M. Leclerc, Electrical and thermoelectric properties of poly(2,7-carbazole) derivatives, *Chem. Mater.* **21**, 751 (2009).

276. C. Yu, Y. S. Kim, D. Kim, and J. C. Grunian, Thermoelectric behavior of segregated-network polymer nanocomposites, *Nano Lett.* **8**, 4428 (2008).

277. D. Kim, Y. Kim, K. Choi, J. C. Grunian, and Ch. Yoo, Improved thermoelectric behavior of nanotube-filled composites with poly(3,4-ethylenedioxythiophene) poly(estirenesulfonate), *ACS Nano* **4**, 513 (2010).

278. K. Chatterjee, M. Mitra, K. Kargupta, S. Ganguly, and D. Banerjee, Synthesis, characterization and enhanced thermoelectric performance of structurally ordered cable-like novel polyaniline-bismuth telluride nanocomposite, *Nanotechnology* **24**, 215703 (2013).

279. O. Bubnova, Z. U. Khan, A. Malti, S. Braun, M. Fahlman, M. Berggren, and X. Crispin, Optimization of the thermoelectric figure of merit in the conducting polymer poly(3,4-ethylenedioxythiophene), *Nature Mater.* **10**, 429 (2011).

280. S. O. Kelley and J. K. Barton, Electron transfer between bases in double helical DNA, *Science* **283**, 375 (1999).

281. M. W. Shiawari, M. J. Deen, E. B. Starikov, and G. Cuniberti, Electrical conductance in biological molecules, *Adv. Funct. Mats.* **20**, 1865 (2010).

282. X. Guo, A. A. Gorodetsky, J. Hone, J. K. Barton, and C. Nuckolis, Conductivity of a single DNA duplex bridging a carbon nanotube, *Nature Nanotechnol.* **3**, 163 (2008).

283. B. Xu, P. Zhang, X. Li, and N. Tao, Direct conductance measurements of single DNA molecules in aqueous solution, *Nano Lett.* **4**, 1105 (2004).

284. H. van Zalinge, D. J. Schiffrin, A. D. Bates, W. Haiss, J. Ulstrup, and R. J. Nichols, Single-molecule conductance measurements of single- and double-stranded DNA oligonucleotides, *ChemPhysChem* **7,** 94 (2006).

285. E. Artacho, M. Machado, D. Sánchez-Portal, P. Ordejón, and J. M. Soler, Electrons in dry DNA from density functional calculations, *Molecular Phys.* **101,** 1587 (2003).

286. M. Taniguchi and T. Kawai, Electronic structure of A- and B-type DNA crystals, *Phys. Rev. E* **70,** 011913 (2004).

287. M. Furukawa, H. S. Kato, M. Taniguchi, T. Kawai, T. Hatsui, N. Kosugi, T. Yoshida, and M. Aida, Electronic states of the DNA polynucleotides poly(dG)-poly(dC) in the presence of iodine, *Phys. Rev. B* **75,** 045119 (2007).

288. K. Schwab, E. A. Henriksen, J. M. Worlock, and M. L. Roukes, Measurement of the quantum of thermal conductance, *Nature (London)* **404,** 974 (2000).

289. N. Nishiguchi, Y. Ando and M. N. Wybourne, Acoustic phonon modes of rectangular quantum wires, *J. Phys. Condens. Matter* **9,** 5751 (1997).

290. I. S. Yang and A. C. Anderson, Specific heat of deoxyribonucleic acid at temperatures below 5 K, *Phys. Rev. B* **35,** 9305 (1987).

291. P. Reddy, S. Y. Jang, R. A. Segalman, and A. Majumdar, Thermoelectricity in molecular junctions, *Science* **315**, 1568 (2007).

292. K. Baheti, J. A. Malen, P. Doak, P. Reddy, S. Y. Jang, T. D. Tilley, A. Majumdar, and R. A. Segalman, Probing the chemistry of molecular heterojunctions using thermoelectricity, *Nano Lett.* **8**, 715 (2008).

293. A. Tan, S. Sadat, and P. Reddy, Measurement of thermopower and current-voltage characteristics of molecular junctions to identify orbital alignment, *Appl. Phys. Lett.* **96,** 013110 (2010).

294. A. Tan, J. Balachandran, S. Sadat, V. Gavini, B. D. Dunietz, S. Y. Jang, and P. Reddy, Effect of length and contact chemistry on the electronic structure and thermoelectric properties of molecular junctions, *J. Am. Chem. Soc.* **133**, 8838 (2011).

295. J. Balachandran, P. Reddy, B. D. Dunietz, and V. Gavini, End-group-induced charge transfer in molecular junctions: Effect on electronic-structure and thermopower, *J. Phys. Chem. Lett.* **3**, 1962 (2012).

296. J. R. Widawasky, P. Darancet, J. B. Neaton, and L. Venkataraman, Simultaneous determination of conductance and thermopower of single molecule junctions, *Nano Lett.* **12**, 354 (2012).

297. E. Maciá, Thermoelectric power and electrical conductance of DNA based molecular junctions, *Nanotechnology* **16**, S254 (2005).

298. E. Maciá, DNA-based thermoelectric devices: A theoretical prospective, *Phys. Rev. B* **75,** 035130 (2007).

299. K. H. Yoo, S. H. Ha, J. O. Lee, J. W. Park, Jinhee Kim, J. J. Kim, H. Y. Lee, T. Kawai, and H. Y. Choi, Electrical conduction through poly(dA)-poly(dT) and poly(dG)-poly(dC) DNA molecules, *Phys. Rev. Lett.* **87,** 198102 (2001).

300. J. C. Poler, R. M. Zimmermann, and E. C. Cox, Scanning thermopower microscopy of guanine monolayers, *Langmuir* **11,** 2689 (1995).

301. E. Maciá, Thermoelectric properties of codon DNA based molecular devices, *Rev. Adv. Mats. Sci.* **10**, 166 (2005).

302. T. M. Tritt, Thermal Conduction: Theory, Properties and Applications, Kluwer Academic, Plenum Publishers, New York (2004).

303. C. Zhou, L. Zhang and J. Sakamoto, Thermoelectric properties of p-type skutterudite nanocomposites, in Nanoscale Thermoelectrics Lecture Notes, X. Wang and Z. M. Wang (Eds.), Nanoscale Sciences and Technology 16, Springer International Publishing, Switzerland 2014.

304. W. Seifert, V. Pluschke, and N. F. Hinsche, Thermoelectric cooler concepts and the limit for maximum cooling, *J. Phys.: Condens. Matter* **26**, 255803 (2014).

305. M. W. Gaultois, T. D. Sparks, C. K. H. Borg, R. Seshadri, W. D. Bonificio, and D. R. Clarke, Data-driven review of thermoelectric materials: Performance and resource considerations, *Chem. Mater.* **25**, 2911 (2013).

306. H. Y. Lv, W. J. Lu, D. F. Shao, and Y. P. Sun, Enhanced thermoelectric performance of phosphorene by strain-induced band convergence, *Phys. Rev. B* **90**, 085433 (2014).

Index